TOPICAL AND TRANSDERMAL DRUG DELIVERY SYSTEMS

Applications and Future Prospects

TOPICAL AND TRANSDERMAL DRUG DELIVERY SYSTEMS

Applications and Future Prospects

Edited by

Nayan A. Gujarathi, PhD

Juliana Palma Abriata, PhD

Raj Kumar Keservani, MPharm

Anil K. Sharma, PhD, MPharm

AAP | APPLE ACADEMIC PRESS

First edition published 2023

Apple Academic Press Inc.
1265 Goldenrod Circle, NE,
Palm Bay, FL 32905 USA

760 Laurentian Drive, Unit 19,
Burlington, ON L7N 0A4, CANADA

CRC Press
6000 Broken Sound Parkway NW,
Suite 300, Boca Raton, FL 33487-2742 USA

4 Park Square, Milton Park,
Abingdon, Oxon, OX14 4RN UK

Library and Archives Canada Cataloguing in Publication

Title: Topical and transdermal drug delivery systems : applications and future prospects / edited by Nayan A. Gujarathi, PhD, Juliana Palma Abriata, PhD, Raj Kumar Keservani, MPharm, Anil K. Sharma, PhD, MPharm.
Names: Gujarathi, Nayan A., 1980- editor. | Abriata, Juliana Palma, editor. | Keservani, Raj K., 1981- editor. | Sharma, Anil K., 1980- editor.
Description: First edition. | Includes bibliographical references and index.
Identifiers: Canadiana (print) 20220422672 | Canadiana (ebook) 20220422702 | ISBN 9781774910702 (hardcover) | ISBN 9781774910719 (softcover) | ISBN 9781003284017 (ebook)
Subjects: LCSH: Transdermal medication. | LCSH: Drug delivery systems. | LCSH: Nanomedicine.
Classification: LCC RM151 .T67 2023 | DDC 615/.6—dc23

Library of Congress Cataloging-in-Publication Data

CIP data on file with US Library of Congress

ISBN: 978-1-77491-070-2 (hbk)
ISBN: 978-1-77491-071-9 (pbk)
ISBN: 978-1-00328-401-7 (ebk)

About the Editors

Nayan A. Gujarathi, PhD

Shri Vile Parle Kelavani Mandal's Institute of Pharmacy, Dhule, Maharashtra, India

Nayan A. Gujarathi, PhD, MPharm, is an Assistant Professor at the Department of Pharmaceutics, Shri Vile Parle Kelavani Mandal's Institute of Pharmacy, Dhule, India. He has 14 years of academic experience in pharmaceutical science. He has published 32 peer-reviewed papers in the field of pharmaceutical sciences in national and international journals with 5 book chapters. He is an approved post-graduate teacher in pharmaceutics by North Maharashtra University, Jalgaon, and Savitribai Phule University, Pune. He is appointed as a referee (evaluator) at the university level for Avishkar 2013 and Avishkar 2017 by North Maharashtra University, Jalgaon. He completed the minor research project sanctioned by BCUD, North Maharashtra University, Jalgaon. He received a Diploma in Pharmacy from the Government Polytechnic, Jalgaon, India, and a Bachelor's degree in pharmacy from the North Maharashtra University, Jalgaon. He completed a Master of Pharmacy degree with a specialization in pharmaceutics from the School of Pharmaceutical Sciences, Rajiv Gandhi Proudyogiki Vishwavidyalaya, Bhopal, India. He accomplished his PhD in pharmaceutical sciences from the Jodhpur National University, Jodhpur, India. He presented his research work at the international conference organized by the Controlled Release Society and delivered four expert talks on the implementation of outcome-based education for the accreditation process. His research interest encompasses novel drug delivery systems, polymer science, nanotechnology, and nutraceuticals.

Juliana Palma Abriata, PhD

School of Pharmaceutical Sciences of Ribeirao Preto, Sao Paulo, Brazil

Juliana Palma Abriata, PhD, is currently working at the School of Pharmaceutical Sciences of Ribeirao Preto, Sao Paulo, Brazil. She received her Master's degree and developed her doctoral research in the same faculty. During this period, she worked on the development and characterization of drug delivery systems for Chagas disease and ovarian cancer treatment. She was a guest PhD student at the College of Pharmacy, The Ohio State University, USA, under the supervision of Dr. Robert Lee. She has publications in the drug delivery field.

Raj Kumar Keservani, MPharm

Faculty of B. Pharmacy, CSM Group of Institutions, Prayagraj, Uttar Pradesh, India

Raj Kumar Keservani, M.Pharm, is a Faculty of B. Pharmacy at CSM Group of Institutions, Prayagraj, Uttar Pradesh, India. He has more than 12 years of academic (teaching) experience from various institutes in Indian pharmaceutical education. He has published 35 peer-reviewed papers in the field of pharmaceutical sciences in national and international journals, 15 book chapters, three co-authored books, and 15 edited books. He is also an active reviewer for several international scientific journals. Mr. Keservani graduated with a pharmacy degree from the Department of Pharmacy, Kumaun University, Nainital (UA), India. He received his Master of Pharmacy (MPharm) (specialization in pharmaceutics) from the School of Pharmaceutical Sciences, Rajiv Gandhi Proudyogiki Vishwavidyalaya, Bhopal, India. His research interests include nutraceutical and functional foods, novel drug delivery systems (NDDS), transdermal drug delivery/drug delivery, health science, cancer biology, and neurobiology.

Anil K. Sharma, PhD

School of Medical and Allied Sciences, G. D. Goenka University, Sohna, Haryana, India

Anil K. Sharma, MPharm, PhD, is an Assistant Professor (Pharmaceutics) at the School of Medical and Allied Sciences, GD Goenka University, Gurugram, India. He has experience of more than 11 years in academics. He graduated with a degree in pharmacy (BPharm) from the University of Rajasthan, Jaipur, India, and received a Master of Pharmacy (MPharm) from the School of Pharmaceutical Sciences, Rajiv Gandhi Proudyogiki Vishwavidyalaya, Bhopal, India, with a specialization in pharmaceutics. He has earned his PhD from the University of Delhi. He has published 30 peer-reviewed papers in the field of pharmaceutical sciences in national/international reputed journals, as well as 16 book chapters and 15 edited books. His research interests encompass nutraceutical and functional foods, novel drug delivery systems (NDDS), drug delivery, nanotechnology, health science/life science, and biology/cancer biology/neurobiology.

Contents

Contributors

Alexsandra C. Apolinário
Department of Pharmacology, Institute of Biomedical Sciences, University of São Paulo, Sao Paulo, Brazil

Tamara Gonçalves Araújo
Federal University of Ceará, Faculty of Pharmacy, Dentistry, and Nursing, Department of Pharmacy, Fortaleza, Brazil

Akshay Babu
Sandip Institute of Pharmaceutical Sciences, Nashik–422213, Maharashtra, India

Akshada A. Bakliwal
Sandip Institute of Pharmaceutical Sciences, Nashik–422213, Maharashtra, India

Maria Vitória Lopes Badra Bentley
School of Pharmaceutical Sciences of Ribeirão Preto-University of São Paulo, Ribeirão Preto, São Paulo, Brazil, E-mail: vbentley@usp.br

Guilherme Dos Anjos Camargo
Pharmaceutical Sciences Department, State University of Ponta Grossa, Ponta Grossa, Paraná, Brazil

Patrícia Mazureki Campos
Pharmaceutical Sciences Department, State University of Ponta Grossa, Ponta Grossa, Paraná, Brazil, E-mail: patriciamazurekicampos@yahoo.com.br

Prashant K. Deshmukh
H.R. Patel Institute of Pharmaceutical Education and Research, Shirpur–425405, Maharashtra, India

Josimar O. Eloy
Federal University of Ceará, Faculty of Pharmacy, Dentistry, and Nursing, Department of Pharmacy, Fortaleza, Brazil

Lillian Maria Uchoa Dutra Fechine
Federal University of Ceará, Faculty of Chemistry, Department of Analytical Chemistry and Physico-Chemistry, Fortaleza, Brazil

Pierre Basílio Almeida Fechine
Federal University of Ceará, Faculty of Chemistry, Department of Analytical Chemistry and Physico-Chemistry, Fortaleza, Brazil

Luiziana Cavalcante Costa Fernandes
Federal University of Ceará, Faculty of Pharmacy, Dentistry, and Nursing, Department of Pharmacy, Fortaleza, Brazil

Priscileila Colerato Ferrari
Pharmaceutical Sciences Department, State University of Ponta Grossa, Ponta Grossa, Paraná, Brazil

Nayan A. Gujarathi
Shri Vile Parle Kelavani Mandal's Institute of Pharmacy, Dhule–424001, Maharashtra, India, E-mail: nayang5678@gmail.com

Gulam Mohammed Husain
National Research Institute of Unani Medicine for Skin Disorders (Formerly Central Research Institute of Unani Medicine) Under CCRUM, Ministry of Ayush, Government of India, Opp. ESI Hospital, AG Colony Road, Erragadda, Hyderabad–500038, Telangana, India

Ivanildo José da Silva Júnior
Federal University of Ceará, Center of Technology, Department of Chemical Engineering, Fortaleza, Brazil

Raj Kumar Keservani
Faculty of B. Pharmacy, CSM Group of Institutions, Prayagraj–212111, Uttar Pradesh, India

Ana Beatriz Klosowski
Pharmaceutical Sciences Department, State University of Ponta Grossa, Ponta Grossa, Paraná, Brazil

Marilisa Guimarães Lara
School of Pharmaceutical Sciences of Ribeirão Preto, University of São Paulo, Ribeirão Preto, São Paulo, Brazil, E-mail: mlara@fcfrp.usp.br

Luciana B. Lopes
Department of Pharmacology, Institute of Biomedical Sciences, University of São Paulo, Sao Paulo, Brazil, E-mail: lublopes@usp.br

Marcela Tavares Luiz
School of Pharmaceutical Sciences of Ribeirão Preto-University of São Paulo, Ribeirão Preto, São Paulo, Brazil

Jéssica Roberta Pereira Martins
Federal University of Ceará, Center of Technology, Department of Chemical Engineering, Fortaleza, Brazil

Mahesh P. More
SRES Sanjivani College of Pharmaceutical Education and Research, Kopargaon–423603, Maharashtra, India; H.R. Patel Institute of Pharmaceutical Education and Research, Shirpur–425405, Maharashtra, India, E-mail: mahesh.more@sanjivani.org.in

Karina Alexandre Barros Nogueira
Federal University of Ceará, Center of Technology, Department of Chemical Engineering, Fortaleza, Brazil

Aldenor Souza de Oliveira
Federal University of Ceará, Faculty of Chemistry, Department of Analytical Chemistry and Physico-Chemistry, Fortaleza, Brazil

Vasim Pathan
Sandip Institute of Pharmaceutical Sciences, Nashik–422213, Maharashtra, India

Pravin O. Patil
H.R. Patel Institute of Pharmaceutical Education and Research, Shirpur–425405, Maharashtra, India

Tulshidas S. Patil
Shri Vile Parle Kelavani Mandal's Institute of Pharmacy, Dhule–424001, Maharashtra, India

Raquel Petrilli
Federal University of Ceará, Center of Technology, Department of Chemical Engineering, Fortaleza, Brazil; Federal University of Ceará, Faculty of Pharmacy, Dentistry, and Nursing, Department of Pharmacy, Fortaleza, Brazil, E-mail: petrilliraquel@gmail.com

Fabíola Garcia Praça
School of Pharmaceutical Sciences of Ribeirão Preto-University of São Paulo, Ribeirão Preto, São Paulo, Brazil

Bhushan R. Rane
Shri D.D. Vispute College of Pharmacy and Research Center, New Panvel–410206, Maharashtra, India

Lívia de Queiroz Carvalho Silva
School of Pharmaceutical Sciences of Ribeirão Preto-University of São Paulo, Ribeirão Preto, São Paulo, Brazil

Ana Vitória Pupo Silvestrini
School of Pharmaceutical Sciences of Ribeirão Preto-University of São Paulo, Ribeirão Preto, São Paulo, Brazil

Meenendra Tripathi
Faculty of B. Pharmacy, CSM Group of Institutions, Prayagraj–212111, Uttar Pradesh, India

Juliana Santos Rosa Viegas
School of Pharmaceutical Sciences of Ribeirão Preto-University of São Paulo, Ribeirão Preto, São Paulo, Brazil

Abbreviations

AAD	adaptive aerosol delivery
ABCs	amphiphilic block copolymers
AD	atopic dermatitis
ALA	5-aminolevulinic acid
BAL	broncho alveolar lavage
BCG	bacillus Calmette-Guérin
CAC	critical association concentration
CMC	carboxymethyl cellulose
CMC	critical micelle concentration
COPD	chronic obstructive pulmonary disease
CPP	critical packing parameter
CPSC	Consumer Products Safety Commission
CTA	clinical trial authorization
CTN	clinical trial notification
CTX	clinical trial exemption
DLS	dynamic light scattering
DPI	dry powder inhalers
DSC	differential scanning calorimetry
DZ	dacarbazine drug
ECGs	electrocardiography
EDTA	ethylene diamine tetra-acetic acid
EE	encapsulation efficiency
EEG	electroencephalography
EMA	European Medicines Agency
EMG	electromyography
EP	epidermis
EPR	enhanced permeability and retention
FDA	Food and Drug Administration
FTIR	Fourier transform infrared spectroscopy
HLB	hybrid lipid bilayer
HLB	hydrophilic-lipophilic balance
HPC	hydroxypropyl cellulose
HPLC	high-performance liquid chromatography

HRT	hormonal replacement therapy
HSV	herpes simplex virus
HSYA	hydroxysaflor A
ICH	International Council for Harmonization
IMPD	investigational medicinal product dossier
IND	investigational new drug
LD	laser diffraction
LDPI	laser doppler perfusion imager
LPHNPs	polymer lipid hybrids NPs
LSE-High	living skin equivalents-high
LTRs	localized transport regions
LUV	large unilamellar vesicles
MDI	metered-dose inhaler
MLV	multilamellar vesicles
MPEG-PLA	methoxy-PEG-PLA
MSN	mesoporous silica NPs
MUNJIs	multi-use-nozzle jet injectors
NDA	pre-new drug application
NLC	nanostructured lipid carriers
NPs	nanoparticles
NRT	nicotine replacement therapy
NSAIDs	nonsteroidal anti-inflammatory drugs
NTA	nanoparticle tracking analysis
O/W	oil in water
PAAm	poly(acrylamide)
PBPK	physiologically based pharmacokinetics
PCA	polyacrylates
PCL	polycaprolactones
PCS	photon correlation spectroscopy
PD	pharmacodynamics
PDT	photodynamic therapy
PEO	polyethylene oxide
PEO–PPO–PEO	poly(ethylene oxide)–poly(propylene oxide)–poly (ethylene oxide)
PET	tomography (positron emission)
PG	poly(glycerol)
PGA	poly-glycolic acid
PGE1	prostaglandin E1
PHPMA	poly[N-(2-hydroxypropyl) methacrylamide]

PII	primary irritation index
PK	pharmacokinetics
PLA	poly-lactic acid
PLA2	phospholipase A2
PLGA	poly-lactic-co-glycolic acid
POxs	poly(oxazolines)
pre-IND	investigational new drug
PTXSLN	paclitaxel was encapsulated in SLN
PVA	poly(vinyl alcohol)
PVP	polyvinyl pyrrolidone
RF	radio frequency
RHE	reconstructed human epidermis
RIPT	repeated insult patch test
ROS	reactive oxygen species
SANS	small-angle neutron scattering
SAXS	small-angle X-ray scattering
SC	stratum corneum
SEM	scanning electron microscopy
SII	secondary irritation index
siRNA	small interfering RNA
SLB	supported lipid bilayers
SLN	solid lipid nanoparticles
SNES	simulated nasal electrolytic solution
ST	subcutaneous tissue
SUV	small unilamellar vesicles
TAA	triamcinolone acetonide
TDDS	transdermal drug delivery system
TEER	transdermal electrical resistance
TER	transcutaneous electrical resistance
TEWL	transdermal water loss
TFTI	thin-film tissue interface
TGA	Therapeutic Goods Association
TGA	thermogravimetric analysis
TTDDS	topical and transdermal drug delivery systems
V	volts
VDC	vertical diffusion cell
W/O	water in oil
WLT	Wen-Luo-Tong
XRPD	X-ray powder diffraction technology

Preface

Oral drug ingestion has been the oldest and the most common methods of drug delivery, yet a number of medicines are known to be less effective or pose bioavailability issues. Several actives are reported to have high first-pass metabolism, cause bowel upset, and decompose in the gastrointestinal tract. Researchers across the globe have consistently endeavored to bring out alternatives to oral drug delivery systems. Though injections have been considered a viable choice over oral formulations, providing the highest bioavailability, injections are invasive, yield hazardous biowaste, as well as possess the danger of spreading ailments via re-use of needles, particularly in developing nations. Transdermal drug delivery is credited to have a significant role in therapeutics by a variety of preparations/approaches designed by formulation scientists. Commencing in the decade of the 1980s, several products have received market authorizations from the FDA for applications to patients. Such formulations may offer liberation of the confined drug for extended periods. In addition, patient compliance is enhanced by such products, which are cost-effective too.

The present book strives to provide requisite information to readers relevant to topical and transdermal drug delivery systems (TDDS). The content of this book has been contributed by highly skilled, experienced, and renowned scientists and researchers across the world. They provide up-to-date information that offers insights to readers, researchers, academicians, scientists, and industrialists worldwide.

The book *Topical and Transdermal Drug Delivery Systems: Applications and Future Prospects* comprises 10 chapters with three parts that discuss the basics of drug delivery systems, strategies to enhance permeation across membranes, formulation, and evaluation of diverse dosages forms.

Part I—Basics of Transdermal Drug Delivery Systems contains three chapters. Chapter 1—*Concepts and Applications in Topical and Transdermal Drug Delivery,* written by Karina Alexandre Barros Nogueira and colleagues, gives an overview of skin structure and focuses on the most used approaches to overcome the stratum corneum (SC), with special attention to topical and transdermal drug delivery, aiming at different health sciences applications.

The text has been enriched with appropriate figures and tables. Finally, some clinical applications are also presented and discussed.

The nuances of micro/nano-TDDS are discussed in Chapter 2—*Micro- and Nano-Systems in Transdermal Drug Delivery System,* written by Nayan A. Gujarathi and associates. The authors have focused on the pros and cons of conventional drug delivery systems against novel ones. The extraordinary packages of nanosystems (vesicular systems, polymeric nanoparticles (NPs), nanoemulsion, and dendrimers) are broadly discussed, and their application is reviewed in this chapter through the transdermal route.

Chapter 3—*Advanced Topical and Transdermal Drug Delivery (TTDS) Used in Skin Diseases and Lesions of the Aging,* written by Ana Vitória Pupo Silvestrini and coworkers, summarizes the main types of nanoparticles used as TTDS, applied in skin ailments such as psoriasis, vitiligo, cancer, lesions of the aging and other.

Part II—Diverse Transdermal Drug Delivery Systems contains five chapters. Chapter 4—*Polymeric Micelles in Topical and Transdermal Delivery,* written by Alexsandra C. Apolinário and Luciana B. Lopes, describes the role of polymeric micelles in topical/transdermal delivery. The authors discuss structural aspects of polymeric micelles, ranging from composition and formation to main characteristics and their application for topical and transdermal delivery.

Chapter 5—*Microneedles: An Innovative Strategy for Skin Drug Delivery, Diagnostic, and Cosmetologic Approaches,* written by Marilisa Guimarães Lara, presents the concept of microneedles and describes the types of microneedles and their respective approaches to application. Aspects related to the fabrication of microneedles as materials forming and requirements for a good performance are also presented. Finally, the applications of microneedles as therapeutic, prophylactic, diagnostic, and cosmetologic tools are discussed, considering the advances in research and scientific development in microneedle technology in each of these areas.

An overview of dispersed systems has been provided in Chapter 6—*Emulsion, Nanoemulsion, and Microemulsion in TDDS,* written by Patrícia Mazureki Campos and colleagues. Colloidal carriers such as emulsions and submicron emulsions (nanoemulsions and microemulsions) have gained considerable interest in the topical application of therapeutic agents over the last few decades, and their fundamentals and applications are described in this chapter.

Chapter 7—*Pulmonary Drug Delivery System,* written by Nayan A. Gujarathi and associates, gives a focused view of drug delivery systems meant for the lungs. The chapter embraces an introduction to the respiratory system, the variety of drug carriers, design, and characterization of drug delivery systems. Applications in diseases relevant to the lungs have been listed.

The nasal application of nano-sized drug carriers has been discussed in Chapter 8—*Nanoencapsulated Nasal Drug Delivery System,* written by Nayan A. Gujarathi and colleagues. The authors have exhaustively described numerous nanomedicines with mention of types, preparation methods, and assessment technologies. Further, the potential future uses are also taken into reckoning.

Part III—Miscellaneous contains two chapters. Chapter 9—*Skin Sensitivity and Irritation Testing for Transposing Transdermal Drug Delivery System,* written by Mahesh P. More and coworkers, nicely listes and discusses the significance of skin sensitivity studies for the development of a viable pharmaceutical product. The recent advances have also found a place in the chapter so as to provide readers with updated learning.

The specialized uses of the science of fragrances to treat diseases are given in Chapter 10—*Regulatory Aspects of Drug Development for Dermal Products,* written by Nayan A. Gujarathi and colleagues. The chapter provides a detailed discussion on over-controlling authority guidelines, which are, in fact, essential for any dermal product for market authorization. The formulators may take cognizance of this and accordingly document formulation design protocols.

PART I

Basics of Transdermal Drug Delivery Systems

CHAPTER 1

Concepts and Applications in Topical and Transdermal Drug Delivery

Karina Alexandre Barros Nogueira,[1]
Luiziana Cavalcante Costa Fernandes,[2]
Jéssica Roberta Pereira Martins,[1] Aldenor Souza de Oliveira,[3]
Lillian Maria Uchoa Dutra Fechine,[3] Pierre Basílio Almeida Fechine,[3]
Tamara Gonçalves Araújo,[2] Ivanildo José da Silva Júnior,[1]
Josimar O. Eloy,[2] and Raquel Petrilli[1,2]

[1]Federal University of Ceará, Center of Technology, Department of Chemical Engineering, Fortaleza, Brazil

[2]Federal University of Ceará, Faculty of Pharmacy, Dentistry, and Nursing, Department of Pharmacy, Fortaleza, Brazil, E-mail: petrilliraquel@gmail.com (R. Petrilli)

[3]Federal University of Ceará, Faculty of Chemistry, Department of Analytical Chemistry and Physico-Chemistry, Fortaleza, Brazil

ABSTRACT

Topical and transdermal drug delivery involves crossing the main barrier of the skin, the stratum corneum (SC). Basically, the SC is the outermost layer of the skin, formed by dead corneocytes immersed in a lipid matrix which limits the skin penetration and permeation of drugs. Some characteristics are needed for the drug to be delivered passively, like adequate lipophilicity and limited molecular weight. These requirements have limited the number of commercially available products based on

Topical and Transdermal Drug Delivery Systems: Applications and Future Prospects. Nayan A. Gujarathi, PhD, Juliana Palma Abriata, PhD, Raj Kumar Keservani, MPharm, Anil K. Sharma, PhD, MPharm (Eds.)

transdermal or topical delivery. Briefly, the transdermal delivery system is capable of delivering the drug through the skin into the blood circulation, preferentially at a constant rate. Topical administration is responsible for delivery in different skin layers, depending on the drug's physicochemical characteristics and desired effect. Nowadays, many strategies have emerged to improve skin delivery to or through the skin, and these can be categorized into passive and physical methods. For passive delivery of drugs, chemical penetration enhancers can be used to improve skin penetration and/or permeation of drugs. Instead, physical methods such as iontophoresis, sonophoresis, and microneedles can be used to overcome the skin barrier. In this chapter, we will present an overview of skin structure and focus on the most used approaches to overcome the SC, with special attention to topical and transdermal drug delivery aiming at different health sciences applications. Finally, some clinical applications will be presented and discussed.

1.1 INTRODUCTION

The skin (Figure 1.1) is the division between the environment and the internal organs, being thus endowed with several functions, such as protection against mechanical shocks, regulation of body temperature, and sensitivity, and is the first line of physical defense against pathogens. It occupies 16% of an individual's body weight, thus being the largest organ in the human body (Venus, Waterman, and McNab, 2011; McLafferty, Hendry, and Farley, 2013). Compared to the oral route, it prevents changes in the gastrointestinal tract and avoids the first-pass metabolism. Dermal applications also have advantages over injections, being less invasive because they do not cause discomfort in the patient and allow self-administration (Münch, Wohlrab, and Neubert, 2017; Petrilli and Lopez, 2018).

The delivery of active compounds in the skin can aim for topical or transdermal absorption. Delivery of compounds aiming at the treatment of conditions in different layers of the skin is named topical delivery, whereas when the active compound is needed for systemic delivery, it is known as transdermal delivery (Lane et al., 2012). All these routes described above must transcend the outermost layer of the skin, the stratum corneum (SC), which is the main barrier of the skin. For transdermal delivery, some requirements are applicable, such as drugs with a molecular weight below 1 kDa and with high lipophilicity, which would restrict the use of transdermal delivery of macromolecules using passive delivery (Margetts and Sawyer, 2007; Paudel et al., 2010; Münch, Wohlrab, and Neubert, 2017).

In this context, different physical methods were studied because they could potentially facilitate the transport of macromolecules and nanoparticles by overcoming the barrier of the SC, allowing higher concentrations of drugs to reach deep skin layers (Gratieri et al., 2008). Among the methods, iontophoresis, low-frequency ultrasound, and microneedles have been widely used for transdermal delivery. More recently, these methods demonstrated an application for topical delivery of drugs, aiming the treatment of dermatological disorders, such as skin tumors and inflammation. Mojeiko et al., for instance, showed an increase in topical delivery of celecoxib using microneedles and microemulsion. The amount of celecoxib in the skin was 30-fold higher than in the receptor phase, suggesting the microneedles as a viable alternative for skin barrier disruption and improved celecoxib penetration for local breast cancer chemoprevention. Also, Herwadkar studied parameters optimization of low-frequency sonophoresis for transdermal and intradermal delivery of ketoprofen used in the treatment of rheumatoid arthritis and osteoarthritis. Some parameters such as application time, coupling medium, duty cycle, and distance of horn from skin were optimized for the higher accumulation of ketoprofen in synovial tissue (Herwadkar et al., 2012; Petrilli and Lopez, 2018; Petrilli et al., 2018b; Mojeiko et al., 2019).

In essence, iontophoresis refers to the use of a mild electric current that is applied, causing the movement of charged and uncharged molecules through the skin where at least two electrodes, the anode and the cathode, are required for electromigration and electroosmosis. In low-frequency ultrasound, cavitation is the main permeabilization mechanism of the skin consisting of the formation of microbubbles that oscillate, collapse, and disorganize the SC, whereas other mechanisms include thermal effects by absorption of ultrasound energy. Microneedles are minimally invasive because they create micrometer-sized pores in the epidermis (EP), with different lengths, materials, and geometry, which unblock the pathway for drug delivery (Park et al., 2014; Petrilli and Lopez, 2018; Byrne, Yeh, and DeSimone, 2018; Petrilli et al., 2018b; Lee et al., 2018).

Therefore, transdermal and topical drug delivery systems are extensively studied in order to overcome the barrier of the skin, the SC, each way made more efficient by applying passive or active methods. Herein, we will describe and discuss the structure of the skin and present passive and physical methods for topical and transdermal delivery and clinically available applications, aiming to highlight the most important aspects in the field.

1.2 SKIN STRUCTURE

The skin (Figure 1.1) consists of a complex organization with different cell layers and a subcutaneous fatty tissue that contains skin appendages and cells able to respond to pain, touch, pressure, and temperature. The EP is the outermost layer of the skin, consisting of the stratified squamous epithelium. The keratinocytes are the most abundant cells of this layer, responsible for the synthesis of keratin. Other important cell types that are present in this layer are melanocytes (responsible for melanin synthesis), Langerhans cells (responsible for the immune response), and Merkel cells (responsible for sensitivity to touch). This same layer presents four sublayers, the basal stratum, spinous stratum, granular stratum, and SC (Venus et al., 2011; McLafferty et al., 2013).

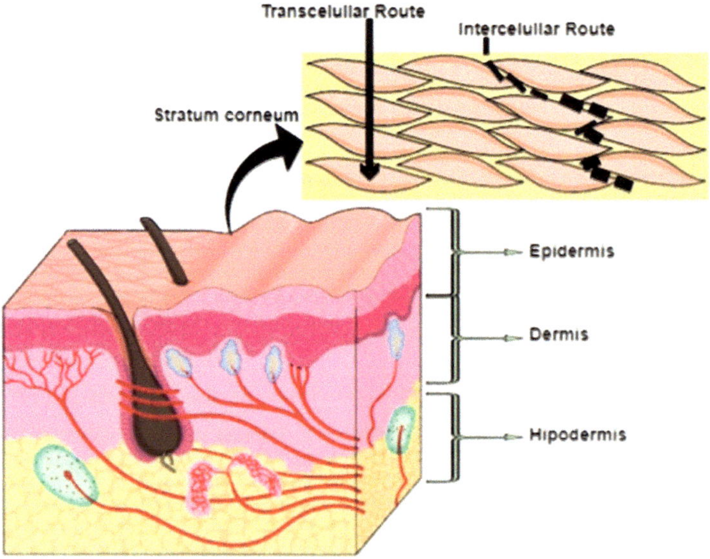

FIGURE 1.1 Skin structure representation.

The basal layer is the deepest layer of the EP, which consists of a layer of keratinocytes, able to differentiate and migrate to the surface of the skin. As these cells move between the other layers of the EP, their differentiation evolves until these cells reach the outermost layer of the EP, called the SC. Currently, the cellular differentiation allows the production of keratin, a protein filament that is associated in the keratinocytes by desmosomes, ensuring that the EP remains healthy and able to develop its defensive processes due to its continuous cellular regeneration. Also, it allows the skin to act as a protective and control barrier against dehydration (Baroli, 2010; Losquadro, 2017).

The SC is characterized as the first physical defense barrier against penetration of invading agents, impact resistance, and permeability barrier for drugs. In its composition, there are the corneocytes, which are the keratinocytes in the last stage of maturation, that basically consist of anucleate cells, rich in proteins and surrounded by a rich lipid matrix. In short, this layer of skin has a composition of around 15% water, so the homeostasis of hydration is fundamental to ensuring the homogeneity of the SC. Hydration agents of the SC are osmotically active amino acids derived from corneocyte proteolysis (Baroli, 2010; Petrilli et al., 2018a).

The EP and dermis are connected by a basal membrane that is composed of two layers, a more superficial, called lamina lucida, and a deeper one, called lamina dense. Both are connected to one another by collagen fibers (Losquadro, 2017).

The dermis is located just below the dermal-epidermal junction. It is a dense vascularized layer that protects the skin from mechanical shocks, and in its structure, there are specialized cells, glands, lymphatic vessels, and sensory nerves. It can be subdivided into two layers: the papillary dermis, which is the first layer of the dermis and the thinnest, and the reticular dermis, the lower and thicker layer (Lai-Cheong and McGrath, 2017; Losquadro, 2017).

The papillary layer is in contact with the dermal-epidermal junction, where are localized nerves and capillaries. The latter is of paramount importance because they allow the exchange of nutrients between the two layers of the skin. The reticular layer is formed of connective tissue and collagen and elastin fibers, blood vessels, and glands. The collagen and elastin fibers give the dermis resistance by stretching and contracting the skin. The collagen corresponds to 70% of the weight of the dermis, mainly types I and II and elastin fibers, which correspond to approximately 5%

of the weight of the dermis and are arranged interlaced in collagen fibers. The structure of these fibers changes over time, becoming less rigid and resistant. The glands present in the dermis can be divided into two types, sweat and sebaceous (Lai-Cheong and McGrath, 2017).

Briefly, the sweat glands are subdivided into eccrine and apocrine. This division takes into account the location of each one, as well as their structure and the composition of their secretion. The sebaceous glands are responsible for the secretion of a mixture of triglycerides, cholesterol, proteins, and salts that covers the surface of the skin, inhibiting excessive evaporation of water (McLafferty, Hendry, and Farley, 2013).

Located below the dermis, the hypodermis is made up of loose connective tissue that forms the adipose tissue that functions as thermal insulation for the body. In this layer are localized cell types, such as fibroblasts, adipose cells (adipocytes), and macrophages, that maintain adipocyte homeostasis and fat thermogenesis stimulation when it is exposed to different climatic conditions and physical stimuli (Wong et al., 2016).

1.2.1 EXPERIMENT SETUP FOR IN VITRO PERMEATION AND PENETRATION STUDIES

Different types of biological models can be prepared for *in vitro* permeation experiments that may influence the results. The full-thickness skin is used after the removal of connective tissue and subcutaneous fat. The variability can be reduced, while maintaining the dermal thickness using dermatomization. Briefly, the term atomization is a process where skin thickness is set from 500 to 750 μm (Abd et al., 2016). It should be performed by a trained person to minimize the variation in thickness obtained, thereby reducing the possible waste of material (Henning et al., 2008). Basically, the skin subcutaneous fat is removed, and the surface is carefully cleaned. The skin is placed in support and dermatomized to the desired thickness. The skin is then stored at –80°C until use (Verbaan et al., 2007). For this purpose, different techniques for skin conservation can be used, and each lab should evaluate the best conditions for their expected use (Praça et al., 2018).

Before the experiments, it is necessary to verify the integrity of the skin samples with a skin barrier integrity test, as well as the visual verification of the skin in order to avoid the use of damaged skin since the evaluation of the penetration of products applied topically in Franz diffusion cells use intact skin (Davies et al., 2015). Commonly verified parameters include tritiated water permeation, transdermal electrical resistance (TEER) of skin conductivity, and transdermal water loss (TEWL) (Davies et al., 2015; Guth et al., 2015; Zhang et al., 2018).

In the tritiated water test, skin integrity is checked based on skin permeability after the application of a small amount of tritiated water in the diffusion cell donor chamber. For skin conductivity testing, skin electrical resistance is measured by the application of small electrical potential through the skin sample providing a rapid assessment of skin integrity (Davies et al., 2004, 2015). This test measures the ionic transport through the skin and therefore is related to the cutaneous permeation of the polar skin pathway, having a reduced correlation with the lipoid pathway of the skin. Pores that have undergone some physical method such as iontophoresis or ultrasound, or chemical treatment have less electrical resistance than others that did not undergo this rupture (Zhang et al., 2018).

TEWL is a useful method because it is non-invasive, and the instrument used in the test can be used for the evaluation of the skin barrier both *in vitro* and *in vivo* (Davies et al., 2015). Knowing that the lipids found in the skin are a barrier to water permeation, TEWL verifies the permeation of water through the skin and is related to the lipoidal pathway of the skin. This method has the disadvantage that for the evaluation of TEWL, a certain period of stabilization is required in an environment with temperature and humidity controlled around the probe; thus, a high number of Franz diffusion cells is not a practical option (Davies et al., 2015; Zhang et al., 2018).

For the *in vitro* experiment setup, after checking the integrity of the skin, usually Franz diffusion cells are mounted with the skin facing the donor compartment, where the formulations are applied. After the required time for the experiment, the skin should be removed and carefully rinsed to remove excessive formulation or drugs. Then, the skin is usually separated into different layers, aiming for the quantification of the drug (Petrilli et al., 2013, 2018b). The technique of tape stripping is used to determination of the amount of drug in the SC. It consists of the sequential removal of layers of cells from the SC by successive application and removal of adhesive tape (Paleco et al., 2014; Praça et al., 2018). Formulation wastes are removed from the surface of the skin by discarding the first tape (Petrilli et al., 2018b). The pressure and the rate of application must be consistent in order to ensure homogeneity, and the samples are then centrifuged, so the drug is recovered in the supernatant of a selected solvent based on the solubility of the drug (Praça et al., 2018).

Finally, the amount of drug retained in viable EP is quantified by skin homogenization with a tissue homogenizer and immersion in an adequate solvent. An additional step of liquid-liquid partition may be required for some drugs (Petrilli et al., 2018a).

Techniques to measure skin penetration and permeation are essential for topical and transdermal delivery. The quantification of drug that penetrated through the donor chamber in the skin and into the receptor solution can be measured by high-performance liquid chromatography (HPLC), spectrofluorimetric analysis, gas chromatography, mass spectrometry, scintillation counting, among others (Touitou, Meidan, and Horwitz, 1998; Moser et al., 2001; Petrilli et al., 2013, 2018a).

1.2.2 APPARATUS AND MODEL SKIN MEMBRANES FOR IN VITRO STUDIES

Regarding the delivery of the drugs and their penetration into the skin layers, following the application of the substance of therapeutic interest, studies that characterize the permeation of the drug, as well as the diffusion process, are required. *In vitro* tests is an important tool, allowing for to evaluation of phenomena that occur between the application of the product and the pharmacologically measured effect, with the relative simplicity of test execution, control of the temperature variable, and low operational cost (Jug et al., 2018).

In order to evaluate and identify the variables for the design of a formulation that can alter the *in vivo* bioavailability of the drug, the diffusion cell is the experimental apparatus used for this evaluation of skin penetration in topical and transdermal systems (Godin and Touitou, 2007; Cevc and Vierl, 2010). The basic structure of the experimental apparatus consists of a donor compartment, where the drug or delivery system is applied. It must penetrate and/or permeate the membrane in order to achieve the receptor chamber in which aliquots can be withdrawn for drug analysis in solution (Leveque et al., 2004). The method approved by the FDA for permeation and dissolution tests of topical preparations uses the Franz diffusion cell (Cal and Centkowska, 2008).

In the mid-1970s, the first studies using diffusion cells were published by Franz (1975). In 1975 the modified Franz cell was presented as a finite-dose static diffusion cell with skin mounted on a vertical diffusion cell (VDC) and the dermis in contact with a receptor solution, then an amount of the formulation under study was applied to the skin in order to mimic *in vivo* conditions (Franz, 1975; Friend, 1992; Moser et al., 2001) The system has its temperature controlled by a thermostatic bath that surrounds the receiving chamber, and that to ensure homogenization of temperature and solution in the receiver compartment, magnetic stirring can be used. The VDC scheme is shown in Figure 1.2 (Ng et al., 2010).

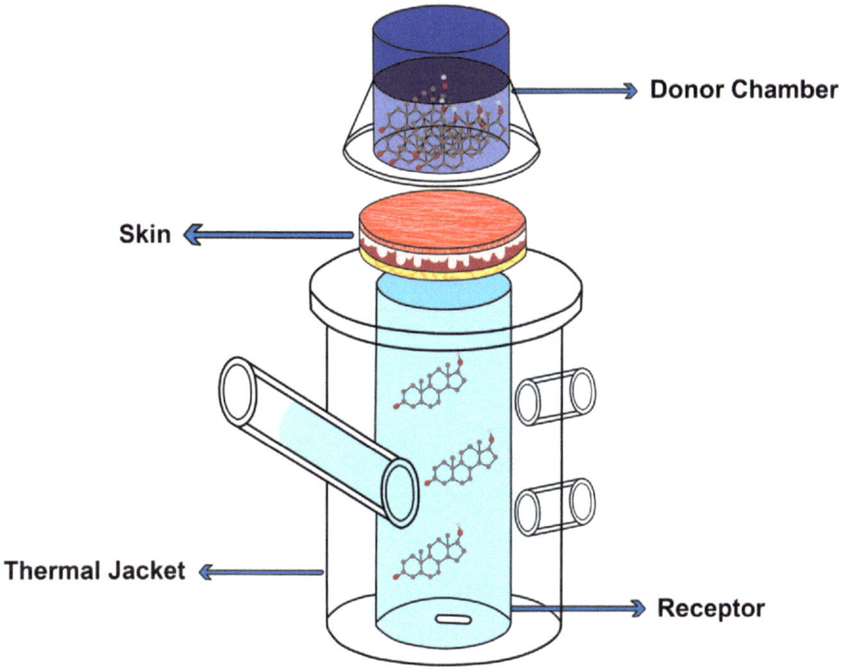

FIGURE 1.2 Apparatus: Franz diffusion cell.

Bronaugh and Steward introduced the continuous flow system, quite similar to the Franz VDC, but with an infusion pump that establishes a continuous flow through the receiver compartment, with the purpose of automating the sampling of cells and thus mimicking the blood flow of the skin, close to reality. In this way, it was possible to monitor the drug release profile by sampling for 24 hours and also to maintain the sink conditions using this system (Bronaugh, 1985).

The methodology applied in the *in vitro* release assay of a drug can be determinant for obtaining results that mimic the *in vivo* effects of the drug release (Praça et al., 2018). In this way, the apparatus to be used and the parameters referring to the applied methodology should be considered. For instance, the composition of the receptor medium, the synthetic membrane used in the *in vitro* release assays, and the agitation speed of the receptor medium should be adjusted according to the study. For the composition of the receptor medium, it should be able to maintain the sink conditions at a temperature of 32 °C (± 0.5), which for hydrophilic drugs can be a buffer at physiological pH, or if the drug is lipophilic, the medium should be hydro-alcoholic or added with surfactants. Thus, when the dissolved drug concentration is less than 10% of the saturation concentration in the release medium, the system operates under sink conditions (Sinko and Gonzales, 2008; Ruela et al., 2014).

Considering the *in vitro* studies of cutaneous permeation and penetration of drugs, human skin obtained from plastic surgeries, biopsies, and human cadavers, is known as the gold standard (Praça et al., 2018). Generally, the skin of the abdominal, breast, or back region is more convenient because there are differences in the absorption of a particular compound in the skin depending on its location in the body, attributed to factors such as the thickness of the SC, hydration, and lipid composition. However, their viability is limited due to the low availability of this type of skin, as well as ethical concerns related to approval of using extensive amounts of skin (Barbero and Frasch, 2009). Other membrane alternatives should be considered in order to replace human skin, among which we can mention animal skin, such as pig, snake, and mouse (Praça et al., 2018), synthetic membranes (Flaten et al., 2015), and skin equivalent or reconstructed EP resulting from three-dimensional cultures (Catarino et al., 2018).

For *in vitro* skin permeation and penetration studies, the porcine skin is recommended because of its similarities with human skin, such as the comparable thickness of the SC of 21–26 μm, the average density of hair follicles, which in the pig ear is 20 cm^2 compared to 14–32 cm^2 in human skin (Wester, 2002; Jacobi et al., 2007). Considering that SC lipids are known for their importance in regulating skin permeability, the organization of lipids in the SC of human skin and pig skin has been analyzed by Fourier transform infrared spectroscopy (FTIR). It was possible to verify marked differences, where the lipids of the porcine SC are predominantly arranged in the form of a hexagonal network, whereas the lipids of the human SC are predominantly packed in a denser orthorhombic network (Caussin et al., 2008). In the human and pig SC, the main classes of lipids are ceramide, cholesterol, and free fatty acid, all of them in an approximately equimolar proportion, but the free fatty acid and ceramide compositions in the two species are different (Bouwstra et al., 1999; Pappas, 2010).

Other species, like the hairless mice, also have similarities with the human skin, with easy handling, small size, and low cost, but its use is limited due to ethical concerns. In contrast, snake's skin is composed of SC devoid of viable EP and hair follicles, which may limit its use. For this membrane model there is no animal sacrifice to obtain the skin because it changes the skin (ecdise) regularly as an adult, every 2 to 3 months (Praça et al., 2018). Other advantages are easy storage and reduced microbiological contamination as they do not contain living tissues (Rigg and Barry, 1990).

Among the disadvantages of the *in vitro* method, the transport of molecules of low solubility in water and especially the absence of living skin metabolism can hamper its comparison to *in vivo* studies. Furthermore, experiments on biological membranes are also subject to alterations, since the desiccated skin presents differences to the living skin, and even the SC, which is a tissue formed by dead cells, can undergo anatomical changes due to degradation that occurs in the viable EP after excision (Plessis et al., 2001).

Considering that biological models such as those described above may have high variability, making statistical significance as well as special storage requirements, high cost, and also a biological risk, it is convenient to consider other membranes for topical and transdermal formulation studies (Barbero and Frasch, 2016). In order to reduce the need for biological membranes in the initial experiments, synthetic or artificial membranes can be used (Haq et al., 2018).

Although synthetic membranes do not model the effects of lipid disturbance that biological samples undergo, conclusions can be made about the partitioning and diffusion phenomena. Synthetic membranes are easier to obtain, and their structure is simpler than the biological membrane, so they may be preferred for large-scale studies as they can be performed with their mechanisms interpreted readily (Ng et al., 2010).

The Food and Drug Administration (FDA, 1997) suggests the application of simple polymer membranes for *in vitro* drugs permeation studies such as cellulose, polysulfone, polyethersulfone, or polydimethylsiloxane, since a suitable synthetic membrane should be inert, does not obstruct drug penetration, and enable high permeability. Thus, depending on the physicochemical characteristics of the synthetic membrane, the Franz diffusion cell assay can only show the release of the drug and not the permeation of the drug through the skin, which limits information on the efficacy of the transdermal method (Simon et al., 2016).

In addition, reconstructed human epidermis kits (RHE) can be used, mimicking the application site of the product, thus allowing the topical application and evaluation of some clinical reactions (Roquel et al., 2000). Other advantages that may be cited are solubility and chemical stability in aqueous solution, which are limiting factors in *in-vitro* methods based on the use of submerged cultures (Gibbs et al., 2013). The cell culture in monolayer and reconstructed EP are alternatives to the biological membranes and the requirements of the regulatory authorities that also provide improvements in the knowledge of the biological processes of the skin (Abd et al., 2016).

In RHE or skin equivalent, epidermal keratinocytes develop in the air-liquid interface on a collagen matrix exposed to the air, generally formed by collagen type I and fibroblast, giving rise to almost normal architecture tissues. These models are interesting approaches for the evaluation of skin corrosion and irritation studies, but they cannot replace the *in vitro* skin penetration studies that will give information about the amount of drug in each skin layer (Catarino et al., 2018). However, the development of RHE within the research laboratory may demonstrate some difficulties during cultivation. Important factors such as cell density, cell viability, enzyme use, cell confluency, culture media, protein extraction, and storage should be controlled (Lemper, De Paepe, and Rogiers, 2014).

1.3 PASSIVE SKIN DELIVERY OF DRUGS

1.3.1 TOPICAL PASSIVE DELIVERY

The skin is being widely studied as an alternative treatment route for diseases, especially for those ones located in specific sites of the skin, such as skin infections, skin cancer, hyperkeratinization, and hyperpigmentation, among others. Some advantages of this treatment route are: it is characterized as a non-invasive treatment, allows for localized action of the drug with the easy application of the drug and consequently leads to better adherence of the patient to the treatment, decreases levels of systemic effects, avoids hepatic metabolism, and may improve the efficacy of the drug, with consequent greater bioavailability. However, drugs generally have low permeability through the skin. Such disadvantage is related to the structure found in the outermost layer of the skin, the SC, a barrier that plays a protective role (Du et al., 2016; Carita et al., 2017).

SC is organized by corneocytes, cells with high keratin concentration, and a lipid matrix that is endowed with unique characteristics, arranged in a single continuous matrix throughout the EP layer, arranged in multilamellar chains. The matrix is composed of ceramides, free fatty acids, and cholesterol. The cells are disposed of in the lipid matrix as a "brick and mortar" arrangement. Due to these characteristics, for a molecule to be permeated through the skin, it must be able to cross the SC, a major barrier for skin drug delivery (Naik et al., 2000; Carita et al., 2017).

Some routes are involved in the penetration/permeation of drugs through the skin layers, including diffusion through the intact EP, and permeation through the EP appendages and hair follicles, and sweat glands. The appendages of the skin are distributed in the epithelial tissue in different locations and generally low indifferent concentrations, occupying 0.1% of the total skin surface, so permeability through these pathways becomes limited. In the case of permeability through the intact EP, we can highlight two routes, the intercellular and the transcellular. The first occurs when molecules cross the EP between the corneocytes, and the second occurs by permeation crossing the existing corneocytes and lipids (Moser et al., 2001; Lane et al., 2013).

Some characteristics must be considered for adequate penetration through the skin, which has been previously reviewed (Baroli, 2010). Among these factors, the most relevant are:

1. **Localization and Skin Conditions:** The distribution of follicles in the extension of epithelial tissue is not uniform; some molecules use these pathways to penetrate SC; some conditions of the individual, such as age, skin conditions, gender, skin photo type, skin hydration condition, may interfere in the lipid composition of SC, acting on the thickness of the epidermal layer, affecting the distribution and concentration of the active substances in the skin layers;

2. **Physicochemical Characteristics of the Molecule:** Parameters such as drug solubility in the vehicle, its pH, pKa, and diffusion factors are relevant. Another important factor is the molecule's oil/water partition coefficient, as the more lipophilic the drug, the easier the permeation in SC. Thus a hydrophilic molecule will have greater penetration difficulties. Compared to nanoparticles, the size of the skin permeation agent is significantly important, so particles smaller than 600 Da have better skin penetrability. However, this does not mean that larger nanomaterials are unable to obtain good penetration. Some characteristics of nanoparticle skin permeators are their dimensions, their ability to penetrate the epithelial layers, the route that will be used for skin permeation and their diffusion coefficient in both vehicle and skin;
3. **Physicochemical Properties of Dispersion Vehicle:** Drugs intended for topical purposes are dispersed in a formulation base, so that the final formulation will result from mixing between the excipients and the active substance. Therefore, all formulation components will come into contact with SC and, different forms will diffuse through the skin layers. High viscosity formulations may decrease the diffusion of the active ingredients in the vehicle, resulting in less penetration into the skin layers; volatile components will evaporate rapidly, nonvolatile components may diffuse through the layers, and formulation components will be mixed with compounds present on the skin, such as sweat, lipid substances. The composition of the final formulation is of great importance as the presence of key substances (solvents, surfactants, absorption promoters) will be able to alter the SC in order to increase the penetration capacity of the drug. Furthermore, when the formulations are very lipophilic, they may face competition with the lipophilicity of SC, negatively influencing the partition of molecules. In contrast, formulations that have an occlusive character may favor the increase in skin absorption.

Skin absorption promoters are widely used to improve drug release by decreasing SC lipid resistance, and some may be used to increase drug concentration in the skin layers. Such absorption promoters should be endowed with good tolerability and without pharmacological activity. They are divided into groups based on their chemistry, structure, and mechanism of action (Table 1.1) (Lane, 2013; Carvajal-Vidal et al., 2017).

TABLE 1.1 Classification and Description of Some Main Absorption Promoters (Lane, 2013)

Classification	Absorption Promoters/Mechanism	Examples
Alcohols	Short-chain: Application as co-solvent and enhancer for cutaneous penetration. Long chains: The permeation of the active is related to the chain size of saturated fatty alcohols Able to modify the function of the skin barrier, such as by modification of lipid order, influence of fluidity of lipids, effects on keratin.	Ethanol and saturated fatty alcohols
Amides	Greater interaction with the lipids present in the skin, increasing the penetration through a greater fluidity of the structure of the skin	Azone
Esters	Ability to interact with the lipid barrier to increase lipid fluidity by means of the negative variation of the lipid transition temperature	Isopropyl myristate
Glycols	Used as a co-solvent in preparations with poorly soluble substances and/or as promoters in increasing the penetration of topical formulations. Mechanism has not yet been fully elucidated	Propylene glycol
fatty acids	Increases the penetration of active compounds by disordering the lipids in the stratum corneum	Oleic acid
Sulfoxides	Co-solvent and facilitator of skin penetration Some are elucidated, such as displacement of keratin water binding, lipid extraction, conformation changes in keratin, interaction with alkyl lipid chains in the stratum corneum	Dimethylsulfoxide
Surfactants	Promoters damage and irritation in the stratum corneum	Sodium lauryl sulfate

1.3.2 TRANSDERMAL PASSIVE DELIVERY

In transdermal delivery, the drug crosses the SC layer by a passive diffusion process that passes through the EP and the dermis by permeation until it is absorbed into the bloodstream (Lai-Cheong and McGrath, 2017). Thus the transdermal system can be described as the system that facilitates the passage of the drug through the skin in order to exert systemic effects because it reaches the blood circulation, providing some advantages over other routes of administration: there are no variations of drug molecules due to gastric pH during gastrointestinal absorption; no interactions with food; reduction in the susceptibility to enzymatic inactivation due to the effect of first-pass metabolism; the prolonged treatment duration in a single application; therapy easily interrupted by simple removal of the transdermal device; lower doses may be administered due to increased bioavailability; maintenance of plasma concentration and lower patient rejection due to easy administration (Allen, Popovich, and Ansel, 2007; Alexander et al., 2012).

Although products have been applied to the skin for centuries, in the relatively short time of around 40 years, in 1979, the Food and Drug Administration (FDA) approved and released the first generation of skin application drugs for the treatment of nausea, vomiting, pain, hormones, and angina in the form of adhesives such as scopolamine, clonidine, testosterone, estradiol, nitroglycerin, fentanyl, and nicotine, the latter being used in the treatment of smoking, which stimulated the development of this technology for the release of transdermal drugs (Allen, Popovich, and Ansel, 2007; Margetts and Sawyer, 2007).

Substances suitable for transdermal administration should have some characteristics: low molecular, usually lower than 1 kDa and adequate hydrophilicity (Münch, Wohlrab, and Neubert, 2017). Good candidate drugs for the transdermal route are potent, non-irritating, have partition coefficients up to three and with short half-life times not undergoing metabolism in the skin. Thus, a condition similar to continuous infusion as a function of slow permeation is generated (Rang and Dale, 2001; Gilman, 2010).

Some of the disadvantages of the transdermal system are the possibility of irritation at the site of administration, the time interval between the desired drug concentration in the plasma and the administration and also that hydrophilic drugs undergo passive diffusion obstruction. The variability among individuals (hydration and skin pH, skin age, area of application, hair distribution, SC integrity) may modify of permeation and metabolism of drugs (Alexander et al., 2012).

There are three distinct types of transdermal systems, also known as patches: the reservoir system, the adhesive system, and the matrix system. The first one has this nomenclature because there is a reservoir with the drug and the release with a controlled rate between the skin and the reservoir occurs through a membrane. The reservoir-like systems have more rigid control over drug release rates, but if membrane rupture occurs, an excessive and abrupt release of drug into the skin may result in over-dosing of the drug. The second has the drug dispersed in adhesive polymers, so this adhesive part allows for greater adhesion to the skin, control of the drug partition in the SC as well as drug storage. The third system incorporates the drug into a polymer matrix, controlling the release of the drug into the skin that is evenly distributed throughout the system, thus avoiding risks of overdosage (Margetts and Sawyer, 2007).

The organization of the transdermal adhesive systems can be divided into some main parts. The coating layer responsible for the storage protection is removed and discarded prior to the application of the adhesive. Pressure-sensitive adhesive that must have the ability to mold under pressure and leave no residue on the skin when removed, maintaining contact between the skin and the transdermal system. The polymer matrix or membrane is the part responsible for controlling the release of the drug, and requires physical, chemical, and biological compatibility. In the system reservoir, the drug is in an inert polymeric matrix or suspended in a viscous liquid medium. Usually, a permeation enhancer is used as a way to increase the skin permeability to the drug. Surrounding the reservoir there is an impermeable membrane and a polymeric membrane responsible for controlling the rate of drug release. This can be microporous or non-porous in nature and is chosen according to the drug. In matrix system, the semisolid matrix contains a drug suspension or solution in the adhesive (Banerjee et al., 2014). The support layer should be chemically inert, in addition to being flexible and occlusive (Kandavilli, Nair, and Panchagnula, 2002; Arunachalam et al., 2010).

Substances applied to the skin must transpose the structure of the SC, from the viable cells of the EP, through the dermis using passive diffusion to reach the bloodstream. The slowest diffusion stage occurs in the SC and is considered the limiting factor for the diffusion of a substance.

Mathematically, the absorption of drugs through the skin can be described by the Fick diffusion second law. In literature, this equation is frequently used to describe the mass and heat diffusion in transient regime. After solved, it is possible to describe how mass and heat are transported through the layer of SC, for example, as a function of position and time (Welty et al., 2008; Bird, Stewart, and Lightfoot, 2002). Considering a very large well-stirred donor phase that is applied in SC and the control surface is represented as a homogeneous membrane (SC), transport of drug is described below (Lane et al., 2012; Jepps et al., 2013):

$$\frac{\partial C(x,t)}{\partial t} = D_{SC} \frac{\partial^2 C(x,t)}{\partial x^2} \tag{1}$$

with the initial and boundary conditions:

$$C(x, 0), C(0, t)\, K_{SC} C_d, C(h_{SC}, t) = 0 \tag{2}$$

where; D_{SC} is the diffusion coefficient; C_d is the concentration in the donor phase; K_{SC} is the partitioning coefficient; $C(x, t)$ is the concentration at depth x and time t; and h_{SC} is the thickness of the SC. Once solved Eqns. (1) and (2) resulted in the amount of solute penetrating in SC (Q) versus time:

$$Q(t) = K_{SC} A C_d h_{SC} \left(\frac{D_{SC}}{h_{SC}^2} t - \frac{1}{6} - \frac{2}{\pi} \sum_{n=1}^{\infty} \frac{(-1)^n}{n^2} exp\left(-t \frac{D_{SC}}{h_{SC}^2} \pi^2 n^2 \right) \right) \tag{3}$$

where; A is the area of application of solute in SC. For systems with long time or $t > h_{SC}^2 / (2D)$, the summation term can be ignored in Eqn. (3) and resulted in the most commonly equation for experimental data in literature for transdermal systems:

$$Q(t) = \frac{K_{SC} A C_d D_{SC}}{h_{SC}} \left(t - \frac{h_{SC}^2}{6 D_{SC}} \right) = k_p A C_d \left(t - t_{lag} \right) \tag{4}$$

where;

$$k_p = \frac{K_{SC} D_{SC}}{h_{SC}} \text{ and } t_{lag} = \frac{h_{SC}^2}{6 D_{SC}} \tag{5}$$

Despite, simple Eqn. (4) is widely used for the solution to diffusion through a homogeneous membrane even for sophisticated SC models (Jepps et al., 2013).

1.4 PHYSICAL METHODS FOR TOPICAL AND TRANSDERMAL DELIVERY

1.4.1 MICRONEEDLES

Over the years, the preference for transdermal delivery of drugs has increased. However, due to the barrier that hampers the penetration of drugs, mainly large molecules present reduced permeability in the whole skin. In this context, physical methods, such as microneedles, can create pores in the skin or temporary pathways that facilitate drug application, with minimal risk of infection compared to conventional injections (Henry et al., 1998; Petrilli and Lopez, 2018).

In addition, the application of microneedles does not need healing and presents less pain than a hypodermic needle. The shape, size, and thickness of the needles change according to the intended application. They usually have a length between 150 and 1,500 μm and diameter varying from 50 to 250 μm (Gill et al., 2008; de Lima, de Souza, and Grignoli, 2015; Waghule et al., 2019).

There are several types of microneedle arrays, produced using biodegradable polymers, silicones, metals, glass, and ceramic with structures designed to pierce the SC and other epithelial tissues. The needles are tested to penetrate the irregularities of the skin surface and break their natural elasticity. In the development of these matrices, several parameters are classified for application, including manufacturing material, applications, and design, making it necessary to know the effects of each factor to achieve the desired result (Fonseca et al., 2019).

There are four groups of microneedles according to their applications. Solid microneedles are used to create micropores in the SC and in the viable EP, which enables the subsequent release of drugs through these pores. The most common model consists of a roller with a rotating drum, which contains the microneedles fixed to its surface. They can be prepared with dried coatings to apply encapsulated polymers in *in-vitro* and *in vivo* tests for delivery of insulin, proteins, and other substances (Banks et al., 2008; Zhou et al., 2010).

The coated microneedles are composed of inert materials and are coated with formulations of different drugs. After the pressure exerted on the SC for application of this device, the material incorporated into the needles is released into the skin and they are withdrawn intact after a certain period (Elbashir et al., 2001; Zhang et al., 2012; Chong et al., 2013; Reis et al., 2014).

The dissolvable or biodegradable microneedles are impregnated with a formulation containing a drug or biopharmaceutical. The microneedles, after insertion, remain in the skin until they are discharged through the interstitial fluids, and their removal is not necessary. Thus, there is very great similarity between such devices and the conventional transdermal patches, as to the mode of use. This mechanism has many advantages, one of which is the risk of infections and local irritations in cases of incomplete breaks and withdrawals (Sullivan et al., 2010; Kommareddy et al., 2012).

Finally, the fourth type of microneedles is hollow, very similar to the traditional hypodermic needles. However, they have a great advantage of the use in micrometric scale being a painless system. Another advantage is it does not require a specialized professional for its application, nor aseptic conditions for that also it avoids the risk of inflammation and microbiological contamination at the device insertion site (Roxhed, Griss, and Stemme, 2008; Wang et al., 2009; Ling and Chen, 2013).

The use of these microneedles has been quite usual in the application of insulin in diabetic patients. They become more effective by leaving exposed pores for less time, in addition to having a low cost of production, and providing more effective immunity. Davis (2005) developed hollow microneedles strong enough to pierce the skin containing 16 microneedles of 500 μm long and 75 μm diameter tested in diabetic and hairless mice. The results obtained by manipulating insulin over a 4-hour period were from blood glucose which dropped to 47%, demonstrating the efficacy of hollow microneedles for *in vivo* insulin administration (Davis et al., 2005; Prausnitz et al., 2009).

Adhesives with dissolvable polymeric microneedles have been widely studied for influenza vaccination (Figure 1.3). Sullivan (2010) evaluated the use of dissolving microneedles fabricated using a biocompatible polymer for insertion and dissolution in the skin in minutes. The study used as a comparative hypodermic injection, both in mice. The results obtained showed a similar immune response between the two mechanisms of drug transport, however, the dissolving microneedles demonstrated an increased production of IL-4 and IFN-γ (Sullivan et al., 2010). In a study developed by Weldon (2011), the delivery of the recombinant H1N1 influenza subunit by coated microneedles was evaluated. The tests were done in mice and the results obtained showed that the microneedles coated with the antigen-induced a systemic hormonal response determined by the ELISA test. Compared to conventional subcutaneous injections, the microscopic needles presented higher efficiency in IgG production. Furthermore, microneedle-immunized mice completely cleared the virus from the lung (Weldon et al., 2011).

Another application with microneedles is in photodynamic therapy (PDT). Briefly, PDT is a non-invasive technique used against bacteria, viruses, fungi, protozoa, and in the treatment of skin cancer lesions or treatment. PDT has similar or superior efficacy to conventional treatments with reduced side effects as it is easy to apply and requires a short recovery time. In the literature, there are already studies of conventional PDT and it is associated with the use of microneedles in the treatment of cancer cells

(Donnelly et al., 2008). For example, Mikolajewska (2010) and collaborators initially evaluated the effectiveness of silicon micro-needles for PDT using rodents; however, the thickness and structure of the SC and hair follicles in the skin of the mouse are different from human skin. Therefore, to gain clinical relevance, the study was performed on 14 healthy volunteers pre-tested with microneedles. The developed creams containing 2%, 8%, and 16% (w/w) of ALA and MAL were applied in areas of 1 cm^2 4 h. In addition, 16% of ALA and MAL creams were applied for 24 h. In the results obtained, the microneedles increased the fluorescence but did not increase the sensations of pain during the exposure to light, nor did it influence the occurrence of erythema (Mikolajewska et al., 2010).

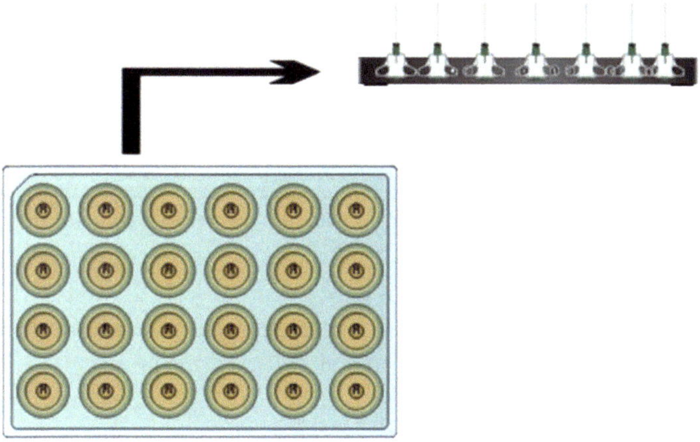

FIGURE 1.3 Microneedle adhesive for the immunization process.

1.4.2 IONTOPHORESIS

In addition to other physical methods, iontophoresis is an efficient way to deliver active substances to localized areas of the skin. Iontophoresis is a non-invasive technique that aims to increase the permeation of drugs through the skin layers to reach the bloodstream. An advantage of this technique is that the rate of drug release depends on the intensity of the current applied and the duration of treatment. The methodology used is the application of low-intensity electric current that assists in the transport of drug molecules through both electroosmosis and electromigration mechanisms (Petrilli and Lopez, 2018).

Electromigration is characterized as the orderly movement of ions that is related to the electric current applied to the charged medium, that is, when the drug is in contact with the electrode of the same polarity, the drug is repelled and the drug is transported through the skin. The flow of each ion is related to the electrical mobility and concentration of high mobility competing molecules in the system (Gratieri et al., 2008; Petrilli and Lopez, 2018). Electroosmosis is characterized by the flow of a solvent from the anode to the cathode, due to the application of electric current to the skin, through which water-soluble neutral molecules diffuse into the skin. In the presence of an electric current, the skin favors the transport of Na^+ over Cl^- (Figure 1.4). The isoelectric point of the skin is between 4.0 and 4.5 when in contact with a solution with physiological pH above its isoelectric point. The transport of cations is favored and makes anions transport more difficult (Kalluri and Banga, 2011; Gratieri et al., 2008). For instance, the effect of the pH on the iontophoretic delivery of 5-fluorouracil (5-FU) was studied by Merino et al. (1999). Basically, 5-FU is a hydrophilic drug with pKa~8 and 5-FU flux was evaluated upon iontophoresis and passive diffusion at different pH. It was observed that at pH 8.5, the drug is negatively charged and thus, the electromigration due to cathodal iontophoresis increases significantly the flux of 5-FU compared to anodal iontophoresis. As expected, at physiological pH 7.4, the amount of drug ionized is near 25% and then both cathodal and anodal iontophoresis result in similar flux. On the contrary, at pH 5, the drug is mostly uncharged resulting in an increase in the contribution of electroosmosis flux. This example illustrates the importance of evaluating drug charges for iontophoresis (Merino et al., 1999). Thus, the pH of the formulation affects the degree of ionization of the drug, therefore, the iontophoretic flow of the formulation must be placed in the compartment near the electrode of equal polarity (Petrilli and Lopez, 2018).

The formulation of the drug is of great importance for the efficiency of the technique. There is a correlation with the iontophoretic flux and the concentration of the drug penetrated into the skin, degree of skin hydration and SC integrity (Gratieri et al., 2008). For the formulation to remain at the site of application may be semi-solid or liquid. If a polymer is used to obtain a gel, it must be compatible with the drug so that it does not interfere with its release. Diffusion of the drug across skin membranes is ensured by increased drug availability. Therefore, the electrical charge of the drug should be taken into account when incorporated into the gel (Huang et al., 2005).

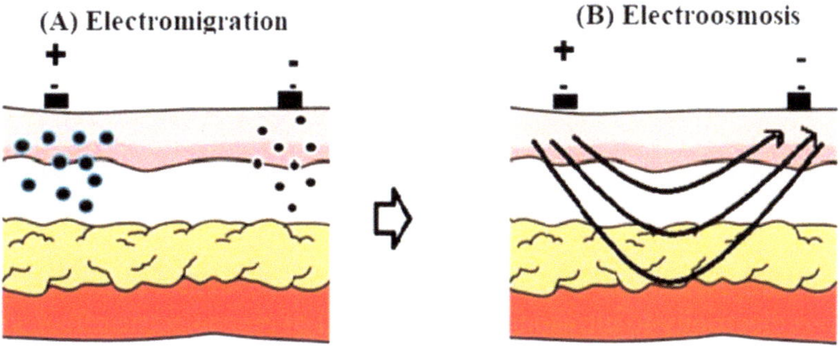

FIGURE 1.4 Illustrative scheme of electromigration and electroosmosis.

Huang (2005) and his collaborators evaluated the effects of iontophoresis and electroporation on transdermal drug release. The hydrogel where the iontophoresis formulation was composed of hydroxypropyl cellulose (HPC) or carboxymethylcellulose (CMC), evaluated for the release of nalbuphine and its prodrugs. Iontophoresis improved drug delivery; however, hydrogels containing low HPC concentration did not affect drug permeation. However, with higher hydrogel concentrations and higher molecular weight of the drug significantly decreased permeation, which can be related to viscosity and ions competition. Finally, the authors concluded that molecular weight and hydrogel compositions, have significant effects on skin permeation by passive diffusion or under iontophoresis (Huang et al., 2005).

The main experimental parameters that guide iontophoresis were revised by Kalia (2004) and collaborators. The authors compiled a series of papers to evaluate the influence of drug concentration, applied current and pH in iontophoretic delivery efficiency, as well as different therapeutic classes and molecules such as peptides and proteins that were investigated as potential for iontophoretic delivery. The study concluded that iontophoresis provides a clear therapeutic benefit over the existing delivery methods in addition to its efficacy in adverse effects. Thus, the main advantage of iontophoresis is the controlled drug entry kinetics it offers, and the ability of patient-optimized adverse effects profile (Kalia et al., 2004).

1.4.3 SONOPHORESIS

Sonophoresis, or phonophoresis, is a powerful technique based on the drug permeation enhancement through the skin under ultrasound irradiation.

As a brief description, sound waves can passes through different media with pressure cycles of compression (high pressure) and rarefaction (low pressure), which its propagation, i.e., speed, depends on the nature of the medium. Sound waves in a frequency >20 kHz are well-known as ultrasound since they are beyond human hearing, which are normally generated by ceramic materials with piezoelectric properties, which are no further covered in this chapter (Ashokkumar, 2018).

Basically, a specific ceramic material, well-known as a transducer, can convert electricity to ultrasound waves by aid of piezoelectric effect, since the ceramic expands or contracts when an alternating voltage is applied through the transducer. This behavior provides a continuous ultrasound irradiation at the same frequency. It is also possible to provide pulsed ultrasound, i.e., "pulse *on*" and "pulse *off*," where time-controlling bursts of alternating voltage is applied through the transducer, generating cycles of ultrasound pulse at the same frequency (Ashokkumar and Cavalieri, 2016; Ashokkumar, 2018).

Ultrasound propagation though the medium is the main responsible factor for all versatilities and innumerous applications of ultrasound irradiation, including skin permeation enhancement and materials synthesis, which is well-called the acoustic cavitation phenomenon. "Cavitation" phenomenon literally means creation of cavities in a liquid, the acoustic cavitation, i.e., under ultrasonic frequency, is related to the growth of bubbles nuclei and their further collapse. That growth is normally generated since all liquids used in lab experiments contain dissolved gas molecules, and also due to negative and positive pressures of the sound waves propagation in the liquid, well-known as "Rectified Diffusion" process, which is deeper described by Ashokkumar (2018). As bubbles surface area expands and compresses, the amount of gas/vapor *in* and *out* of the bubbles also changes, where the diffusion into the bubble overcome the diffusion out, resulting in an interesting and specific ratio between ultrasonic frequency (F, Hz) and bubble radius (R, m), as seen in Eqn. (5).

$$F \times R \sim 3 \tag{5}$$

As the ultrasound frequency decreases, the radius of the cavitation bubbles tends to grow, resulting in the inertial collapse, and then generating extreme high temperatures, from 1,000 to 5,000 K, and pressures around 100 atm. Such extreme environments and physical phenomena are responsible for ultrasound irradiation versatility, that has been used, for example, in drug skin permeation enhancement, well-called sonophoresis, which is described in the next section (Meidan and Michniak-Kohn, 2010)

As briefly discussed, ultrasound waves propagate through a medium, inducing a compression and expansion area. For all different media, their resistance to propagation depends on the acoustic impedance (Z) (Eqn. (6)).

$$Z = \rho \times C \tag{6}$$

where; ρ is the mass density of the medium; and C is the ultrasound speed propagation. For instance, for high acoustic impedance values, ultrasound waves do not propagate well. Herein, for skin ultrasound-applied method, a coupling liquid is required, which can effectively transfer the ultrasound irradiation from the transducer to the skin surface. Generally, coupling liquids are emulsions, gel, or aqueous solutions, or even oils, and these media act as "smart" drug delivery systems once they provide both: (i) topical or transdermal drug delivery; and (ii) efficient US propagation interface (Meidan and Michniak-Kohn, 2010).

Once in a biological environment, ultrasound irradiation now leads to body tissue, and some physical effects may be evidenced if the released energy was absorbed by the skin. Hence, for each body tissue is established an absorption coefficient (a), as shown in the following equation.

$$I(x) = I_0 \times e^{-ax} \tag{7}$$

where; considering perpendicular US incidence, $I(x)$ is the intensity at depth x, I_0 is the intensity at the surface, and a is the absorption coefficient (Meidan and Michniak-Kohn, 2010).

From the interaction between ultrasound waves and body tissues arises two "new" phenomena, once cavitation was already discussed in a previous section: heating and microstreaming. Under *in vivo* environment, temperature must be a controlling parameter since any temperature variation might lead to a cell disruption, i.e., any absorption of the US released energy will increase the temperature of the exposed tissue. According to Eqn. (7), tissues with high absorption coefficient resist higher and longer heating periods in comparison to low ones, such as bone and skin, respectively. In this case, as absorption coefficient directly increases with ultrasonic frequency, fixing frequency regime, the tissue heating directly varies with acoustic intensity and time regime (Polat, Blankschtein, and Langer, 2010; Petrilli and Lopez, 2018).

In biological tissues, it is also observed the microstreaming phenomenon that occurs in liquid and semiliquid media mainly due to acoustic impedance variation between adjacent sonicated tissues, and also under continuous US irradiation. The microstreaming itself consists of acoustically driven unidirectional current flow that is produced by ultrasonic reflections and other distortions, and is a cavitation secondary phenomenon (Polat, Blankschtein, and Langer, 2010).

During 50 decades, the sonophoresis was studied and also improved in order to be applied as an effective physical method for drug absorption through skin barriers. However, the great success of sonophoresis was just proved at 90s, when scientists reported that low-frequency applied ultrasound (20–500 kHz) effectively enhances drugs transdermal permeation and absorption (Meidan and Michniak-Kohn, 2010; Petrilli and Lopez, 2018).

In topical applications, the coupling medium is placed on the skin, and the skin permeability changes due to the formation of low and high pressures areas under US irradiation, where frequency and amplitude are key parameters. Amplitude refers to the ultrasound horn displacement during each half cycle, and frequency is related to the number of times that the tip is moved per second (Petrilli and Lopez, 2018).

On the skin, the expected mechanism of ultrasound cavitation is that bubbles oscillate and collapse, which generates localized shock waves and liquid microjets directed at the SC (Figure 1.5) (Azagury et al., 2014). From the Eqn. (5), it is possible to observe why cavitation mechanism is responsible for increasing drug permeability. Once bubbles size is inversely proportional to frequency-applied US, low frequencies provide larger bubbles in the coupling medium. For instance, Polat (2011) reported too large bubbles with a radius around 150 μm under US irradiation between 20 and 100 kHz, resulting in no penetrability by cavitation-formed bubbles through the SC, and causing a more heterogeneity in the skin disorganization. The cavitation indeed plays as a predominant mechanism for enhancing drug skin permeation, since it facilitates the penetration of even macromolecules mainly by localized transport regions (LTRs) (Polat et al., 2011).

Mathematic models propose that during acoustic cavitation, large bubbles continue to enlarge, and this results in channels though SC, forming small routes for transportation of macromolecules or even drug nanocarriers. The cavitation indeed plays as a predominant mechanism for enhancing drug skin permeation, since it facilitates the penetration of even macromolecules. However, other mechanisms also contribute to enhancing skin permeabilization such as heating, as already reported (Meidan and Michniak-Kohn, 2010; Polat et al., 2011).

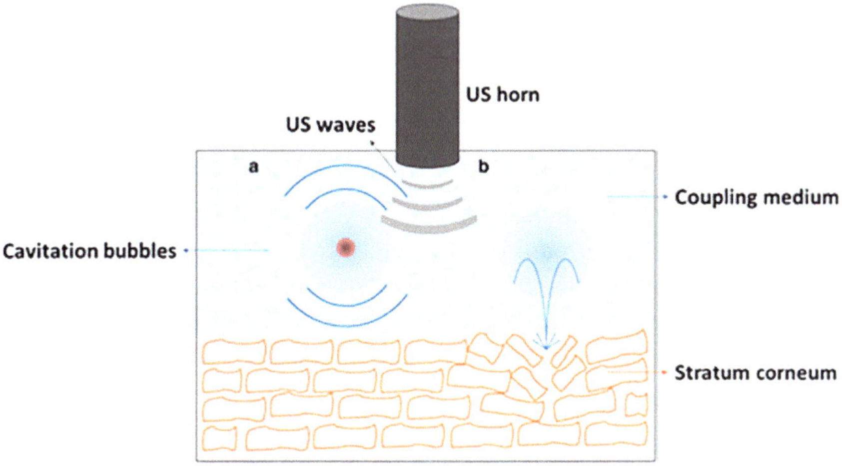

FIGURE 1.5 Cavitation mechanism at low frequency ultrasound and further inertial bubbles collapse – (a) shock waves; and (b) microjets.

Source: Reprinted with permission from: Azagury et al. (2014). © Elsevier.

The first sonophoresis device was FDA-approved only in 2004, used in the transdermal delivery of lidocaine. Much effort has been taken in the improvement of ultrasound devices and also accomplish with other enhancement methods for transdermal drug delivery. Most of the researches use simultaneous treatment or pretreatment, where the US irradiation is applied simultaneously or previously to drug application, respectively. The pretreatment method has shown great success over simultaneous one, since it kept the skin permeability for some hours (Azagury et al., 2014; Petrilli and Lopez, 2018). In pretreated skin the drug penetration occurs through the well-called LTRs, in which the molecules can permeate though hydrophilic porous region (Polat et al., 2011; Petrilli and Lopez, 2018). However, most promise route involves the synergistic effect between low-frequency ultrasound and other enhancer methods or chemicals, such as perpendicular US-applied waves and surfactants in the coupling medium on the skin (Meidan and Michniak-Kohn, 2010; Azagury et al., 2014), and also combining them with other physical methods such as microneedles and iontophoresis (Petrilli and Lopez, 2018).

Besides sonophoresis has been studied over decades, a well understanding about ultrasound and its bioeffects is missing, as well as the chemical effect of acoustic cavitation in the drug molecules, and it can be expected that in the near future, replacement of needle vaccines and immunization might be considered with skin permeabilization by ultrasound.

1.5 CLINICAL APPLICATION

Topical and transdermal drugs available on the market have limited numbers because they require correct physical and chemical properties and pharmacokinetics (PK) capable of effectively delivering the drug to its therapeutic target. Clinical studies have been conducted for decades and are in various stages of study. There are examples of studies for various diseases in different phases of the clinical study, as presented in Table 1.2.

TABLE 1.2 Clinical Trials for Topical and Transdermal Applications

Description	Delivery System	Disease	Phase	References
Transdermal continuous oxygen therapy	Transdermal	Anorectal surgeries	2	https://clinicaltrials.gov/ct2/show/NCT02617706#studydesc
Transdermal testosterone nanoemulsion	Transdermal	Menopause	2	https://clinicaltrials.gov/ct2/show/NCT02445716
Transdermal estrogen in patch	Transdermal	Anorexia nervosa Osteoporosis	2	https://clinicaltrials.gov/ct2/show/NCT02475265
Topical steroid ointment	Topical	Actinic keratoses	4	https://clinicaltrials.gov/ct2/show/NCT03279328
Topical Centella Asiatica ointment	Topical	Xerosis cutis diabetes mellitus, type 2	4	https://clinicaltrials.gov/ct2/show/NCT03815305
Topical cyanoacrylate adhesive	Topical	Wounds	1	https://clinicaltrials.gov/ct2/show/NCT01514383

Thus, drug delivery systems have been extensively researched and developed for better delivery of compounds of therapeutic interest leading to novel systems that make use of micro/nanoparticles, adhesives, gels, and solutions as well as combinations with physical methods such as iontophoresis. These systems are effective and already use nanotechnology to expand results as presented by Botelho and collaborators when using a nanostructured formulation combined with the transdermal system can verify effective restoration of estradiol and follicle-stimulating hormone levels and consequent relief of menopausal symptoms (Botelho et al., 2014).

Also, studied devices can promote the healing of sickle ulcers, chronic wounds, and cutaneous necrosis through continuous transdermal oxygen therapy by the mechanism of oxygen supply of the air by a thin flexible tube that is in the transdermal system in contact with the surface of the wound, facilitating the treatment of the wound for the patient and the doctor (Zhang and Barralet, 2017).

Although increased research activity in this field shows a proliferation of such drug delivery systems there must be a corresponding clinical and commercial relationship for the future of this field of interest. In Table 1.3, it is possible to observe the more recent formulations approved by the FDA. Herein we highlight the plurality of compounds used in the elaboration of new topical and transdermal drugs, showing that there is still much to be developed in the field of administration of drugs from topical and transdermal.

Currently, a few devices are available in the market aiming at the application of physical methods. For instance, LidoSite® Topical System from Vyteris is a device for iontophoresis that uses both lidocaine and epinephrine dermal analgesia. Other devices available are and Iogel® from Iomed Inc and Ionsys®. For sonophoresis, some examples of devices are PainShield®, WoundShield TM, UroShield TM and NG-Shield TM. Microneedles are currently available for vaccination, such as Micronjet® (Nanopass Technology Ltd.) (Petrilli and Lopez, 2018).

1.6 CONCLUSION

Skin structure acts as a protection layer limiting the contact of the internal organs and the environment. However, usually topical and transdermal delivery of drugs are desired as they can provide drug delivery directly to the target site, such as deep skin layers or prevent the drug from bypassing the hepatic metabolism. In this context, it is possible to consider both

TABLE 1.3 Examples of Topical and Transdermal Drug Products Approved by FDA (2019)

Proprietary Name	Active Ingredient	Strength	Route	Applicant Holder	Approved Date
BUTRANS	Buprenorphine	Film, extended-release	Transdermal	Purdue Pharma LP	Jun. 30, 2010
SANCUSO	Granisetron	Film, extended-release	Transdermal	Kyowa Kirin Inc.	Sep. 12, 2008
DAYTRANA	Methylphenidate	Film, extended-release	Transdermal	Noven Pharmaceuticals Inc.	Apr. 6, 2006
EVAMIST	Estradiol	Spray	Transdermal	Perrigo Pharma International DAC	Jul. 27, 2007
ACZONE	Dapsone	Gel	Topical	Allergan Inc.	Jul. 7, 2005
FLECTOR	Diclofenac epolamine	Patch	Topical	Institut Biochimique SA	Jan. 31, 2007
JUBLIA	Efinaconazole	Solution	Topical	Bausch Health Americas Inc.	Jun. 6, 2014
PICATO	Ingenol mebutate	Gel	Topical	Leo Laboratories Ltd.	Jan. 23, 2012

the passive and active delivery methods described in this chapter. Also, both passive and physical methods can be used for topical or transdermal delivery of drugs. Models for predicting the effect of drug administration in or through the skin were developed over the years and were summarized in this chapter. Finally, clinical application is already a reality, but just a few devices are currently available for physical methods application. It is expected that the growing research interest in the field will introduce new products to the market over the next few years.

KEYWORDS

- **drug delivery**
- **high-performance liquid chromatography**
- **reconstructed human epidermis**
- **transdermal electrical resistance**
- **transdermal water loss**
- **vertical diffusion cell**

REFERENCES

Abd, E.; Yousef, S. A.; Pastore, M. N.; Telaprolu, K.; Mohammed, Y. H.; Namjoshi, S.; Grice, J. E.; Roberts, M. S. Skin models for the testing of transdermal drugs. *Clinical Pharmacology: Advances and Applications*, **2016**, *8*, 163–176.

Alexander, A.; Dwivedi, S.; Ajazuddin; Giri, T. K.; Saraf, S.; Saraf, S.; Tripathi, D. K. Approaches for breaking the barriers of drug permeation through transdermal drug delivery. *Journal of Controlled Release*, **2012**, *164*(1), 26–40.

Allen, L. V.; Popovich, N. G.; Ansel, H. C. *Formas Farmacêuticas e Sistemas de Liberação de Fármacos [Pharmaceutical Forms and Drug Delivery Systems]* (8th edn.); Artmed, Ed.; Porto Alegre, **2007**.

Arunachalam, A.; Karthikeyan, M.; Kumar, D. V.; Prathap, M.; Sethuraman, S.; Ashutoshkumar, S. Review article current pharma research transdermal drug delivery system : A review. *Current Pharma Research*, **2010**, *1*(1).

Ashokkumar, M. Introductory text to sonochemistry. *ChemTexts*, **2018**, *4*(2), 7.

Ashokkumar, M.; Cavalieri, F. *Handbook of Ultrasonics and Sonochemistry*. Springer: Singapore, **2016**.

Azagury, A.; Khoury, L.; Enden, G.; Kost, J. Ultrasound mediated transdermal drug delivery. *Advanced Drug Delivery Reviews*, **2014**, *72*, 127–143.

Banerjee, S.; Chattopadhyay, P.; Ghosh, A.; Datta, P.; Veer, V. Aspect of adhesives in transdermal drug delivery systems. *International Journal of Adhesion and Adhesives*, **2014**, *50*, 70–84.

Banks, S. L.; Pinninti, R. R.; Gill, H. S.; Crooks, P. A.; Prausnitz, M. R.; Stinchcomb, A. L. Flux across of microneedle-treated skin is increased by increasing charge of naltrexone and naltrexol *in vitro*. *Pharmaceutical Research*, **2008**, *25*(7), 1677–1685.

Barbero, A. M.; Frasch, H. F. Effect of frozen human epidermis storage duration and cryoprotectant on barrier function using two model compounds. *Skin Pharmacology and Physiology*, **2016**, *29*(1), 31–40.

Barbero, A. M.; Frasch, H. F. Pig and guinea pig skin as surrogates for human *in vitro* penetration studies: A quantitative review. *Toxicology in Vitro*, **2009**, *23*(1), 1–13.

Baroli, B. M. Penetration of nanoparticles and nanomaterials in skin: Fiction or reality? *Journal of Pharmaceutical Sciences*, **2010**, *99*, 21–50.

Bird, R. B.; Stewart, W.; Lightfoot, E. *Transport Phenomena* (2nd edn.); John Wiley & Sons, Inc., **2002**.

Botelho, M.; Queiroz, D.; Barros, G.; Guerreiro, S.; Fechine, P.; Umbelino, S.; Lyra, A.; Borges, B.; Freitas, A.; Queiroz, D.; Ruela, R.; Almeida, J.; Quintans Jr, L. Nanostructured transdermal hormone replacement therapy for relieving menopausal symptoms: A confocal Raman spectroscopy study. *Clinics,* **2014**, *69*(2), 75–82.

Bouwstra, J. A.; Dubbelaar, F. E. R.; Gooris, G. S.; Weerheim, A. M.; Ponec, M. The role of ceramide composition in the lipid organization of the skin barrier. *Biochimica et Biophysica Acta – Biomembranes*, **1999**, *1419*(2), 127–136.

Bronaugh, L. Methods for *In Vitro* Percutaneous absorption studies IV: The flow-through diffusion cell. *Journal of Pharmaceutical Sciences*, **1985**, *74*(1).

Byrne, J.; Yeh, J. J.; DeSimone, J. Use of iontophoresis for the treatment of cancer. *Journal of Controlled Release*, **2018**, *284*, 144–151.

Cal, K.; Centkowska, K. Use of cyclodextrins in topical formulations: Practical aspects. *European Journal of Pharmaceutics and Biopharmaceutics*, **2008**, *68*(3), 467–478.

Carita, A. C.; Eloy, J. O.; Chorilli, M.; Lee, R. J.; Leonardi, G. R. Recent advances and perspectives in liposomes for cutaneous drug delivery. *Current Medicinal Chemistry*, **2017**, *25*(5), 606–635.

Carvajal-Vidal, P.; Mallandrich, M.; García, M. L.; Calpena, A. C. Effect of different skin penetration promoters in halobetasol propionate permeation and retention in human skin. *International Journal of Molecular Sciences*, **2017**, *18*(11).

Catarino, C. M.; Do Nascimento, P. T.; Pennacchi, P. C.; De Assis, S. R.; Gimenes, F.; Consolaro, M. E. L.; De Moraes, B. S. B.; Maria-Engler, S. S. Skin corrosion test: A comparison between reconstructed human epidermis and full-thickness skin models. *European Journal of Pharmaceutics and Biopharmaceutics*, **2018**, *125*, 51–57.

Caussin, J.; Gooris, G. S.; Janssens, M.; Bouwstra, J. A. Lipid organization in human and porcine stratum corneum differs widely, while lipid mixtures with porcine ceramides model human stratum corneum lipid organization very closely. *Biochimica et Biophysica Acta - Biomembranes*, **2008**, *1778*(6), 1472–1482.

Cevc, G.; Vierl, U. Nanotechnology and the transdermal route. A state-of-the-art review and critical appraisal. *Journal of Controlled Release*, **2010**, *141*(3), 277–299.

Chong, R. H. E.; Gonzalez-Gonzalez, E.; Lara, M. F.; Speaker, T. J.; Contag, C. H.; Kaspar, R. L.; Coulman, S. A.; Hargest, R.; Birchall, J. C. Gene silencing following SiRNA delivery to skin via coated steel microneedles: *In Vitro* and *in Vivo* proof-of-concept. *Journal of Controlled Release*, **2013**, *166*(3), 211–219.

Davies, D. J.; Heylings, J. R.; McCarthy, T. J.; Correa, C. M. Development of an *in vitro* model for studying the penetration of chemicals through compromised skin. *Toxicology in Vitro*, **2015**, *29*(1), 176–181.

Davies, D. J.; Ward, R. J.; Heylings, J. R. Multi-species assessment of electrical resistance as a skin integrity marker for *in vitro* percutaneous absorption studies. *Toxicology in Vitro*, **2004**, *18*(3), 351–358.

Davis, S.; Martanto, W.; Allen, M.; Prausnitz, M. Hollow metal microneedles for insulin delivery to diabetic rats. *IEEE Transactions on Biomedical Engineering*, **2005**, *52*(5), 909–915.

De Lima, A.; De Souza, T.; Grignoli, L. The benefits of microneedling in the treatment of aesthetics. *Revista Científica Da FHO/Uniararas*, **2015**, *3*(1), 92–99.

Donnelly, R. F.; Morrow, D. I. J.; McCarron, P. A.; Woolfson, A. D.; Morrissey, A.; Juzenas, P.; Juzeniene, A.; Iani, V.; McCarthy, H. O.; Moan, J. Microneedle-mediated intradermal delivery of 5-aminolevulinic acid: Potential for enhanced topical photodynamic therapy. *Journal of Controlled Release*, **2008**, *129*(3), 154–162.

Du, F.; Hönzke, S.; Neumann, F.; Keilitz, J.; Chen, W.; Ma, N.; Hedtrich, S.; Haag, R. Development of biodegradable hyperbranched core-multishell nanocarriers for efficient topical drug delivery. *Journal of Controlled Release*, **2016**, *242*, 42–49.

Elbashir, S. M.; Harborth, J.; Lendeckel, W.; Yalcin, A.; Weber, K.; Tuschl, T. Duplexes of 21-nucleotide RNAs mediate RNA interference in cultured mammalian cells. *Nature*, **2001**, *411*(6836), 494–498.

FDA. SUPAC-SS: Nonsterile Semisolid Dosage Forms; *FDA-Guidance for Industry: Scale-Up and Post-Approval Changes: Chemistry, Manufacturing and Controls*; *In Vitro Release Testing and In Vivo Bioquivalance Documentation*, **1997**.

Flaten, G. E.; Palac, Z.; Engesland, A.; Filipović-Grčić, J.; Vanić, Ž.; Škalko-Basnet, N. *In vitro* skin models as a tool in optimization of drug formulation. *European Journal of Pharmaceutical Sciences*, **2015**, *75*, 10–24.

Fonseca, D. F. S.; Vilela, C.; Silvestre, A. J. D.; Freire, C. S. R. A compendium of current developments on polysaccharide and protein-based microneedles. *International Journal of Biological Macromolecules*, **2019**, *136*, 704–728.

Franz, T. J. Percutaneous absorption. On the relevance of *in vitro* data. *Journal of Investigative Dermatology*, **1975**, *64*(3), 190–195.

Friend, D. R. *In vitro* skin permeation techniques. *Journal of Controlled Release*, **1992**, *18*, 235–248.

Gibbs, S.; Corsini, E.; Spiekstra, S. W.; Galbiati, V.; Fuchs, H. W.; Degeorge, G.; Troese, M.; Hayden, P.; Deng, W.; Roggen, E. An epidermal equivalent assay for identification and ranking potency of contact sensitizers. *Toxicology and Applied Pharmacology*, **2013**, *272*(2), 529–541.

Gill, H. S.; Denson, D. D.; Burris, B. A.; Prausnitz, M. R. Effect of microneedle design on pain in human subjects. *Clinical Journal Pain*, **2008**, *24*(7), 585–594.

Gilman, G. *As Bases Da Farmacologia Terapêutica* (11th edn.); Laurence L. Bruton: Porto Alegre, **2010**.

Godin, B.; Touitou, E. Transdermal skin delivery: Predictions for humans from *in vivo*, *ex vivo* and animal models. *Advanced Drug Delivery Reviews*, **2007**, *59*(11), 1152–1161.

Gratieri, T.; Martins, G.; Fonseca, R.; Lopez, V. Princípios básicos e aplicação da iontoforese na penetração cutânea de fármacos [Basic principles and application of iontophoresis in the cutaneous penetration of drugs]. *Química Nova*, **2008**, *31*(6), 1490–1498.

Guth, K.; Schäfer-Korting, M.; Fabian, E.; Landsiedel, R.; Van Ravenzwaay, B. Suitability of skin integrity tests for dermal absorption studies *in vitro*. *Toxicology in Vitro*, **2015**, *29*(1), 113–123.

Haq, A.; Dorrani, M.; Goodyear, B.; Joshi, V.; Michniak-Kohn, B. Membrane properties for permeability testing: Skin versus synthetic membranes. *International Journal of Pharmaceutics*, **2018**, *539*(1, 2), 58–64.

Henry, S.; Mcallister, D. V.; Allen, M. G.; Prausnitz, M. R. Microfabricated microneedles: A novel approach to transdermal drug delivery. *Journal of Pharmaceutical Sciences*, **1998**, *87*(Volume 2), 1–3.

Herwadkar, A.; Sachdeva, V.; Taylor, L. F.; Silver, H.; Banga, A. K. Low frequency sonophoresis mediated transdermal and intradermal delivery of ketoprofen. *International Journal of Pharmaceutics*, **2012**, *423*(2), 289–296.

Huang, J. F.; Sung, K. C.; Hu, O. Y. P.; Wang, J. J.; Lin, Y. H.; Fang, J. Y. The effects of electrically assisted methods on transdermal delivery of nalbuphine benzoate and sebacoyl dinalbuphine ester from solutions and hydrogels. *International Journal of Pharmaceutics*, **2005**, *297*(1, 2), 162–171.

Jacobi, U.; Kaiser, M.; Toll, R.; Mangelsdorf, S.; Audring, H.; Otberg, N.; Sterry, W.; Lademann, J. Porcine ear skin: An *in vitro* model for human skin. *Skin Research and Technology*, **2007**, *13*(1), 19–24.

Jepps, O. G.; Dancik, Y.; Anissimov, Y. G.; Roberts, M. S. Modeling the human skin barrier - towards a better understanding of dermal absorption. *Advanced Drug Delivery Reviews*, **2013**, *65*(2), 152–168.

Jug, M.; Hafner, A.; Lovrić, J.; Kregar, M. L.; Pepić, I.; Vanić, Ž.; Cetina-Čižmek, B.; Filipović-Grčić, J. An overview of *in vitro* dissolution/release methods for novel mucosal drug delivery systems. *Journal of Pharmaceutical and Biomedical Analysis*, **2018**, *147*, 350–366.

Kalia, Y. N.; Naik, A.; Garrison, J.; Guy, R. H. Iontophoretic drug delivery. *Advanced Drug Delivery Reviews*, **2004**, *56*, 619–658.

Kalluri, H.; Banga, A. K. Transdermal delivery of proteins. *AAPS PharmSciTech*, **2011**, *12*(1), 431–441.

Kandavilli, S.; Nair, V.; Panchagnula, R. Polymers in transdermal drug delivery systems. *Pharmaceutical Technology*, **2002**, No. May, 62–80.

Kommareddy, S.; Baudner, B. C.; Seajin, O.; Kwon, S. Y.; Singh, M.; O'Hagan, D. T. Dissolvable microneedle patches for the delivery of cell-culture-derived influenza vaccine antigens. *Journal of Pharmaceutical Sciences*, **2012**, *101*(1), 322–332.

Lai-Cheong, J. E.; McGrath, J. A. Structure and function of skin, hair and nails. *Medicine (United Kingdom)*, **2017**, *45*(6), 347–351.

Lane, M. E. Skin penetration enhancers. *International Journal of Pharmaceutics*, **2013**, *447*, 12–21.

Lane, M. E.; Santos, P.; Watkinson, A. C.; Hadgraft, J. Enhancement. In *Transdermal and Topical Drug Delivery: Principles and Practice*; Benson, H. A. E., Watkinson, A. C., Eds.; John Wiley & Sons, Inc., **2012**; Vol. 1.

Lee, H.; Song, C.; Baik, S.; Kim, D.; Hyeon, T.; Kim, D. H. Device-assisted transdermal drug delivery. *Advanced Drug Delivery Reviews*, **2018**, *127*, 35–45.

Lemper, M.; De Paepe, K.; Rogiers, V. Practical problems encountered during the cultivation of an open-source reconstructed human epidermis model on a polycarbonate membrane and protein quantification. *Skin Pharmacology and Physiology*, **2014**, *27*, 106–112.

Leveque, N.; Makki, S.; Hadgraft, J.; Humbert, P. Comparison of Franz cells and microdialysis for assessing salicylic acid penetration through human skin. *International Journal of Pharmaceutics*, **2004**, *269*(2), 323–328.

Levintova, Y.; Plakogiannis, F. M.; Bellantone, R. A. and improved *in vitro* method for measuring skin permeability that controls excess hydration of skin using modified Franz diffusion cells. *International Journal of Pharmaceutics*, **2011**, *419*(1, 2), 96–106.

Ling, M. H.; Chen, M. C. Dissolving polymer microneedle patches for rapid and efficient transdermal delivery of insulin to diabetic rats. *Acta Biomaterialia*, **2013**, *9*(11), 8952–8961.

Losquadro, W. D. Anatomy of the skin and the pathogenesis of nonmelanoma skin cancer. *Facial Plastic Surgery Clinics of North America*, **2017**, *25*(3), 283–289.

Margetts, L.; Sawyer, R. Transdermal drug delivery: Principles and opioid therapy. *Continuing Education in Anaesthesia, Critical Care and Pain*, **2007**, *7*(5), 171–176.

McLafferty, E.; Hendry, C.; Farley, A. The integumentary system: Anatomy, physiology and function of skin. *Nursing Standard*, **2013**, *27*(3), 35–42.

Meidan, V.; Michniak-Kohn, B. B. Ultrasound-Based Technology for skin barrier permeabilization. *Handbook of Non-Invasive Drug Delivery Systems*, **2010**, 119–133.

Merino, V.; Lópes, A.; Kalia, Y. N.; Guy, R. H. Electrorepulsion versus electroosmosis: Effect of PH on the iontophoretic flux of 5-fluorouracil. *Pharmaceutical Research*, **1999**, 16.

Mikolajewska, P.; Donnelly, R. F.; Garland, M. J.; Morrow, D. I. J.; Singh, T. R. R.; Iani, V.; Moan, J.; Juzeniene, A. Microneedle pre-treatment of human skin improves 5-aminolevulinic acid (ALA)- and 5-aminolevulinic acid methyl ester (MAL)-induced PpIX production for topical photodynamic therapy without increase in pain or erythema. *Pharmaceutical Research*, **2010**, *27*(10), 2213–2220.

Mojeiko, G.; De Brito, M.; Salata, G. C.; Lopes, L. B. Combination of microneedles and microemulsions to increase celecoxib topical delivery for potential application in chemoprevention of breast cancer. *International Journal of Pharmaceutics*, **2019**, *560*, 365–376.

Moser, K.; Kriwet, K.; Naik, A.; Kalia, Y. N.; Guy, R. H. Passive skin penetration enhancement and its quantification *in vitro*. *European Journal of Pharmaceutics and Biopharmaceutics*, **2001**, *52*(2), 103–112.

Münch, S.; Wohlrab, J.; Neubert, R. H. H. European journal of pharmaceutics and biopharmaceutics dermal and transdermal delivery of pharmaceutically relevant macromolecules. *European Journal of Pharmaceutics and Biopharmaceutics*, **2017**, *119*, 235–242.

Naik, A.; Kalia, Y. N.; Guy, R. H.; Guy, R. H. Rationale for transdermal drug delivery. *Pharmaceutical Science & Technology Today*, **2000**, *3*(9).

Ng, S. F.; Rouse, J. J.; Sanderson, F. D.; Meidan, V.; Eccleston, G. M. Validation of a static Franz diffusion cell system for *in vitro* permeation studies. *AAPS PharmSciTech*, **2010**, *11*(3), 1432–1441.

Paleco, R.; Vučen, S. R.; Crean, A. M.; Moore, A.; Scalia, S. enhancement of the *in vitro* penetration of quercetin through pig skin by combined microneedles and lipid microparticles. *International Journal of Pharmaceutics*, **2014**, *472*(1, 2), 206–213.

Pappas, A. Epidermal surface lipids. *Dermato-Endocrinology*, **2010**, *1*(2), 72–76.

Park, D.; Park, H.; Seo, J.; Lee, S. Sonophoresis in transdermal drug delivery. *Ultrasonics*, **2014**, *54*(1), 56–65.

Paudel, K. S.; Milewski, M.; Swadley, C. L.; Brogden, N. K.; Ghosh, P.; Stinchcomb, A. L. Challenges and opportunities in dermal/transdermal delivery. *Therapeutic Delivery*, **2010**, *1*(1), 109–131.

Petrilli, R.; Eloy, J. O.; Paschoal, J.; Lopez, R. F. V. Quantification of 5-FU in skin samples for the development of new delivery systems for topical cancer treatment. *Pharmazie*, **2018**, *3*, 133–138.

Petrilli, R.; Eloy, J. O.; Saggioro, F. P.; Chesca, D. L.; De Souza, M. C.; Dias, M. V. S.; daSilva, L. L. P.; Lee, R. J.; Lopez, R. F. V. Skin cancer treatment effectiveness is improved by iontophoresis of EGFR-targeted liposomes containing 5-FU compared with subcutaneous injection. *Journal of Controlled Release*, **2018b**, *283*, 151–162.

Petrilli, R.; Lopez, R. F. V. Physical methods for topical skin drug delivery: Concepts and applications. *Brazilian Journal of Pharmaceutical Sciences*, **2018a**, *54*(Special Issue), 1–19.

Petrilli, R.; Praca, F.; Carollo, A. R.; Medina, W.; Oliveira, K. T.; Fantini, M.; Neves, M. Da G.; Cavaleiro, J.; Serra, O.; Iamamoto, Y.; Bentley, M. V. Nanoparticles of lyotropic liquid crystals: A novel strategy for the topical delivery of a chlorin derivative for photodynamic therapy of skin cancer. *Current Nanoscience*, **2013**, *9*(4), 434–441.

Plessis, J.; Goosen, C.; Venter, J. P.; Muller, D. G. A comparative study of an in situ adapted diffusion cell and an *in vitro* Franz diffusion cell method for transdermal absorption of doxylamine. *European Journal of Pharmaceutical Sciences*, **2001**, *13*, 169–177.

Polat, B. E.; Blankschtein, D.; Langer, R. Low-frequency sonophoresis: Application to the transdermal delivery of macromolecules and hydrophilic drugs. *Expert Opinion on Drug Delivery*, **2010**, *7*(12), 1415–1432.

Polat, B. E.; Figueroa, P. L.; Blankchtein, D.; Langer, R. Transport pathways and enhancement mechanisms within localized and non-localized transport regions in skin treated with low-frequency sonophoresis and sodium lauryl sulfate. *Journal of Pharmaceutical Sciences*, **2011**, *100*(2), 512–529.

Praça, F. S. G.; Medina, W. S. G.; Eloy, J. O.; Petrilli, R.; Campos, P. M.; Ascenso, A.; Bentley, M. V. L. B. Evaluation of critical parameters for *in vitro* skin permeation and penetration studies using animal skin models. *European Journal of Pharmaceutical Sciences*, **2018**, *111*, 121–132.

Prausnitz, M. R.; Mikszta, J. A.; Cormier, M.; Andrianov, A. K. Microneedle -based vaccines. In *Vaccines for Pandemic Influenza*; Compans, R. W., Orestein, W. A., Eds.; Springer: Berlin, **2009**; pp. 369–393.

Rang, H. P.; Dale, M. M. *Farmacologia* (4th edn.); Guanabara Koogan: Rio de Janeiro, **2001**.

Reis, T. A.; Barreto, L. C.; Cunha, F. M.; Gratieri, T.; Gelfuso, G. M. Microagulhas: Estado da arte e aplicações médicas. *Brasília Medica*, **2014**, *51*(2).

Rigg, P. C.; Barry, B. W. Shed snake skin and hairless mouse skin as model membranes for human skin during permeation studies. *The Society for Investigative Dermatology*, Inc, **1990**, 90.

Roquel, R.; Cohen, C.; Leclaire, J.; Tessonneaud, T.; Gagne, C.; Teissier, M. Use of a standardized reconstructed epidermis kit to assess *in vitro* the tolerance and the efficacy of cosmetics. *International Journal of Cosmetic Science*, **2000**, *6*, 409–419.

Roxhed, N.; Griss, P.; Stemme, G. Membrane-sealed hollow microneedles and related administration schemes for transdermal drug delivery. *Biomedical Microdevice*, **2008**, *10*(2), 271–279.

Ruela, A. L. M.; Perissinato, A. G.; Lino, M. E. De S. L.; Mudrik, P. S.; Pereira, G. R. P. Evaluation of skin absorption of drugs from topical and transdermal formulations. *Brazilian Journal of Pharmaceutical Science*, **2014**, *52*(3).

Simon, A.; Amaro, M. I.; Healy, A. M.; Cabral, L. M.; De Sousa, V. P. Comparative evaluation of rivastigmine permeation from a transdermal system in the Franz cell using synthetic membranes and pig ear skin with *in vivo-in vitro* correlation. *International Journal of Pharmaceutics*, **2016**, *512*(1), 234–241.

Sinko, P. J.; Gonzales, O. *Físico Fármacia e Ciências Farmacêuticas* (5th edn.); Artmed, Ed.; Porto Alegre, **2008**.

Sullivan, S. P.; Koutsonanos, D. G.; Del Pilar Martin, M.; Lee, J. W.; Zarnitsyn, V.; Choi, S. O.; Murthy, N.; Compans, R. W.; Skountzou, I.; Prausnitz, M. R. Dissolving polymer microneedle patches for influenza vaccination. *Nature Medicine*, **2010**, *16*(8), 915–920.

Touitou, E.; Meidan, V. M.; Horwitz, E. Methods for quantitative determination of drug localized in the skin. *Journal of Controlled Release*, **1998**, *56*(1–3), 7–21.

Venus, M.; Waterman, J.; McNab, I. Basic physiology of the skin. *Surgery*, **2011**, *29*(10), 471–474.

Verbaan, F. J.; Bal, S. M.; Van Den, B. D. J.; Groenink, W. H. H.; Verpoorten, H.; Lüttge, R.; Bouwstra, J. A. Assembled microneedle arrays enhance the transport of compounds varying over a large range of molecular weight across human dermatomed skin. *Journal of Controlled Release*, **2007**, *117*(2), 238–245.

Waghule, T.; Singhvi, G.; Dubey, S. K.; Pandey, M. M.; Gupta, G.; Singh, M.; Dua, K. Microneedles: A smart approach and increasing potential for transdermal drug delivery system. *Biomedicine and Pharmacotherapy*, **2019**, *109*, 1249–1258.

Walters, K. *Dermatological and Trandermal Formulations*, Marcel Dek.; New York, **2002**.

Wang, P. C.; Wester, B. A.; Rajaraman, S.; Paik, S. J.; Kim, S. H.; Allen, M. G. Hollow polymer microneedle array fabricated by photolithography process combined with micromolding technique. *Proceedings of the 31st Annual International Conference of the IEEE Engineering in Medicine and Biology Society: Engineering the Future of Biomedicine EMBC*, **2009**, 7026–7029.

Weldon, W. C.; Martin, M. P.; Zarnitsyn, V.; Wang, B.; Koutsonanos, D.; Skountzou, I.; Prausnitz, M. R.; Compans, R. W. Microneedle vaccination with stabilized recombinant influenza virus hemagglutinin induces improved protective immunity. *Clinical and Vaccine Immunology*, **2011**, *18*(4), 647–654.

Welty, J.; Wicks, C.; Wilson, R.; Rorrer, G. *Fundamentals of Momentum, Heat and Mass Transfer,* (5th edn.); John Wiley & Sons, Inc., **2008**.

Wester, R. Maibach, H. *In vivo* methods for percutaneous absorption measurement. *Journal of Toxicology: Cutaneous and Ocular Toxicology*, **2002**, 411–422.

Wong, R.; Geyer, S.; Weninger, W.; Guimberteau, J. C.; Wong, J. K. The dynamic anatomy and patterning of skin. *Experimental Dermatology*, **2016**, *25*(2), 92–98.

Zhang, H.; Barralet, J. E. Mimicking oxygen delivery and waste removal functions of blood. *Advanced Drug Delivery Reviews*, **2017**, *122*, 84–104.

Zhang, Q.; Murawsky, M.; LaCount, T.; Kasting, G. B.; Li, S. K. Transepidermal water loss and skin conductance as barrier integrity tests. *Toxicology in Vitro*, **2018**, *51*, 129–135.

Zhang, Y.; Brown, K.; Siebenaler, K.; Determan, A.; Dohmeier, D.; Hansen, K. Development of lidocaine-coated microneedle product for rapid, safe, and prolonged local analgesic action. *Pharmaceutical Research*, **2012**, *29*(1), 170–177.

Zhou, C. P.; Liu, Y. L.; Wang, H. L.; Zhang, P. X.; Zhang, J. L. Transdermal delivery of insulin using microneedle rollers *in vivo*. *International Journal of Pharmaceutics*, **2010**, *392*(1, 2), 127–133.

CHAPTER 2

Micro- and Nano-Systems in Transdermal Drug Delivery

Nayan A. Gujarathi,[1] Tulshidas S. Patil,[1] Bhushan R. Rane,[2] Akshay Babu,[3] and Raj Kumar Keservani[4]

[1]Shri Vile Parle Kelavani Mandal's Institute of Pharmacy, Dhule–424001, Maharashtra, India, E-mail: nayang5678@gmail.com (N. A. Gujarathi)

[2]Shri D.D. Vispute College of Pharmacy and Research Center, New Panvel–410206, Maharashtra, India

[3]Sandip Institute of Pharmaceutical Sciences, Nashik–422213, Maharashtra, India

[4]Faculty of B. Pharmacy, CSM Group of Institutions, Prayagraj–212111, Uttar Pradesh, India

ABSTRACT

The extensively used novel methods such as micro- and nano-systems for transdermal administration of the medication are hypodermic needles, topical creams, vesicular system, and transdermal patches. The influence of the majority of the therapeutic marketers is confined because of the stratum corneum (SC) layer of the skin, functions as an obstruction for the active constituents and therefore just a few active constituents are able to attain the movement across the SC. Microneedles enable transport of the drug through this pathway and overcome different issues related to the traditional dosage forms. The main principle involves the distraction

Topical and Transdermal Drug Delivery Systems: Applications and Future Prospects. Nayan A. Gujarathi, PhD, Juliana Palma Abriata, PhD, Raj Kumar Keservani, MPharm, Anil K. Sharma, PhD, MPharm (Eds.)

of the skin layer, and thereby developing micron length pathways that lead the drug immediately to the epidermis (EP) or higher EP vicinity from wherein the drug can immediately move into the systemic flow without facing the barrier. This describes the diverse capability of the microneedles. The dimensions of nano-systems usually range from 1 nm to a 100 nm. The nano-systems enriched with various advantages such as increased drug permeability, drug stability, and drug retention make them suitable for transdermal drug delivery. The extraordinary packages of nano-systems (vesicular systems, polymeric nanoparticles, nanoemulsion, and dendrimers) were widely studied and its application is reviewed in this chapter through transdermal route.

2.1 INTRODUCTION

Transdermal therapeutic system is characterized as independent dosage form when spread on or bring into contact with the skin, transports the medication across the epidermis (EP) at a predetermined rate to the systemic circulation and keep the drug concentration inside the therapeutic window for prolonged time frame (Pawar et al., 2018). Transdermal drug delivery system (TDDS) is used to attain the localized as well as systemic therapeutic effect. The TDDS offers a means of sustaining the release of drugs and reducing the intensity of drug action, thereby reducing the adverse effects associated with oral treatment. Transdermal absorption governs through a moderate procedure of dispersion driven by the gradient between the high concentration of drug in the delivery system and the zero concentration prevailing in the skin. Subsequently, the delivery system must be kept in touch with the skin for a significant time. TDDS is an informal way for transporting active pharmaceutical ingredient primarily by applying a drug formulation onto flawless and sound skin. The medication at first enters through the stratum corneum (SC) and after that goes through the more profound EP and dermis devoid of medication aggregation in the dermal layer. At the point when medication reaches the dermal layer, it becomes accessible for systemic absorption by means of the dermal microcirculation. It can give a non-intrusive option in contrast to parenteral routes. In this way overcome the associated issues, for example, needle fear and inconvenience. A large surface area of the skin and its ease of access offers several choices to deliver drug through the skin for transdermal absorption. Moreover, the pharmacokinetic profiles of medications are progressively uniform with fewer pinnacles, along these lines limiting the danger of lethal reactions. It can improve patient compliance because of the decrease of dosing frequencies and are additionally appropriate for patients who are oblivious or spewing, or the individuals who depend on self-administration. TDDS avoids pre-systemic metabolism, along these lines improving bioavailability. Transdermal delivery provides an improved way to deal with the administration of medications by maintaining the therapeutic concentration of active ingredients in the blood for predetermined timeframe. TDDS escape gastrointestinal absorption, which is associated with the enzymatic and pH-related deactivation. This strategy likewise takes into consideration for decrease in pharmacological dosing for drug because of the reduced biotransformation/stability issue associated with medication when administered through gastrointestinal tract versus transdermal route.

TDDS releases the incorporated or encapsulated active ingredient through the patient's skin giving a uniform blood level. Furthermore, the drug moiety moves on with the circulatory system to the targeted site, required less effective dose of drug than that in the traditional medication type (Lee et al., 2018). The TDDS permits steady dosing as opposed to the peaks and valleys in drug level associated with orally administered drug. Transdermal method of administrating drug is frequently utilized route in the management of different fundamental disorders, for example, hypertension, joint inflammation, diabetes, malignancy, and so forth (Chanda et al., 2011; Keservani et al., 2020). It overcomes disadvantages related with oral and intravenous courses. Dermal and transdermal routes offer bigger surface area accessible for medication absorption, ease in availability and end of treatment at whatever point required. Drug delivery through skin helps in the management of systemic as well as topical infections. It is without pain technique for administration, encourages self-administration of drug in the patients, favored in long term management of the afflictions like chronic pain and stays away from hepatic first pass metabolism. These routes are affordable on a month-to-month cost premise, as these are intended to convey drugs from days to months. Drug delivery through topical route supports pain free self-medication in the patients. In contrast with the oral or parenteral medication administration, topical route helps reduced frequent dosing and keep up steady medication plasma level for expanded time frame.

- **Advantages (Tyagi et al., 2011):**
  - Hepatic first pass digestion, salivary digestion and intestinal digestion are evaded.
  - The simplicity of use makes it convenient for patients to self-administer such type of drug delivery systems.
  - If there is any occurrence of an emergency then removing patch anytime during the treatment can right away stop medication delivery to systemic circulation.
  - Furthermore, the structure of the skin is fundamentally and organically same in almost every people so there is no significant variation observed in results for patients.
  - Drug demonstrating gastrointestinal disturbance and retention can be appropriately delivered through the skin.
  - The non-invasive technique can be utilized for medications with short half-lives, which would some way or another, require continuous dosing.
  - Due to reduced recurrence of dosing, the increase in patient compliance reported (Uchechi et al., 2014).
  - The antagonistic impacts are limited because of an enduring and ideal blood concentration time profile.
  - The pain and inconvenience related with parenteral treatment are avoided.
  - The medication release is with the end goal that there is an anticipated and extended length of activity.
- **Disadvantages (Alkilani et al., 2015; Tyagi et al., 2011):**
  - There is probability of skin aggravation.
  - Dose dumping may occur due to burst release.
  - It cannot be utilized for chronic conditions where medication treatment is wanted for an extensive period of time, e.g., hypertension, angina, diabetes, and so on.
  - Lag time is inconstant and may be shift from few hours to day for various system.
  - Cutaneous degradation of drug will influence therapeutic result of the medication.
  - Transdermal treatment is plausible for certain potent medications only.
  - Transdermal treatment isn't achievable for ionic medications.
  - It can't convey drug in pulsatile style.

- Erythema, itching, tingling, local edema can be brought about by the medication, the adhesive or different excipients in the formulation.

2.2 TRANSDERMAL PATCHES (SHARMA, 2019)

A transdermal patch is a type of medicated adhesive patch that is set on the skin to convey a particular portion of medicine through the skin and into the circulation system. Often, this elevates healing to an injured region of the body. A wide range of pharmaceuticals can be delivered through transdermal patch. The most significant problem in the fabrication of novel system meant for transdermal delivery is to modify the delivery of drug through the skin on interest. Skin patches hold guarantee for transdermal administration of a broad extent of medicinal therapy. Patches regulate the release of medications and keep away from pinnacles and valleys related with multiple-dose oral drug, consolidating broadened span of delivery with patient solace, while fundamentally improving patient compliance. Patch delivery is simpler than infusion, and eliminates the danger of infection.

2.2.1 TYPES OF PATCHES

1. **Single-Layer Drug-in-Adhesive:** The drug is incorporated into an adhesive layer of transdermal patch. The adhesive layer not only hold the different layers together with the whole structure to the skin, but also to release the drug. A new cover and a backing surround the adhesive layer.
2. **Multi-Layer Drug-in-Adhesive:** The adhesive patch of multi-layer drug is comparable to the single-layer system. The only difference is that multi-layer system comprise of an extra layer of drug-in-adhesive, normally (but not in all cases) separated by a membrane (Pawar et al., 2018). The release of the drug is the responsibility of both adhesive layers. This patch has a temporary liner layer as well as a permanent backing.
3. **Reservoir System:** The reservoir transdermal structure has a distinct drug layer unlike the single-layer and multi-layer drug-in adhesive structures. The discharge rate in this sort of system is

zero order. The drug reservoir is maintained in this system between the rate controlling membrane and the back layer. Drug releases through membrane that is microporus and at a controlled rate. In the reservoir compartment, the drug may be incorporated in the form of a drug solution, drug suspension, or drug is dispersed in a solid polymer matrix or converted into gel.

4. **Matrix System:** This appears as a rigid network-like structure made up of polymer in which fine particles of drug moiety is uniformly distributed (Pawar et al., 2018). The drug is distributed homogenously in a polymeric matrix of lipophilic-type or hydrophilic-type known as matrix-dispersion system. Polymer is then attached together with drug to an occlusive base plate in a compartment made from a drug-impermeable back layer. Instead of applying adhesive on the drug reservoir face it is distributed along the circumference.
5. **Micro-Reservoir System:** This type of transdermal patch is the combination of both the reservoir systems and the matrix dispersion-type. Here the drug is suspended in a water-soluble polymer solution and then dispersed homogeneously into a lipophilic polymer to create thousands of un-bleachable, microscopic drug reservoir areas.
6. **Vapor Patch:** The adhesive layer serves not only to hold the different layers together in this category but it will also release the essential oil vapors. The vapor patches are recently introduced in pharma market, releases the essential oils for as long as 6 hours. The vapors patches are primarily used in decongestion cases. Other type of vapor patch available in the market are enhances the sleep quality. Also accessible on the market are vapor patches that decrease the amount of cigarettes one smokes in a month.

2.2.2 EVALUATION OF TRANSDERMAL PATCH (UCHECHI ET AL., 2014; PAWAR ET AL., 2018)

1. **Interaction Studies:** The compatibility of drugs and polymers was described through FTIR spectroscopy. Compatibility was verified by creating physical drug and polymer combination (1:1) and then the mixture's FTIR assessment was performed. The peaks

in the FTIR spectra of mixtures should not be changed, and they should be similar to the FTIR spectra of pure drug and polymer. There are three categories for evaluation of interaction studies: physicochemical assessment, *in-vitro* studies, and *in vivo* research.

Physical or physicochemical evaluation of transdermal system:

i. **Film Thickness:** The instruments used for measurement of film thickness are micro-meter, electronic vernier calipers, with a least count of 0.01 mm, dial gauge, or screw gauge. Thickness is measured at five different points on the film and average of five readings is taken.
ii. **Percentage Flatness:** Strips are selected as the average percent of length calculated from the 7 cm strips.

    If the constriction of the film due to applied pressure is 0% then it indicates 100% flatness

$$\% \, Constriction = (initial\ length - final\ length)/initial\ length \times 100$$

iii. **Folding Endurance:** It is estimated by constantly folding the transdermal system at same location on a film strip (2 x 2 cm) until it breaks. The number of times the film can be folded at the same location without breaking is recorded as the folding endurance value for the strip.
iv. **Tensile Strength:** A modified pulley system is used to determine tensile strength. The tensile strength of the film is the magnitude of force applied to breakdown the film. The unit of the tensile strength for the film is kg/cm^2.
v. **Patch Thickness:** It is determined by using digital micrometer screw gauge, by placing it at three different points and calculating mean value of all the value obtained.
vi. **Elongation Break Test:** The elongation break can be calculated by the formula:

$$Elongation\ break = (final\ length - initial\ length)/initial\ length.$$

vii. **Weight Uniformity:** The transdermal patches required for weight uniformity test are selected randomly. A given patch region is to be cut at separate areas and weighed in a digital balance. Calculate the average weight for the individual patch and determine the standard deviation value.

viii. **Drug Content:** A film is cut from the obligatory region. Place small pieces of film in a buffer solution up to 100 ml (pH 7.4 or 6.8 or as prescribed) and shake for 24 hours continually. Then the entire solution will be ultrasonic for 15 minutes. Filtration is performed followed by spectrophotometric estimation of the drug is carried out and then drug content is determined.

ix. **Percentage of Moisture Content:** Individual films are weighed and placed in a desiccator for 24 hours to be maintained at room temperature. The desiccator contains anhydrous calcium chloride or activated silica. They are separately weighed until constant weight is shown. The percentage of moisture is calculated as the difference between the initial and the final weight.

$$\%\ moisture\ Content = [initial\ weight - final\ weight] / final\ weight * 100.$$

x. **Percentage of Moisture Uptake:** Accurately weighed film is kept in a desiccator. The relative humidity (a saturated potassium chloride solution) is maintained at 84% for next 24 hours at room temperature. The films were exposed for sufficient time to attained constant weight. The proportion of absorption of moisture is calculated by original weight as the distinction between the final and initial weight.

$$\%\ moisture\ uptake = [final\ weight - initial\ weight] / initial\ weight * 100.$$

xi. **Water Vapor Transmission Rate:** For the research of transmission, glass vials of approximately 5 ml capability of equal diameter are taken. All vials are carefully cleaned and fully dried in an oven. Weigh approximately 1 gm of anhydrous/fused calcium chloride and are stored in all vials taken. Films are attached to the border of the vials and weighed individually, then stored in a closed desiccator containing a saturated potassium chloride solution to maintain approx. humidity 84%. The vials were weighed respectively at different intervals after 6 h, 12 h, 24 h, 36 h, 48 h and 72 h. Water vapor transmission rate is calculated at every 24 h expressed in gm/m^2.

$$\%\ Transmission\ rate = [final\ weight - initial\ weight] / area * time] * 100.$$

2. **Adhesive Studies:**
   i. **Shear Adhesion Test:** The cohesive strength of polymer and transdermal patch is estimated by shear adhesion test. The resistance value is influenced by the polymer structure, degree of cross-linking, the quantity of tackifiers incorporated and the weight of the molecules. A coated adhesive patch is stacked between the stainless steel plate and the specified weight suspended parallel to this plate from the patch. The cohesive power is directly proportional to the time required to take this patch off.
   ii. **Peel Adhesion Test:** Measuring the power of the patch between an adhesive and a substratum is described as adhesion. The force needed is used as a test substrate, which is intended to remove adhesive coating from the steel. The adhesive characteristics are determined by the type and quantity of polymer, the molecular weight, and the structure of polymers. The single patch is pasted to test substratum (Steel) and pulled at an angle of 180° from the substratum. Adhesive failure is reported with no substrate residue.
   iii. **Tack Properties:** Tack is a polymer's capacity to adhere to a low contact pressure substrate. In transdermal structures, application with low finger pressure is essential. Tack depends on the molecular weight as well as the polymer and tackling resin composition used in the polymer.

   Tests for tack include:
      a. Thumb tack test;
      b. Rolling ball tack test;
      c. Peel tack or quick stick test;
      d. Probe tack test.

3. ***In Vitro* Drug Release Studies:** The paddle over disk method (USP device V) can be used to evaluate the release of the drug from the patches. The film of specified thickness and shape is accurately weighed and fixed with adhesive on the glass plate. It is held in 500 ml of dissolution media (phosphate buffer solution – pH 7.4) and set the device at 32 ± 0.5°C. The distance between glass tray and paddle is 2.5 cm and it is operated at a velocity of 50 rpm. The 5 ml of sample is withdrawn after specific time interval given

in monograph or type of patch. The sample is analyzed by UV or HPLC.

4. ***In Vitro* Skin Permeation Studies:** Permeation studies is an *in-vitro* test performed through the skin by diffusion cell. The male Wistar rat weighing 200–250 gm were generally utilized to perform the test. The hair is removed from the surface of rat's abdominal skin using electric clipper. Distilled water is used to clean the skin's dermal side to remove adhesive tissue and then stored for 1 h in phosphate buffer solution (pH 7.4) or suitable dissolution media. Before beginning the experiment, it was put on a small magnetic needle type magnetic stirrer for standardized diffuser distribution. Using thermostatically controlled heaters, cell temperature was maintained at 32 ± 0.5°C. Rat skin is put in the donor compartment between the diffusion cell compartments with an EP facing upwards. Specific volume is removed at a particular time interval from the receptor compartment and an equivalent volume of new sample is introduced. The sample withdrawal is filtered and analyzed by UV or using HPLC. Flux can be determined by plotting the slop between stable state values of the quantity of mg/cm^2 vs. permeated by the drug. By separating the flux by the original drug load mg cm^2 the time in hours and permeability coefficient was deduced.
5. **Skin Irritancy Studies:** Skin irritation study is conducted on healthy rabbit/mice albino/rats. The transdermal system potential can be accessed through altered Draize testing. The hair is removed from the test animal's dorsal surface and properly clean. Then rectified spirit was added to the test surface. Apply transdermal formulation for 24 hours. The adverse effect on the surface of skin is reported to conclude the result of the skin irritation test.
6. **Stability Studies:** These have been conducted in accordance with ICH guidelines where TDS samples are stored for the minimum period of 6 months. The temperature and humidity level is maintained at 40 ± 0.5°C and 75 ± 5% RH. The drug content present in the sample patch was determined by suitable analytical method at the end of 30 days, 60 days, 90 days, and 180 days appropriately.

2.3 MICROSYSTEMS IN TRANSDERMAL DRUG DELIVERY

2.3.1 MICRONEEDLE

Microneedle unit comprises of micron size needles that are organize properly on a small patch. In view of the hypodermic needle issues and the transdermal patch, the microneedle drug delivery system was designed, believed that it is the hybrids of both systems. The recognition that small needles is adequate for transportation across 10–20 m thick SC was first suggested in the 1970s, but progress was mainly postponed owing to absence of methods for manufacturing such small constructions. The first research on transdermal drug delivery using microneedles was revealed in the early 1990s. Microneedles have the ability to be used as an alternative to subcutaneous and hypodermic needle techniques. Microneedles combine ease of use of a transdermal patch with delivery effectiveness obtained using standard hypodermic needles and syringes. Microneedles are numerous microscopic projections typically assembled on one side of a supporting foundation or patch, normally varying from 25 to 2,000 μm in height, 1 to 25 μm in tip diameter and the base width is 50 to 250 μm. When inserted into skin layers, the needles should be of suitable size, shape, and width in order to circumvent the nerve contact. Usually, they are intended in arrays to enhance skin surface contact and promote therapeutic molecules penetration into the skin. Microneedles are intended to produce temporary aqueous conduits across the skin, improving pain-free flux of molecules from small hydrophilic molecules such as alendronate to macromolecules, including low molecular weight heparins, insulin, and vaccines. Apart from the pain-free delivery side, there are many other benefits of microneedle technology, such as: the reality that they do not cause bleeding; the elimination of transdermal dose variation of tiny molecules; the minimal introduction of pathogens by microneedle-induced holes; the ability for self-administration; the ability to overcome and decrease accidental needle-stick injuries and the risk of transmitting infections, in addition to the ease of microneedle waste disposal (Waghule et al., 2019).

During design and manufacturing, the shape and geometry of MN is critical. The needles must be capable of inserting into the skin without breaking and the needles should be of appropriate length, shape, and width to circumvent the nerve contact. The needle also create an efficient pathway for the delivery of small drugs, macromolecules, and nanoparticles as well as for the extraction of fluids, depending on the rational of individual device. During implementation, especially in the event of blunt and short MN, the elastic properties of human skin can stop efficient MN penetration by twisting the skin fibers around the needles. Based on the type of the material used and the microneedle type it can be developed into different dimensions. Since the thickness of EP is up to 1,500 μm, the length of the needle is also adjusted to 1,500 μm sufficient enough to release the drug across the EP. Longer and thicker the diameter of needles will penetrate deeply into the dermis. Thus, produces harm to the nerves and cause pain. They are mostly 150–1,500 microns long, 50–250 microns broad, with a tip thickness is about 1–25 microns. Based on earlier discussion, the objective of microneedle device is to generate a transportation pathway of micron size, the needle diameter is should be range between few microns. The microneedle tips are accessible in many more forms and can be cylindrical, triangular, pointed, pentagonal, and octagonal.

Types:

1. **Solid Microneedles:** Most of the solid microneedles are used by the formation of pores on the surface of skin to treat the disorders. The pointed tip of the needle is pierce into the skin; generate micron size channels through which the drug enters the straight away the skin layers when a drug patch is applied, thereby improving the permeation. The capillaries take the drug to demonstrate a systemic effect. It may also be used for a local impact. Solid microneedles supply the drug to skin layers with passive diffusion. Bioavailability, permeation, and mechanical strength were discovered to improve in different research.
2. **Coated Microneedles:** The microneedles are enclosed by the layer of dispersion of the drug solution or drug suspension (Waghule et al., 2019). The drug is successively liquefied from the coating and released promptly. The quantity of drug that can be loaded relies on the layer density and the needle size that is generally very small. Studies indicate higher stability.
3. **Dissolving Microneedles:** These are manufactured with biodegradable polymers in which already the drug molecule is encapsulated. Dissolution occurs after inserting the microneedle into the skin. Inside the skin, the polymer is degraded and regulates the release of drugs. The bio-acceptability and dissolution of the polymer within the skin create one-off finest long-term treatment

options with enhanced patient compliance. A significant factor facing issues while developing dissolving microneedles is effective needle drug distribution. Polymer-drug mixing is therefore a critical step in this manufacturing process. The polymer dissolve slowly distributed evenly on the surface of microneedles. Studies have shown fast and effective drug delivery, no skin irritation, enough mechanical strength.

4. **Hollow Microneedles:** The empty area is hold by a hollow microneedle provide space for the dispersion or solution of the drug to be filled. They've got the tips holed. The drug is placed straight into the EP or upper dermis layer when it is inserted into the skin. It is mainly used for compounds of elevated molecular weight such as proteins, vaccines, and oligonucleotides. The rate of drug flow and release pressure can be adapted if a fast bolus injection is to be used to administer the drug. The large amount of drug can be administered by microneedles as it is feasible to place within the needle in the empty area. It is vital to maintain a steady flow rate here. Increasing the bore of the microneedle may improve the flow rate but result in lower strength and sharpness. A metal coat is sometimes applied on the microneedle to boost the microneedles power; however, it may sharpen the needle.
5. **Hydrogel-Forming Microneedles:** This sort of microneedle has been created lately. Microneedles are made with super-swelling polymers. The polymers are the hydrophilic structure that enables them to absorb a large quantity of water into their polymeric three-dimensional network. These polymers swell owing to the existence of the interstitial fluid when inserted into the skin. This results in channels being formed between the drug patch and the capillary circulation. These microneedles are only used to interrupt the skin barrier before needling. They act as a rate-controlling membrane when swelling. They have size and shape flexibility. The distinctive characteristics of such microneedles are easily sterilized and completely removed from the skin (Waghule et al., 2019).

2.3.1.1 METHODS OF DRUG DELIVERY

- "Poke and flow" method; which involves inserting the MN into the tissue and then transporting the drug solution through the MN bore to a hypodermic needle in similar fashion, but hollow MN usually requires very precise and cost-effective manufacturing technology. Passive drug solution diffusion may happen through the MN, but active delivery enables faster delivery rates. Active delivery needs a driving force, a syringe can be used to push the solution into the tissue through the MN, but some studies have combined the MN systems with a pump or pressurized gas.

- "Poke and patch" primarily for strong MN by piercing the upper parts of the skin with solid MN and producing micro-channels followed by the use of drug formulations (e.g., patch, gel) at the piercing site. Pretreatment of the skin produces micro-conducts in the skin, improving the molecular flux through the skin.
- "Poke and coat" by penetrating drug-coated solid MN into the skin that solves the two-step implementation issue and provides incredibly fast drug delivery. It has been discovered that delivery from covered MN is appealing, particularly for molecules of elevated molecular weight. Due to the small size of the MN shaft and tip, however, drug delivery is restricted.
- "Poke and release" involve release of drug into systemic circulation from the dissolvable/porous/hydrogel-forming MN. The materials from which the MN is manufactured act as drug depots holding the drugs until the release trigger happens, i.e., dissolution. This approach eliminates the need for sharp disposal and the ability to reuse MN accidentally. In addition, dissolvable MN patches in "poke and release" methods were revealed to produce both small (MW 500 Dalton) and macromolecules (MW 500 Dalton) effectively.

The drug delivery through the topical path follows the mechanism of diffusion. The skin is momentarily interrupted in the microneedle drug delivery scheme. A microneedle device is formed by organizing hundreds of microneedles in groups on a small patch (same as a normal transdermal patch available on the market) to deliver enough drug to provide the required therapeutic response. The SC is pierced by it thereby bypass the barrier layer. The drug is placed straight in the EP or upper dermis layer, which then enters the systemic circulation and demonstrates a therapeutic reaction when the site of action is reached (Waghule et al., 2019). As a novel and minimally invasive strategy, for small drugs, macromolecules, nanoparticles, or fluid extractions, MN is capable of generating superficial pathways across the skin to obtain increased transdermal drug delivery. Their sharp tips are brief enough to restrict contact with skin nerves, thus preventing pain sensation and are small enough to cause minimal trauma and decrease the possibilities for post-insertion diseases to grow. This technique combines the effectiveness of standard needles for injection with the comfort of transdermal patches, while minimizing the disadvantages of these techniques of administration. In addition; MN can be produced using different material kinds such as polymers, metal, or silicon (Donnelly et al., 2010). Biocompatible and biodegradable polymers can be applied securely to the skin and are cost-effective in general. Different polymeric materials such as poly-L-lactic acid, poly-glycolic acid (PGA), poly-carbonate, poly-lactic-co-glycolic acid (PLGA), polydimethylsiloxane, poly-methyl vinyl ether-alt-maleic anhydride copolymer, carboxymethyl cellulose (CMC), maltose, dextrin, and galactose were all used to make MN. MN can also supply a broad variety of drugs from low molecular weight such as ibuprofen to elevated molecular weight such as ovalbumin compounds. In developing nations, immunization programs through MN could be implemented with minimal medical training and reduced related expenses. Moreover, MN arrays have lately been used as an alternative strategy to minimally invasive sampling of patient liquids, without causing pain or bleeding in advancing new therapeutic drug monitoring systems. It is also used for delivery of oligonucleotides, vaccine therapy, peptide delivery, delivery of hormones, cosmetics, delivery of lidocaine, pain therapy, ocular delivery, cancer therapy, etc. (Yang et al., 2019).

2.3.2 MICROEMULSION

The word microemulsion is used for a clear, thermodynamically stable, or kinetically stable dispersion of two liquid phases in which one is water and one is oil. It is stabilized by a surfactant and co-surfactant interfacial film. "Microemulsion is a water, oil, and amphiphilic system that is thermodynamically stable and an optically isotropic fluid solution," according to Danielsson and Lindman (Kogan and Garti, 2006). The microemulsion is a versatile carrier with its numerous notable characteristics such as improved poorly soluble drug bioavailability, elevated absorption, and permeation due to tiny droplet size and extremely low surface tension as well as cost-effective approach. The microemulsion is basically emulsion of less than 0.1 μm of droplet size. Because of their small size (much lower than the length of the visible light wavelength of 400–800 nm), these droplets are invisible. They are unable to reflect the light and are not noticeable through the optical microscope, making the microemulsion system transparent. Microemulsion is an appropriate drug carrier mechanism for nearly all types of drugs, including both lipophilic and hydrophilic moods. Usually, the microemulsion was characterized by the ternary phase diagram. The elements representing three corners of the ternary phase diagram are oil, water, and surfactants. Co-surfactant used in microemulsion is combined at a set proportion with surfactant and handled as a pseudo-component. Microemulsion appear to be perfect for drug delivery as they provide all possible liquid system requirements including thermodynamic stability (extended shelf life), easy to prepare (No interfacial tension and practically extemporaneous formation), low Newtonian conduct viscosity, elevated surface area (high solubilizing ability) and very small droplet size. Small droplets have better chances of adhering to membranes and more regulated transportation of bioactive molecules. The microemulsion vehicle is used for to solubilized water-insoluble and oil-soluble parts from various plant extracts to achieve synergistic impact for a particular therapeutic objective.

Microemulsion can be orally, topically on the skin, or nasally brought into the body as an aerosol for direct pulmonary entry. Due to their excellent potential in many applications, microemulsion has been subject to countless research over the past centuries. Many scientists have created important attempts to gain a better knowledge of these microstructures due to their rather complex phase conduct and the intriguing microstructures found in microemulsion forming structures. Comprehensive reviews have been published on the microstructures found in stages of microemulsion. It is also important to evaluate the microstructure of the drug in the charged microemulsion to explore the potential of microemulsion as delivery cars. In some cases, the microstructure of the system may be altered due to molecular interactions between the loaded drug and the microemulsion. The evaluation of the system is relatively a difficult assignment due to the complexity of the microemulsion and the several types of constituents added in the manufacture of the microstructure, as well as the restriction allied with individual method. Viscosity and conductivity measurements, as well as pulsed gradient spin echo (self-diffusion) NMR, dynamic light scattering (DLS), cryo-TEM, small-angle neutron scattering (SANS), and small-angle X-ray scattering (SAXS), remain some of the major advanced methods relevant for characterizing microemulsion (Kogan and Garti, 2006). Routinely the portion of the microemulsion includes oil phase, surfactant phase, co-surfactant phase and phase aqueous.

Due to their elevated solubilizing ability reported for both type– hydrophilic and lipophilic active ingredients, the implementation of microemulsion vehicles for cutaneous drug delivery is becoming increasingly common. It has been shown that microemulsion permeation rates were substantially greater than standard emulsions. The strong interactions between the surfactants that occur in the interfacial membrane film in emulsions (creams, lotions, etc.), limit the drug's mobility within the formulation between the internal and external phases. Furthermore, emulsions are not stable formulations and powerful bioavailability fluctuations have been identified. In microemulsion, the co-surfactant helps to lower the surfactant film's interfacial tension, leading in a microemulsion forming spontaneously and encouraging its thermodynamic stability. The thermodynamic process of drug diffusion between the microemulsion phases through the flexible interfacial surfactant film may improve partitioning and diffusion into the SC. The microemulsion droplets formulation leads to enhanced surfactant concentration in distributed structures that will reduce its permeation to the skin. Some performance of microemulsion in cutaneous drug application is repaglinide charged microemulsion gel, Minoxidil charged microemulsion, Dexibuprofen loaded transdermal microemulsion patch, Methyl dihydrojasmonate loaded transdermal microemulsion, T4 microemulsion charged bacteriophage, Zaltoprofen charged microemulsion, Celecoxib charged microemulsion gel, Amphotericin B (Amp B) charged microemulsion (Shukla et al., 2018).

2.3.3 MICROPORATION (SHARMA ET AL., 2019)

Microporation-based drug delivery systems used to deliver drug molecules across biological obstacles have been defined as one of the few improvement techniques of the third generation that will have a major effect on medicine. Microporation systems were also suggested to collect biological fluid samples from the body in relation to delivering drugs to the body. Numerous skin microporation techniques have been explored, including laser or thermal ablation, iontophoresis, electroporation, radiofrequency, microneedle, ultrasound/phonophoresis, and fluid or powder microporation. Skin interstitial fluid sampling methods include subcutaneous sensor implantation, inverse iontophoresis, microdialysis, and wick and microcannula implantation. The biological barrier in the microporation technique is utilized to develop aqueous pathways, which enhances transdermal delivery.

2.3.3.1 RADIOFREQUENCY MICROPORATION (BHOWMIK ET AL., 2010)

The percutaneous diffusion can be facilitated by using an alternative electrical current at a radio frequency (RF) of 100 to 500 KHz leads to the removal of an external strata of the skin. The flow of the current across the cells in the uppermost layer of the skin through a variety of microelectrodes, positioned on the surface of skin, circulates ionic vibrations across the skin cells leading to localized heating, liquid evaporation, and cell removal. As a result, transient aqueous micro-channels are created through SC and EP, called RF-micro-channels, enabling, or increasing the efficient transport of water soluble drug or excipient across the skin. The micro-channels created do not reach the underneath blood capillaries and nerve endings, thus minimizing the danger of skin trauma and cellular stimulation. Some of the devices that use this microporation form are as follows:

- A patented transdermal patch consisting of a plurality of microelectrodes coupled with a control unit that applies electrical energy between the electrodes, creating micro-channels throughout the skin.
- Patented electronic pharmacological delivery card using RF technology and extra characteristics to improve active substance administration.
- Patented device creating micro-channels in SC by producing sparks that, in turn, remove the skin's exterior layer in a much localized area.
- Patented handheld device that allows micro-channel generation while the device is rolled over or slid across chosen area of the skin.
- A patented technique for transdermal water-insoluble drug delivery, preferably steroids, using the ViaDerm scheme.

2.3.3.2 THERMAL MICROPORATION

The flux of the drug across the EP is well recorded and interpreted as temperature susceptible. Consequently, this factor has been utilized in the recent time to modulate transdermal and subsequent distribution systems for drug delivery. It is initiated with patented transdermal patch systems that applied heat to manufacture the patch instead of increasing drug flux across the skin. In comparison, biological membrane heat microporation was used to improve drug delivery. Thermal skin microporation includes applying fast, regulated heat power pulses to a specified site on the skin surface by means of miniaturized resistive components to generate macropores. For several milliseconds, the thermal energy is distributed over the set of tiny components, causing flash vaporization of SC cells in an area about a human hair's width. Devices that used heat straight to boost drug flux across the skin in the 1990s became more common. Many new instruments and techniques that use the microporation impacts that are accomplished when thermal energy is focused on the skin and other biological membranes have been researched over the past 20 years. More sophisticated devices have been created in latest years, some of which have reached complete clinical path and mass production. Some of the devices are as follows:

- A standard transdermal patch consists of a conductive top back layer next to a drug-permeable adhesive polymer layer. Before use, a release layer was removed and the patch was applied to the skin. Then heat generation of the device and skin could happen either by applying an external heat source or by instigating an exothermic reaction inside the real device. As such, this heat generation would result in a sufficient temperature rise (although below 45°C) to permit increased conveyance of active pharmaceutical ingredient across the skin.
- A patented device for the painless creation of micropores (1 to 1,000 μm) in biological membranes, such as skin outer layers or mucosal linings, facilitates the trans-membrane transport of substances to the body. This was accomplished by vaporizing water and other vaporizable substances on the skin or mucosal linings by putting a heated probe on the target region to generate micropores.
- A patented method where a thin film tissue interface (TFTI) between the biological membranes and the controller portion of the device was used to achieve microporation. This invention provided an enhanced and low-cost TFTI device that created micropores using thermal energy generated by passing through resistive components of electrical current.

- A device consisting of a micro-heater with a base end and an interior volume to house a material for ablation. The activation of the microheater linked to a power supply triggers the development of the ablation material that reacts by puncturing the SC mechanically to generate a first-depth microporus.
- Patented technique for improving biological tissue microporation using thermal energy in conjunction with other methods such as sonic, electromagnetic, and mechanical methods.

2.3.3.3 LASER MICROPORATION

Skin ablation or other biological membranes can be accomplished by using a laser emitting at a specified wavelength that is directly absorbed into micropores by the tissue. Laser energy absorption creates immediate vaporization of the tissue owing to flash water evaporation. The resulting micro-explosion leads to ablation of tissue. This fast loss of energy from the ablated site protects the adjacent tissue from harm caused by heat. The skin ablation is accomplished by the two ideal wavelengths are ultraviolet and mid-infrared short-wavelengths, which are respectively captured by tissue proteins and tissue water. The region of SC removed can be regulated effectively by changing the quantity of energy supplied to the skin, particularly when applied at reduced concentrations of energy. Some devices:

- A patented portable microporator device. The device consists of a laser beam shaping component that reshapes the distribution of the laser beams energy intensity, in particular the focal range, focal depth and focus on the biological membrane to create multiple uniform micropores.
- A patented biological membrane microporation device consisting of a feedback mechanism for scanning the created pores and possessing the ability to reshape the micropores, so that the initial and final pores created remain similar.
- A gadget called Epiture Easy touch appealed to be a revolutionary drug delivery tool that utilizes a laser to remove the SC painlessly, allowing for fast percutaneous absorption of topical local anesthetics (Bhowmik et al., 2010).

2.3.3.4 ULTRASOUND (PHONOPHORESIS/SONOPHORESIS) MICROPORATION

Ultrasound is described as noise with 0.02–10.0 MHz frequency and 0.0–3.0 W/cm^2 intensity range. Sonophoresis, also known as phonophoresis, explains ultrasound impacts on drug motion via unbroken skin into soft tissue. Ultrasound-enhanced medication delivery has a variety of major advantages as it is non-invasive, can be closely monitored and can penetrate into the body to the required depths. The early use of ultrasound in transdermal drug delivery as a physical enhancer was created almost 50 years ago. Exposure to ultrasound of a biological membrane creates sonoporation that is the cell membrane provisional perforation. This transient state increases the permeability in cells and tissues of therapeutic agents. A transducer, generates the ultrasound waves, composed of a piezoelectric crystal that describes the frequency of radiated waves, converts electrical energy in the form of oscillations into mechanical energy, generating acoustic waves. A wave is partly dispersed and absorbed by the medium during the propagation through a specified medium, leading to diminution of the transmitted wave with the energy lost being transformed into heat. You can emit ultrasound either continually (constant mode) or sequentially (pulsed mode). There is no specific interpretation about the mechanism of ultrasound impacts on skin in drug delivery. However, various mechanisms were suggested.

2.3.3.5 ELECTROPORATION MICROPORATION

Electroporation or electropermeabilization is a temporary disturbance of the anatomical phospholipid bilayer of physiological membrane triggered by applying elevated voltage pulses (almost 50 V) and may allow the membrane to be permeable to macromolecular organisms such as DNA. When a brief pulse is applied to create a potential for supra-breakdown across the membrane, the membrane breakdown will result in microchannels being created (Bhowmik et al., 2010).

2.3.4 LIQUID JET INJECTORS

A high-speed jet is used in liquid jet injection, to pierce the EP and transport the drug devoid of a needle. In the early 1930's, research on jet injectors started by Arnold Sutermesiter who remarked fortuitous diesel oil injections into workers hands when insignificant leaks happened in high-pressure lines (Subedi et al., 2010). Since then, two primary types of liquid jet injectors being classified. It includes single-dose jet injectors, well-known as DCJIs and MUNJIs (multi-use-nozzle jet injectors). Some DCJIs can only be partially removed while others can be fully removed. MUNJIs had no disposable components and were implemented for rapid mass immunization However, owing to their use; their use was discontinued in the aftermath of reports of hepatitis B spread in the 1980s. Cross-contamination was believed to be the cause of the epidemic by splashing back interstitial liquid from the skin to the nozzle. Commercial liquid jet injectors have a fundamental design consisting of drug-loaded compartment, piston, an energy source (spring or compressed gas) and orifice nozzle typically varying from 150 to 300 μm. The energy source pushes the piston that affects the drug-loaded compartment after triggering the actuation mechanism, resulting in a rapid rise in pressure. This forces the drug solution as a liquid jet with velocity ranging from 100 to 200 m/s through the nozzle orifice. In fact, the jet is turbulent and the jet's diameter is related to that of the orifice but rises with the traveled range. The jet punctures through the skin when it impinges on the skin and initiates the formation of the hole. Because of a mixture of skin erosion and fracture, the formation of a hole is thought to be finished during the first few 100 microseconds (Subedi et al., 2010). As the jet advances deeper into the skin, speed reduces until there is not enough energy to continue the development of the hole. This finishes the first injection stage, i.e., unidirectional puncture of the skin, and follows the second phase, multidirection jet dispersion from the endpoint of permeation. Furthermore, from this phase, the diffusion of liquid appears to be approximately hemispheric, the shape of which is controlled by jet power. MUNJIs were used for disease mass immunization programs, including measles, smallpox, malaria, hepatitis B, influenza, and polio (Subedi et al., 2010). Several proteins were delivered using DCJIs.

2.3.5 POWDER JET INJECTORS

Powder jet injectors bring dry powdered vaccines or drugs into the skin's shallow layers. Also frequently used for these injectors were the words biolistic injectors and gene weapons, the latter being used solely for DNA transmission (Subedi et al., 2010). Early research on injecting strong micro-particles into biological samples was documented, showing transfection of plant cells using nucleic acid-coated tungsten particles using DNA and RNA. Researchers have since studied the potential of this method for protein delivery, gene therapy, and traditional and DNA vaccination applications. Basic design of strong jet injectors includes compressed gas as the source of energy, a drug section comprising drug particulates formulation, and a particle flow nozzle (Subedi et al., 2010). The drug compartment is closed on either side with diaphragms that are usually few microns dense. Compressed gas from a storage canister extends and pushes against the diaphragms when the actuation mechanism is triggered, breaking them sequentially. The gas stream brings with it the particles of the drug. The particles then exit and impinge on the skin through a nozzle. Particles puncture micron-sized holes into SC due to their momentum when they impact the skin. More or less few particles were retained in SC whereas substantial percent of them cross the EP for the required therapeutic effect. Another structure used to study powder injection processes is a light gas piston that utilizes an accelerating piston to deliver required particle velocity (Arora et al., 2008). The piston accelerates and takes the particles with it when the actuation mechanism is triggered. A mechanism of deceleration forces the piston to slow down, causing the particles to leave the piston surface. The particles are expelled and affect the surface of the target tissue. The particle size, density of particle and velocity at which it impacts are key parameters in determining particle distribution across the SC. Particles are powdered drug or vaccine preparation and range from 10 to 20 μm. Solid jet injectors were studied for the provision of DNA coding for bacterial and viral antigens by means of coated gold microparticle. Mice showed induction of humoral and cell-mediated immune reaction to malaria, hepatitis B, and rabies. Tumor protection was also proved by injecting in polymeric particles DNA-coated gold micro-particles and DNA encapsulated.

2.4 NANOSYSTEMS IN TRANSDERMAL DRUG DELIVERY

2.4.1 LIPOSOMES

Liposomes are hollow cholesterol and phospholipid spherical vesicles. Alice Bengham introduced these for the first time (Bibi et al., 2017). Bilayers form phospholipids that contain water in their middle and dissolve it in an aqueous medium. The polar head of the phospholipid (lecithin) is an interface to the aqueous medium. Inside the vesicle, hydrophilic agents are incorporated into the water phase, while lipophilic agents are loaded into the bilayers. The hydrophilic agent, however, can also be loaded to the outer surface in a high amount. Liposomes have excellent ability to permeate drugs and other cargo through the skin's epidermal barrier as liposome. Liposome structure is comparable to EP lipids. Currently, two popular techniques, thin film hydration and solvent injection, prepare the liposomes. Most liposomes are hydrated through thin film. Liposomes can be categorized into categories depending on the preparation method and the technique used: small unilamellar vesicles (SUVs), large unilamellar vesicles (LUVs), and multilamellar vesicles (MLVs). SUVs and LUVs are made up of one lipid bilayer only, whereas multilamellar vesicles are made up of many bilayers. The dimensions of SUVs, LUVs, and MLVs, respectively, are 10–100 nm, 100–500 nm, and > 500 nm. The physicochemical characteristics of liposomes influence the drug's skin-wide penetration. Size is the most significant among them. The skin has tiny pores of 0.3 nm, which can be opened up to 20–40 nm without significant skin illnesses. The liposome size influences the delivery of transdermal drugs via intercellular and transcellular processes. By raising the size of the liposome, the penetration of the encapsulated drug is reduced. If the liposome invades from the skin appendage unit, then the size does not affect the penetration. The thermodynamic condition of liposomes also plays a crucial part in penetration in relation to size. It relies on the cholesterol structure and proportion and the nature of the used phospholipids.

In transdermal drug delivery, the primary aim of the liposome is to generate localized or systemic impacts. Much research has been carried out to study possible interactions between keratinocytes and liposomes using human skin recreated. The mechanism for delivering liposomal drugs to the skin is still not evident. Several processes have been demonstrated in the literature, including: (i) adsorption of the SC liposomes and junction with lipid domain of the SC where lipid exchange occurs between the liposomal phospholipid and the SC; (ii) disintegration of liposomes on the surface of the skin by disrupting the cell packing and fluidity of the lipid bilayer of the SC; (iii) Liposomes that may act as a drug reservoir by penetrating into pilo-sebaceous units; and (iv) Liposomes that may be penetrated into the SC in intact form. Despite all their distinctive characteristics, however, the main issues that limit liposome formulation in transdermal drug delivery are stability, low drug loading, sterilization complexity, and batch-to-batch reproducibility. Because of aggregation, fusion, and vesicle leakage, oxidation, and ester hydrolysis, they undergo a shift in vesicle size. The liquid state of liposomes produced from non-saturated phospholipids is usually connected with these issues. Using hydrogenated phospholipids in the formulation, these are problems that can be resolved and incorporated into the gel matrix at the end. The use of such phospholipid kinds, although issues linked to physical stabilization owing to viscous nature have been fixed, but have a small penetration capacity compared to liquid liposome. In relation to instability, the formation of a reservoir in the SC to create a localized impact of drugs and cosmetics is another liposome-related issue. New classes of lipid vesicles are explored to solve this disadvantage, including transferosomes, ethosomes, invasosomes, and propylene glycol-containing liposomes. They show pronounced elasticity and superior penetration of the skin through the transdermal pathway.

The first liposomal formulation marketed in 1988 is Econazole (Pevaryl 1) Lipogel (Bibi et al., 2017). It was in place for topical dermatoses mycosis therapy. The list of drugs delivered in the form of liposome via skin includes lignocaine, ketoprofen, melatonin, clindamycin hydrochloride, amphotericin B, methotrexate, and estradiol. The transfersomes encapsulate the following drugs include corticosteroid, insulin, triamcinolone-acetonide, diclofenac, superoxide dismutase, tetanus toxoid, and DNA.

2.4.2 NIOSOMES (BIBI ET AL., 2017)

Niosome, also known as elastic vesicles based on nonionic surfactants, are molecular mass developed in an aqueous phase by the self-association of nonionic surfactants, as well as cholesterol or its derivatives in many cases. Niosomes are commonly explored as nonionic surfactant vesicles as an option to liposomes. Niosome-based formulations have been shown to be superior in structure and stability to liposomes in a number of cases, and they are economical to produce. These vesicles are biocompatible, non-immunogenic, and biodegradable. Niosomes are used to deliver drugs aimed and monitored. Their manufacturing techniques include reverse-phase evaporation, ether injection, and hand-shaking method (they enhance drug permeation across the skin by disrupting the SC lipids due to the existence of a surfactant and the vesicle's versatile nature. Another method for altering the barrier function is adsorption and fusion with the skin surface, followed by adsorption and fusion with the skin surface. On the basis of size of vesicles, the niosome referred as second generation elastic vesicles classified into three class: (a) SUV (10–100 nm); (b) LUV (100–3,000 nm); and (c) multi-lamellar vesicles (MLV) – bilayer vesicles are entrapped in large vesicles. Niosomes include proniosomes, surfactant ethosomes, elastic niosomes, polyhedral niosomes, discosomes (disk-shaped vesicles), aspasomes (ascorbylpalmitate vesicles), and so on, without considering the size. Phospholipid will not serve as a carrier material for niosome compared to liposomes, which prevents phospholipid oxidative degradation resulting in elevated chemical stability. There are many kinds of surfactants that are readily derivative and offer greater versatility to the niosomes, no unique circumstances are required for their manufacturing and storage, and the method is streamlined with a reduced price to accomplish large-scale production. The solubility and bioavailability of poorly soluble drugs is enhanced by niosomes which are osmotically active. In addition, trapping lipophilic drugs into bilayer membranes of vesicular structure and hydrophilic drugs in the aqueous compartment, accessibility of prepared products in pure form, ease of handling are several benefits of niosomes.

Niosomes can boost drug residence in both the SC and the EP, while decreasing the drug's systemic absorption. The niosomal permeation mechanism is realized primarily through nonionic surface-active agents. Niosomal transdermal characteristics are comparable to liposome characteristics. The close contact between the skin's niosomes and the SC and subsequent adsorption or fusion on the skin's surface leads to a rise in the drug's gradient of thermodynamic activity at the interface, leading to a rise in lipophilic drug permeation. Second, niosomes enhance the property of SC by decreasing the loss of trans-epidermal water, which can boost the hydration of SC and loosen its closely packed cellular structure and enhance smoothness by replenishing lost skin lipids. Third, niosomes lipid bimolecular layers act as a rate-limiting membrane and play a slow-release impact at the local level. The quantity of drug permeation also has to do with the characteristics of the drug encapsulated. The systemic circulation time in the body for drug may be extended if the drug is encapsulated by the niosomes, thus enhancing the drug's penetration into the target organ, and reducing the curative impact. Transdermal delivery effectiveness relies on the nature and concentration of the nonionic surfactant and cholesterol content in niosomes. Niosomes are the successful drug delivery carriers who discover their application in the delivery of dermal and transdermal drugs. In the therapy of a number of dermatological illnesses, they play an important part in transdermal drug delivery. Ellagic acid and transdermal formulation of minoxidil niosomal are now accessible on the market. Minoxidil collected in niosomes is a promising strategy in the therapy of hair loss. Other study on the niosomal formulation of span 60 and tween 60 has shown that these are efficient carriers of ellagic acid dermal delivery. Due to the flexible nature of the niosomes, the transdermal formulation of enoxacin containing span 80 as a surfactant showed high skin penetration compared to its free form and liposomal formulation. Anti-acne agent (e.g., tritinone) when trapped in niosomes showed higher capsulation efficiency and also protected labile agents against degradation.

2.4.3 ETHOSOMES

Ethosomes are novel nanostructures consisting of phospholipids, ethanol, and water. Elka Touitou first created these nanostructures in 1997. These are soft and flexible vesicles can be prepared from the tenths of a nanometer to the range of a micrometer. They are the modified liposome form with high ethanol content (Bibi et al., 2017). Phospholipids in ethosomes may be phosphatidic acid, phosphatidylserine, or phosphatidylcholine, and may be ethanol or isopropyl alcohol. It has been stated that ethosomes improve the delivery of drugs into the systemic circulation through the profound layers of the skin and permeate the SC intercellular route by disturbing the lipid bilayers organization. Due to the existence of ethanol in large concentration the lipid membrane is less tightly packed, has a more malleable structure and allows for better distribution of the drug through the SC lipids. It has been researched that ethosomes improve drug delivery *in vivo* and *in vitro* through the skin under both occlusive and non-occlusive circumstances (Bibi et al., 2017). The mechanism for entry of the incorporated drug from ethosomes is based on synergistic action between skin lipid parts, ethanol, and vesicles. The precise interaction mode between ethosomes and skin lipids is not yet apparent. This nanostructure, however, is supposed to deliver the encapsulated drug on the grounds of two impacts: ethanol and ethosome. The ethanol effect includes the interaction of ethanol with the lipid molecules 'polar head area, which causes their enhanced fluidity and reduced multilamellar layer density by decreasing the lipid transition temperature in the SC. Due to their malleable nature, the ethosomes fuse with the skin lipids at different points. This results into the opening of new paths leads to diffusion of the encapsulated drugs through the different coats of the skin.

Ethosomes are multilayer vesicles that contain elevated alcohol levels added to the prescription of standard liposomes. The ethosome mechanism that enhances permeability is associated with its structure and elevated alcohol concentration (Zhou et al., 2018). The elevated alcohol concentration improves the mobility of the lipid molecules polar lipid heads, resulting in enhanced lipid fluidity and membrane flexibility. It also reduces the intercellular lipid domains density, causing deformation of the ethosomes during transmission. They communicate with the troubled SC (changing the intercellular lipid lamella) as well as with the deranged SC, creating their own routes to cross the SC and reach the profound layers of the skin to increase their permeability. Second, the elevated alcohol level improves the lipid layers drug solubility as it also dehydrates and the fats-free SC, which encourages long-term permeation. At the same time, it can change the tight arrangement of the SC lipid bilayers, increases the union of ethosome with lipids of SC, decrease the phase-transition temperature, and also promote percutaneous drug absorption. The higher will be the content of alcohol in ethosomes, the greater the flux of permeation. The release and permeation of the drug in the profound skin is results from the fusion of ethosomes with the skin lipid. Many cosmetic and pharmaceutical industries are engaged in successful studies in formulating ethosome-based drug delivery systems for market use. Examples of drugs available in transdermal formulations of ethosomes are tacrolimus, an immunosuppressant used for atopic dermatitis (AD), clotrimazole, trihexyphenidyl HCl, methotrexate, rhodamine, 5-aminolevulinic acid (ALA), ketoprofen, and testosterone.

2.4.4 LIPID NANOPARTICLE (BIBI ET AL., 2017)

Lipid nanoparticles are innovative drug delivery systems that have been created for dermatological and cosmetic reasons because of their notable characteristics such as physical stability, controlled release, and outstanding tolerability. Two kinds of lipid nanoparticles are available: SLN (solid lipid nanoparticles and NLC (nanostructured lipid carriers). Two scientists, Rainer H. Muller, and Maria Rose, established SLNs in 1990 as an appealing alternative to emulsion, liposomes, and polymeric nanoparticles. SLN are produced by solid lipids or a combination of solid and liquid lipids for the liquid emulsion. They are composed of three ingredients: (i) solid lipid (0.1%–30% w/w); (ii) water; and (iii) emulsifier (surfactants 0.5%–5% w/w). In the aqueous stage, which is preferably stabilized with surfactants, the strong lipids are spread. An SLN is a colloidal carrier system that combines the benefits of emulsion, liposomes, and polymeric nanoparticles in a single system for controlled drug delivery. The SLNs are able to modulate active compound release features and also boost skin hydration relative to liposomes and emulsion. However, during storage an SLN has some limits of low ability for drug loading and uncontrolled drug leaching from the carrier system. The lipid nanoparticles of second-generation well-known as NLCs has been created to tackle these problems. Compared to SLNs, they have greater load ability and less leaching characteristics. NLCs consist of solid and liquid lipids where a strong lipid matrix is placed in a liquid lipid or where the liquid lipid is located on the surface of a strong lipid matrix with a surfactant coating. Cosmetics and pharmaceuticals can be loaded into NLCs and SLNs. They are frequently used in topical drug delivery due to protection from the enzymatic degradation and transepidermal water loss of the encapsulated agent and controlled release of the therapeutic agent for extended span of time. Thereby increasing the overall therapeutic action of active compounds. In TDDS, there are several advantages for SLNs and NLCs (e.g., they can be given directly to the skin in sufficient quantities with fewer side effects). The biodegradable as well as physiological lipids are utilized to prepare SLNs and NLCs. Hence, they are nonirritant and non-toxic and can be applied to skin that is harmed and inflamed. The enhanced skin permeation is defined as a mixture of three processes: adhesiveness, occlusive, and hydration of the skin. These lipid nanoparticles of less than 200 nm leads to formation of a monolayer on the EP displaying (i) adhesiveness and occlusive; (ii) avoidance of transepidermal water loss; and (iii) opening of intersection between corneocytes, facilitating the penetration of therapeutic agent deep into skin layers. When a SLN loses water content, its strong matrix is altered by crystallization that results into expulsion of drug encapsulated in the carrier system subsequently leads to its penetration into profound skin layers. The occlusive property of lipid nanoparticles is influenced by a several variables such as particle size, crystalline nature, lipid concentration and applied quantity. Nevertheless, these particles are usually deemed not to penetrate the corneocytes, so one of the paths is permeation through hair follicles (Uchechi et al., 2014).

Possible lipid nanoparticles mechanisms promoting percutaneous drug permeation are as follows: lipid nanoparticles have excellent skin adhesion, encouraging contact between drugs and SC, and forming a monolayered lipid film. Through capillary force, the lipid nanoparticles fuse and deform which results in the formation of hydrophobic film onto the skin surface, both resulting in an occlusion impact. This impact reduces the water evaporation occurs naturally from the skin surface, improves skin hydration, reduces in keratinocytes packing, and reduces the inter-keratinocytes gap, which improves the penetration of drugs deeper into the skin. Second, the lower the particles size of system, the faster the penetration of the skin. The tiny particle size may provide a greater particular region of the surface and may improve the occlusive impact' for transdermal penetration. Third, nonionic surfactants also play a part in encouraging the permeability in the composition of lipid nanoparticles as they improve the thermodynamic stability of NLC and SLN. This causes the disturbance of the skin structure. The concentration of lipid and its type, the carrier for the system and the location of the drug in the nanoparticle framework may also affect the system. Fourth, as far as SLN is concerned, the reservoir effect is existed on the skin surface and SC, because they have been found on the skin surface and in the SC leads to formation of the drug reservoir into the bottom layers of the skin. In a biphasic regimen, they release the active substance – an initial burst release from the particle surface and an aqueous phase, followed by an effect of reservoir. The drug in the NLC continues in the liquid lipid surrounded by the strong lipid, according to some reports. This arrangement provides the drug some mobility, even if the strong lipid undergoes polymorphic modifications, it provides stability. In short, the lipid nanoparticles skin adhesion and 'occlusion impact' is the primary mechanisms that promote permeability (Rane et al., 2018).

A number of techniques for preparing lipid nanoparticles have been widely documented in literature. These include the technique of microemulsion, evaporation of emulsification solvents, method of diffusion of emulsification solvents, method of injection of solvents or displacement of solvents, technique of various emulsions, technique of homogenization of high pressure, technique of contractor of membranes and technique of ultrasonication. The warm and cold homogenization method can also prepare lipid nanoparticles, particularly SLNs. In cosmetics, SLNs are commonly used for the topical therapy of skin infections in particular the acne, psoriasis, and atopic eczema, as well as for rheumatoid arthritis. The transdermal flurbiprofen formulation prepared by the aqueous dispersion technique in SLNs and NLCs has enhanced permeation through the SC and overcomes the side effects of the gastrointestinal tract connected with oral medicine. Examples of drugs encapsulated delivered topically as well as by transdermal route in the form of SLN and NLC are celecoxib, coenzyme Q10, podophyllotoxin, artemether, econazole nitrate, and cyclosporine A.

2.4.5 NANOEMULSION

Nanoemulsion is low viscosity, isotropic, thermo-dynamic, and kinetically stable (without any obvious long-term storage flocculation or coalescence). The blend comprises of transparent or translucent petroleum globules, dispersed in aqueous stage, stabilized by an interfacial film of exceptionally small droplet-sized surfactant or co-surfactant molecules (mean droplet-size generally less than 500 nm). As for the particle size of the discussed nanoemulsion; the upper particle size limits of 1,000 nm, 500 nm, 200 nm, and 100 nm were proposed (Rai et al., 2018). It is not necessary to confuse the term nanoemulsion with microemulsion. Despite the fact that nanoemulsion has almost the identical constitution, approximate size of dispersed phase, and same look as that of microemulsion but the thermodynamic stability and structural aspects differ tremendously. Nanoemulsion very small particle size, big specific surface area and low surface tension offer them excellent wettability allow the system to remain in close proximity with the skin. Nanoemulsion also provides many other benefits including greater solubilization ability, physical stability, enhanced bioavailability, ease of preparing, less energy input, and longer shelf life. The transdermal time of nanoemulsion is shorter compared to frequently used topical skin preparations and the impact of percutaneous absorption is better. Nanoemulsion can be categorized on the basis of the parts into three kinds: oil in water (O/W: oil phase dispersed in constant aqueous stage), water in oil (W/O: water phase dispersed in constant oil stage), bi-continuous/multiple emulsion (micro oil and water phase microphases are inter-dispersed in the scheme). Reports show that O/W nanoemulsion are increasingly being used in pharmaceutical products as a delivery system to encapsulate lipophilic parts. The oil phase plays a crucial role in the formulation of nanoemulsion as it solubilizes lipophilic drugs intended to be used to treat a variety of disorders. In o/w type nanoemulsion, depending on the site of administration, the quantity of oil may differ from 2 to 20% w/w. Based on their solubility and emulsification capacity, the finest emulsifier devices are chosen from a pool (Rai et al., 2018). Nonionic surfactants are used in many cases because they are less toxic and irritating than their anionic counterparts and their cationic counterparts in particular. Selection of emulsifiers is based on their oil and aqueous phase solubility, HLB value, and lower toxicity profile, etc. Nonionic surfactants with HLB value 8–16 are preferred in the preparing of o/w nanoemulsion. The sole use of single chain emulsifiers is less likely to decrease the o/w interface tension to an appreciable extent and therefore the emulsifier system uses a co-emulsifier of amphiphilic nature. The penetration of co-emulsifiers into the emulsifier interface film at the o/w interface enables to further reduce the interface fluidity and thus boost the entropy of the whole colloidal system. It is possible to formulate hydrophobic and hydrophilic drugs in nanoemulsion. They are non-toxic and non-irritating substances and can generally be applied onto the skin or mucous membranes and also have been used in the cosmetic sector. Nanoemulsion can be prepared primarily by three techniques: homogenization of high pressure, microfluidization, and temperature of phase-inversion. Nanoemulsion transdermal delivery has reduced owing to the stability issues intrinsic in this type of dosage (Rai et al., 2018).

Nanoemulsion promotes drug transdermal penetration by means of the following processes: First, the structure of nanoemulsion such as the surface active agent, co-surfactant, and petroleum phase may play a part in encouraging permeability by disturbing lipid arrangement in the SC. This leads to a rise in the quantity of drug transdermal permeation. Secondly, due to the distinctive nanoemulsion composition, both lipid soluble and water-soluble medications (particularly lipid soluble drugs) can be considerably solubilized. This results in a rise in the semipermeable concentration of drug on either side of the skin (i.e., concentration gradient), which in fact increases drug penetration. Third, the internal phase storage feature retains a steady driving force of the diffusion of drugs from the peripheral part towards the skin and extends the absorption of drugs. Fourth, by using the strong mixture with water to create a quicker percutaneous passage through the keratinocytes, Nanoemulsion extends the links between SC. They also release some level of the collagen fiber appears in the dermal part of the skin. Fifth, the nanoemulsion has a tiny particle size, a big specific surface area and a low surface tension that allows excellent wettability and enhanced skin contact, thus benefiting the drug's penetrability. Some examples of drugs deliver transdermally in the form of nanoemulsion are nimesulide, plasmid DNA, caffeine, methyl salicylate, gamma tocopherol, insulin, and aspirin. Transdermal formulations of nanoemulsion are currently not prepared in an extent to that of liposome or nanoparticle owing to the issues of stability intrinsic in this type of dosage. However, the use of these nanocarriers to produce/delivered corticosteroid, analgesic, anticancer agent, etc., is extremely essential because they can behave instantly because they do not need to cross additional obstacles (Rai et al., 2018).

2.4.6 POLYMERIC NANOPARTICLE (ZHOU ET AL., 2018)

Polymeric nanoparticles are strong colloidal carriers made from a polymer matrix comparable to synthetic, semi-synthetic or natural materials of 10 nm to 1,000 nm size. In the discipline of TDDS particularly the polymeric nanoparticle has attracted more attention because it is able to rout the restrictions of additional lipid systems thereby protecting the labile drug from deterioration and denaturation. They constantly releasing the active ingredient thereby increasing the concentration gradient to improve the percutaneous permeation of drugs. The polymeric nanoparticle can be categorized into polymeric micelles, nanosphere, and nanocapsule based on their preparation method and architecture. The most commonly used polymers include the synthetic-type polyesters namely polylactic acid, poly (DL-lactide-co-glycolide) (PLGA), polycaprolactones (PCL), poly-acrylates (PCA), carbomer, PACA, PMMA, and artificial polyesters such as sodium alginate, deacetylchitin, gelatin, or other polymer carriers. These polymer chains consist of two or more covalently bound homopolymer units and can be formulated into di-, tri-or multi-block, random, star, and graft. Under certain environment, they may construct a complex structure in particular synthetic membrane which imitate the cell membrane's lipid bilayer. However, the polymer chain possess high molecular weight that causes the structure of the polymeric membrane becomes highly inter-twined, which is the major difference in comparison to the lipid bilayer. Because of this, elevated mechanical strength and non-deformability of polymeric nanoparticle reported. Nanoparticles maintain the concentration of drug for prolonged period and spreading them into profound skin layers.

This path of drug permeation is of excellent significance as the storage time of the drug integrated into carriers only lasts 10 days. Sometimes the trans-appendageal pathway is negligible as it constitutes only 0.1% of the complete skin surface so that the drug can diffuse below the hair follicles through the unfinished corneocyte layer and generate its systemic effect.

The drug permeation processes promoted by polymeric nanoparticles are as follows: First, full polymeric nanoparticles are too big to spread straight into the profound place of the skin passively through the SC. However, for slow and steady drug released from nanoparticles, they conceivably arranged themselves onto the skin surface or hair follicles and subsequently face towards drug reservoir. This results in a greater local drug concentration in the polymeric nanoparticles as the drug diffuses at the concentration gradient into the active layer of the skin. Second, skin integrity and nanoparticles physicochemical characteristics influence the permeability of polymeric nanoparticles. Therefore, if the skin's integrity can be properly altered, it could promote the drug's percutaneous penetration in the nanoparticles. Third, their permeability is affected by the surface charge of polymeric nanoparticles. The extremely cationic elements boost the surface charge, so there is also a much greater binding takes place with negatively charged surface of the skin resulting in increased drug release from the nanoparticles. Fourth, the permeation of polymeric nanoparticles across the skin is further improved via the hair follicle. Transdermal drug delivery polymeric nanoparticles are less studied than lipid nanoparticles are. Research on developing a stable nanoparticulate structure and investigating pathways of permeation through the skin is still underway. For topical drug delivery, polymeric nanoparticles are frequently used. As a carrier in cosmetics, a biodegradable polymer like PLA is used as a sunscreen agent where it remains on the SC by casting a preventive film over the surface of the skin. Polymeric nanoparticles are used to supply therapeutic agents to different tumor kinds, bone healing therapies, and vaccinations. For example, 5-fluorouracil, flufenamic acid, chlorohexidine, and insulin are transdermal drugs that use nanoparticles.

2.4.7 TRANSFERSOMES

Transfersomes are a fresh class of liposomes also known as vesicles or liposomes that are deformable, ultra-flexible or elastic (Bibi et al., 2017).

They were first launched by GregorCevc in 1990 and have since been the topic of countless studies on patents and literature (Bibi et al., 2017). These vesicles have an internal aqueous compartment that is surrounded by an external lipid bilayer complex. The word transfersome was recorded as the trademark of its proprietary drug delivery technology by the German business IDEA AG. Because of the existence of an edge activator in the lipid bilayer membrane, they are elastic in nature. Transfersomes shape and structure assemble them into self-optimizing and self-regulating nanostructures that allow them to effectively cross different transport obstacles. These are non-invasive nanostructure for targeted delivery of drugs and sustained impact of release. Typically, transfersomes consist of two components. In an aqueous setting, amphiphilic agents such as phospholipids (phosphatidylcholine) self-assemble into lipid bilayers and form a vesicle nearby. The second agent is an edge activator that can be Span 80, Tween 80, sodium cholate, deoxycholate, or potassium glycyrrhizinate, also recognized as an edge activator (Carter et al., 2019). The edge activator in the membrane of the transfersomes destabilizes the lipid bilayers and reduces the vesicle's deformability. Phospholipids (egg phosphatidylcholine, soy phosphatidyl choline), surfactants (sodium cholate, Tween 80) and alcohol (ethanol, methanol) are the main ingredients required for the formulation of transfersomes. They are designed to prevail over standard liposomal constraints. These vesicles compress to one-tenth of their diameter to squeeze in comparison with liposomes through the SC intercellular lipid bilayers (Rane and Gujarathi, 2017).

Transfersomes are created by enhancing the liposome element. The surfactants integrated in the membrane of the vesicle promote a large degree of deformation in order to penetrate the pores several times larger than their own size. The transfersomes mechanism by which they enter the SC effectively can be summarized in three ways. First, surfactants can enter the SC and interact with it causes swelling. Consequently, the keratin becomes stiff and lipids get extracted from the SC completely disturb its arrangement. The crystalline and gel state of long-chain molecules enhances lipid fluidity and produces or expands hydrophilic voids, thereby improving the permeability of SC to water-loving substances. The transfersomes may penetrate the so-called "virtual channels" created by intercellular lipid lamellae with structural irregularities because of the "edge activators." Second, in the SC, the water content (15%) is smaller compare to other EP layers (75%), leads to water gradient. The hydration gradient for the skin is naturally created under non-occlusive circumstances. Transfersome gets shrink when enters in the opening of SC because of the elevated deformability. The study community embraced the concept of osmotic gradient and water mixture. Third, recent research has suggested that transfersomes are primarily passed through the path of hair follicles. When these vesicles are applied under non-occlusive circumstances, following the natural water gradient in the EP, they join the skin as a consequence of the surrounding mechanical stress in a self-adaptive way. Transfersomes function as great nanostructures for transdermal delivery of small and high molecular weight medicines such as sex hormones, gap protein, insulin, albumin, corticosteroids, and anesthetics, 5-fluorouracil, di-potassium glycyrrhizinate, ketotifen fumarate, interleukin-2, ketoprofen, triamcinolone acetonide (TAA), DNA, superoxide dimusate, tetnus oxide, and diclofenac.

2.4.8 INVASOMES (ZHOU ET AL., 2018)

Invasomes, presented by the Verma and Fahr associates, are the vesicles comprising of phosphatidylcholine, ethanol, and terpenes mixture. The bilayer matrix of invasome is prepared by soybean lecithin. Lysophosphatidyl choline is an edge activator that gives the phosphatidylcholine bilayers flexibility. Both ethanol and terpenes function as drug penetration enhancers as well as providing the phospholipid bilayers with fluidity or flexibility. Invasomes for both hydrophilic and lipophilic agents have been shown to be effective drug delivery systems. Invasomes could be a helpful carrier system for transdermal delivery of many drugs. Invasomes enhance hydrophilic and lipophilic drug penetration of the skin. The invasomes transdermal mechanism is found as follows. During their entry into the upper layers of the skin, one portion of the vesicles is divided. The terpenes and phospholipids produced may boost the penetration of many medicines by disrupting the SC tight lipid packing, which acts as penetration enhancers. Invasomal deformation is also linked with ethanol. They are able to accomplish this by transporting through the compact hydrophilic channel available in intercellular space of the SC. As conveyed the invasome also permeate through the hair follicle transport pathway. Invasome are used for transdermal delivery of Isradipine, Olmesartan, etc.

KEYWORDS

- **dynamic light scattering**
- **jet injectors**
- **microneedle**
- **microporation**
- **nanoemulsion**
- **poke and flow method**
- **radio frequency**

REFERENCES

Alkilani, A. Z.; McCrudden, M. T.; Donnelly, R. F. Transdermal drug delivery: Innovative pharmaceutical developments based on disruption of the barrier properties of the stratum corneum. *Pharmaceutics*, **2015**, *7*(4), 438–470.

Arora, A.; Prausnitz, M. R.; Mitragotri, S. Micro-scale devices for transdermal drug delivery. *International Journal of Pharmaceutics*, **2008**, *364*(2), 227–236.

Bhowmik, D.; Bhattacharjee, C.; Chandira, M.; Jayakar, B.; & Sampath, K. P. Recent advances in transdermal drug delivery system. *International Journal of PharmTech Research*, **2010**, *2*, 68–77.

Bibi, N.; Ahmed, N.; Khan, G. M. Nanostructures in transdermal drug delivery systems. In *Nanostructures for Drug Delivery;* Andronescu, E., Grumezescu, A., Eds.; Elsevier Inc., **2017**; pp. 639–668.

Carter, P.; Narasimhan, B.; Wang, Q. Biocompatible nanoparticles and vesicular systems in transdermal drug delivery for various skin diseases. *International Journal of Pharmaceutics*, **2019**, *555*, 49–62.

Chanda, S.; Bagga, M.; Tiwari, R. K. Microneedles in transdermal drug delivery: An unique painless option. *International Research Journal of Pharmacy*, **2011**, 72–78.

Donnelly, R. F.; Singh, T. R. R.; Woolfson, A. D. Microneedle-based drug delivery systems: Microfabrication, drug delivery, and safety. *Drug Delivery*, **2010**, *17*(4), 187–207.

Keservani, R. K.; Bandopadhyay, S.; Bandyopadhyay, N.; Sharma, A. K. Design and fabrication of transdermal/skin drug-delivery system. In *Drug Delivery Systems*; Elsevier, **2020**; pp. 131–178.

Kogan, A.; Garti, N. Microemulsions as transdermal drug delivery vehicles. *Advances in Colloid and Interface Science*, **2006**, *123*, 369–385.

Lee, H.; Song, C.; Baik, S.; Kim, D.; Hyeon, T.; Kim, D. H. Device-assisted transdermal drug delivery. *Advanced Drug Delivery Reviews*, **2018**, *127*, 35–45.

Pawar, P. M.; Solanki, K. P.; Mandali, V. A. Recent advancements in transdermal drug delivery system. *International Journal of Pharmaceutical and Clinical Research*, **2018**, *10*(3), 65–73.

Rai, V. K.; Mishra, N.; Yadav, K. S.; Yadav, N. P. Nanoemulsion as pharmaceutical carrier for dermal and transdermal drug delivery: Formulation development, stability issues, basic considerations and applications. *Journal of Controlled Release*, **2018**, *270*, 203–225.

Rane, B. R.; Gujarathi, N. A. Transfersomes and protransfersome: Ultradeformable vesicular system. In *Novel Approaches for Drug Delivery*; Keservani, R. K., Sharma, A. K., Eds; IGI Global Inc., **2017;** pp. 149–169.

Rane, B. R.; Jain, A. S.; Gujarathi, N. A.; Keservani, R. K. Aquasomes: A nanocarrier system. In *Nanoconjugate Nanocarriers for Drug Delivery;* Keservani, R. K., Sharma, A. K., Eds; Apple Academic Press, **2018;** pp. 263–302.

Sharma, M. Transdermal and intravenous nano drug delivery systems: Present and future. In *Applications of Targeted Nano Drugs and Delivery Systems*. Elsevier Inc., **2019**, pp. 499–550.

Shukla, T.; Upmanyu, N.; Agrawal, M.; Saraf, S.; Saraf, S.; Alexander, A. Biomedical applications of microemulsion through dermal and transdermal route. *Biomedicine & Pharmacotherapy*, **2018**, *108*, 1477–1494.

Subedi, R. K.; Oh, S. Y.; Chun, M. K.; Choi, H. K. Recent advances in transdermal drug delivery. *Archives of Pharmacal Research*, **2010**, *33*(3), 339–351.

Tyagi, R. K.; Chandra, A.; Singh, D.; Rahman, M. A. Transdermal drug delivery system (TDDS): An overview. *International Journal of Pharmaceutical Sciences and Research*, **2011**, *2*(6), 1379–1388.

Uchechi, O.; Ogbonna, J. D.; Attama, A. A. Nanoparticles for dermal and transdermal drug delivery. *Application of Nanotechnology in Drug Delivery*, **2014**, *4*, 193–227.

Waghule, T.; Singhvi, G.; Dubey, S. K.; Pandey, M. M.; Gupta, G.; Singh, M.; Dua, K. Microneedles: A smart approach and increasing potential for transdermal drug delivery system. *Biomedicine & Pharmacotherapy*, **2019**, *109*, 1249–1258.

Yang, J.; Liu, X.; Fu, Y.; Song, Y. Recent advances of microneedles for biomedical applications: Drug delivery and beyond. *Acta Pharmaceutica Sinica B*, **2019**, *9*(3), 469–483.

Zhou, X.; Hao, Y.; Yuan, L.; Pradhan, S.; Shrestha, K.; Pradhan, O.; Li, W. Nano-formulations for transdermal drug delivery: A review. *Chinese Chemical Letters*, **2018**, *29*(12), 1713–1724.

CHAPTER 3

Advanced Topical and Transdermal Drug Delivery (TTDS) Used in Skin Diseases and Lesions of the Aging

Ana Vitória Pupo Silvestrini, Juliana Santos Rosa Viegas, Marcela Tavares Luiz, Lívia de Queiroz Carvalho Silva, Fabíola Garcia Praça, and Maria Vitória Lopes Badra Bentley

School of Pharmaceutical Sciences of Ribeirão Preto-University of São Paulo, Ribeirão Preto, São Paulo, Brazil,
E-mail: vbentley@usp.br (M. V. L. B. Bentley)

ABSTRACT

Topical and transdermal drug delivery systems (TTDDS) are important tools to improve the treatments of several skin disorders. The application of products on the skin tissue avoids many adverse effects and also contributes to patient adhesion to the management of diseases. The uses of nanoparticulated systems to perform topical or transdermal release of drugs represent the main innovation in the concern of drug delivery into the skin. These nanoplatforms expand the idea of target delivery in the skin, acting in the controlled release of molecules. Also, the nanoplatforms protect the skin in front of irritation, increase the stability of the drugs, and decrease the drug dosage that may reduce the local and systemic adverse effects. The skin tissue presents some challenges, such as permeability in *stratum corneum* (SC), that intensify treatment efficacy. Regarding SC permeation the use of nanoparticles have advantages when compared

Topical and Transdermal Drug Delivery Systems: Applications and Future Prospects. Nayan A. Gujarathi, PhD, Juliana Palma Abriata, PhD, Raj Kumar Keservani, MPharm, Anil K. Sharma, PhD, MPharm (Eds.)

to the conventional treatments, because the nanometric size increase the membranes and cellular internalization of active molecules, perform the controlled and/or sustained release and also can be designed to targeted specific type of cells. In a nutshell, this chapter summarizes the main nanoparticles used as TTDS, applied in psoriasis, vitiligo, cancer, lesions of the aging, and others.

3.1 INTRODUCTION

Topical and transdermal drug delivery systems (TTDDS) are the well-known terms used for localized and systemic treatment of biological conditions, respectively, using the skin as administration. Topical delivery therapy examples include dermatological conditions treatments like eczema or psoriasis in which the drug diffuses into the skin to therapeutic response. Whereas, transdermal delivery therapy includes the treatment of hormones, inflammatory, and other conditions in which the drug crosses the whole skin to reach the bloodstream and consequently induces the therapeutic response (Münch, Wohlrab, and Neubert, 2017).

Conventional topical dosage forms are marketed as ointments, creams, gels, pastes, lotions, emulsion, suspension, or solution used for: (i) surface skin effects including water loss prevention, microbial penetration protection, cleansing, and cosmetic effect; (ii) effects of the stratum corneum (SC), including photoprotection, and keratolytic activity (helpful in the treatment of psoriasis), and (iii) viable epidermal effect in which several classes of drugs such as anti-inflammatory, antihistamine, anesthetic, and others, can penetrate to epidermal layers with local effect without successive uptake by the blood (Kaur et al., 2016).

Transdermal delivery systems called "patches" were first demonstrated in the 1970s and FDA approved as a medicine for motion sickness treatment using the scopolamine drug in 1979, followed by nitroglycerin in 1981. Nowadays, transdermal patches are commercially available for several drugs, including nicotine, clonidine, fentanyl, estradiol, testosterone, lidocaine, and others. For the drug from the transdermal patch to cross different layers of the skin until to find the bloodstream, three different types of drug delivery systems which are matrix system containing a drug rate-controlling membrane, matrix systems without drug rate-controlling membrane and reservoir systems are used (Pastore et al., 2015; Patel et al., 2012; Ruela et al., 2013).

Considering that transdermal patches present a linear layer (a protective film that must be removed prior to use), drug (active agent), adhesive layer (with or without the drug), membrane (controls the drug release from reservoir and matrix systems) and backing layer that protects the patch from the outer environment (Singh, Naik, and Sawant, 2010). In matrix systems without drug rate-controlling membrane, the adhesive layer, which is surrounded by linear and backing layers, it is responsible for the releasing of the drug. Moreover, in the matrix system containing a drug rate-controlling membrane it adds another layer of drug-in-adhesive responsible to controlled drug delivery. Unlike matrix systems, for reservoir systems, a liquid compartment containing the drug is present (Pastore et al., 2015; Patel et al., 2012).

TTDDS advantages and limitations (Table 3.1) were exhaustively discussed in the literature. In order to overcome their limitations, several strategies have been researched in the last decades to improve skin delivery of active agents applied on the skin. Regarding the advent of the nanotechnology, effective formulations at the nanoscale have been considered approaches to enhance passive cutaneous drug absorption leading to release drug at the target site (topical or transdermal action) and providing new hope to treatment and diagnosis of several diseases (Kaur et al., 2016). Jotinder Kaur and colleagues (2016) reported in their review the efforts to produce topical drug delivery systems incorporating novel carriers as nanoparticles (NPs). The authors showed that compared to other conventional systems, these NPs demonstrated high efficacy, improved stability with extended shelf life and patient compliance (Kaur et al., 2016).

3.2 CHALLENGES TO TOPICAL AND TRANSDERMAL DELIVERY

Medicines applied through the skin have increased its use over the years due to the nanotechnology evolution. Since the skin is the largest barrier of penetration of substances from the external environment to the interior of the body, researchers have been working aiming the development and validate tools to overcome the skin barrier and deliver the active agent in the deeper layers of the skin reaching a topical or systemic drug effect. Among the many approaches used for this purpose, NPs demonstrate application success (Carter, Narasimhan, and Wang, 2019) and will be the subject of study in this chapter.

TABLE 3.1 Summary Main of TTDDS Features

TTDDS/ Features	Topical	Transdermal
Popular use	Anti-inflammatory, antihistamine, anesthetic cosmetic, anti-aging, and other applications.	Nicotine replacement therapy, anti-inflammatory, pain relief, menopausal symptoms, postmenopausal osteoporosis, contraceptive patch, anti-hypertensive drug and other applications.
Advantages	Increases bioavailability avoids the first-pass metabolism, minimizes side effects, favors safe and effective drug delivery with lower doses, is easy to apply, and increases patient compliance as compared to the conventional system.	Like topical routes, the transdermal drug delivery also avoids the first-pass metabolism, improves drug bioavailability, promotes reduction in dosing frequency, minimizes side effects, avoids gastric emptying time, and increases patient compliance as compared to the conventional system.
Limitations	Drug physicochemical features are required for passive penetration across the stratum corneum. Skin irritation or dermatitis may occur due to the drug or formulation components such as enhancers of dermal drug delivery used to increase percutaneous absorptions can be present.	Ionic drugs and macromolecules have a limited passive skin penetration, and consequently, low drug level in plasma is present. Limited to plasma drug therapeutic values higher than 10 mg/day and use of non-skin compatible components.

Source: Patel et al. (2012).

A major challenge to TTDDS is the slow penetration of active agents through the skin considering that it has extremely low permeability and large molecules cannot readily pass through as a skin barrier function (Mithun Bhowmick, 2012). Despite, human skin is a complex tissue comprising three main regions such as epidermis (EP), dermis (D), and subcutaneous tissue (ST) (Matsui and Amagai, 2015).

The EP is a very complex layer subdivided into four layers: (i) corneal, (ii) translucent, (iii) granular, and (iv) basal. The SC is the outmost layer of the EP and is composed mostly of anucleate, flattened, and keratin-rich cells called keratinized dead cells or corneocytes, forming a resistant barrier that has often been compared to a "brick wall" by several researchers (Kiselev, 2007; Wickett and Visscher, 2006). This superficial layer comprises up to 20 micrometers of thickness composed of 25 to 30 corneocytes cells deep arranged in a lipid matrix which is critical to the barrier function of the SC (Depieri et al., 2015). The layers below SC are arranged within a 50 to 100 micrometers of thickness composed of keratinocytes in varying differentiation degrees, melanocytes (melanin-producing cells), Merkel cells (sensory perception cells) and Langerhans cells (immune function cells) (Bolzinger et al., 2012).

The vascularized dermis layer separates the EP from the ST and it is composed of collagen, elastin fibrous proteins, interfibrillar gel of glycosaminoglycans with salts as well as blood and lymph vessels, nerves, sweat glands and hair follicles within 1 to 4 mm thick (Depieri et al., 2015). Finally, the ST, which is also known as the hypodermis, is related to the maintenance of body temperature due to the presence of fat and connective tissue that house blood vessels and act as thermal insulation (Amjadi, Mostaghaci, and Sitti, 2017).

Thus, NPs to be effective need to cross the SC and penetrate through cutaneous tissue until reaching the target cells or systemic effect. The mechanisms by which the NPs penetrate the skin are directly interconnected by three different routes: intracellular, intercellular, and transfolicullar (Figure 3.1). In the intracellular pathway, the NPs can interact and pass through the corneocytes while in the intercellular pathway they pass through the lipid matrix layers. In addition, the transfolicullar route is a passive penetration route that delivers the NPs to the deeper layers of the skin by accumulating them in the follicles and glands openings (Roberts, 1997; Salimi and Mohammad Soleymani, 2018). In fact, the hair follicles are reported as a promising target for transdermal delivery of different drugs, gene therapy, and vaccination (Carolina et al., 2019; Lademann et al., 2015; Patzelt and Lademann, 2013; Radtke et al., 2017).

Although all the NPs applied on the skin are typically administrated by definition, the term "topical delivery system" refers to a delivery system which promises a local effect on the skin while "transdermal delivery system" are those aimed at a systemic effect (Kulkarni, 2010; Münch et al., 2017). NPs applied to the skin surface for topical effect are challenged to have penetration into the skin layers and remain where their effects are expected. On the other hand, NPs directed for transdermal effect need to pass across the whole cutaneous tissue, ensuring that significant drug amount is absorbed to reach effective plasma concentrations until reaching the target cells where their effect is expected. Some examples of NPs applied for transdermal effects are related to hormonal replacement, inflammatory diseases, nicotine replacement therapy (NRT), vaccines, and gene therapy (Carvalho et al., 2016; Jain et al., 2016).

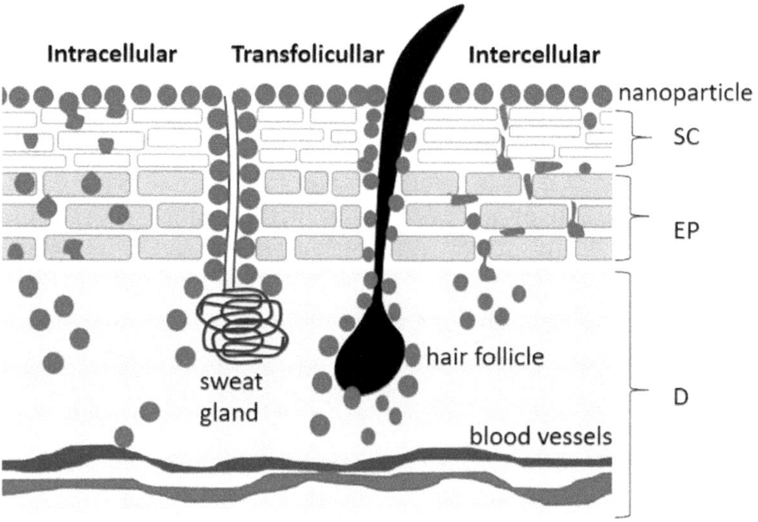

FIGURE 3.1 Representation of nanoparticles penetration throughout the skin from different pathways, which are intracellular, transfolicullar, and intercellular routes. The pink circles are representations of nanoparticles [SC: stratum corneum; EP: epidermis; and D: dermis].

3.3 NANOPARTICLES TO PASSIVE ENHANCEMENT STRATEGIES IN TTDDS

As previously mentioned, nanotechnology has been largely studied for improve passive penetration of both topical and transdermal drug delivery into the skin. The advance of this technology becomes an important role to reach molecules in different skin layers, from the most superficial layer to the deepest, being able to reach the systemic pathway. For these purposes, several organic and inorganic NPs have been studied, and some of the most used will be addressed in the following sessions of this chapter.

3.3.1 LIPIDS-BASED NANOPARTICLES

Lipid-based NPs are widely used in cutaneous applications for their compatibility with the skin, since the lipids used in the NPs composition have physical and structural similarity to those from the skin layers. The main characteristics of lipid NPs is that they are biodegradable and this is an important characteristic that given this system low toxicities and easy to scale up due to established preparation protocols (Desmet, Van Gele, and Lambert, 2017; Jain et al., 2017; Md et al., 2017). Several advantages of encapsulating drugs in lipid-based NPs include the ease of stabilizing the encapsulated drug, protecting the drug against chemical degradation, and improving TTDDS stability. Moreover, for topical delivery, they can have their lipid bilayer functionalized with specific molecules resulting in specific targeting of the disease site and overcoming off-target toxicity (Petrilli et al., 2016, 2017).

3.3.1.1 LIPOSOMES

There are several types of NPs composed of lipids, the first to be discovered were the liposomes, introduced in the 60's, when the phospholipids were described. The liposomes are versatile, and can encapsulate several molecules, lipophilic, hydrophilic, and amphiphilic. Liposomes are mainly known for their characteristic lamellar, the phospholipid bilayer. They can be classified as small, large, giant, oligolamellar, and multilamellar vesicles according to their measurements between 20 and 100 nm, > 100 nm, > 1,000 nm, 100 to 100 nm, and > 500 nm, respectively. The choice of liposome type depends on the purpose, components, and preparation methodologies (Hua, 2015). The most common liposome preparation techniques are related to reversed-phase evaporation, thin-film hydration, and solvent injection (Jain et al., 2017). Currently, there are several variations of the conventional liposomes, the ethosomes, known to be malleable due to the addition of ethanol in their formulation; the transfersomas, which are ultra-formable and designed for greater penetration; the exosomes, which are derived from endosomes, among others (Ascenso et al., 2014; Desmet et al., 2017; Hua, 2015).

3.3.1.2 MICROEMULSIONS AND NANOEMULSIONS

The micro and nanoemulsions are dispersions composed of lipid phase, surfactants, and aqueous phase. There are some similarities between them, but the differences are also pronounced. Recent reviews on this topic have shown that they are both low-viscosity dispersions with nanoscale particle size. Considering the micro means 10^6 and nano refers to 10^9, this could suggest that nanoemulsion has nanoscale particles size while microemulsion has microscale particle size, but they are both dispersions with an internal droplet size smaller than 250 nm (Lopes, 2014). The difference between them is related to the surfactants amount or mixture of surfactants used. Microemulsions use high proportions of surfactants and when formed are thermodynamically stable and present reduced sizes (~ 100 nm). On the other hand, nanoemulsions are advantageously formed using smaller proportions of surfactants, and because of this, they are kinetically, not thermodynamically, stable, and are larger in size than microemulsions (Md et al., 2017).

3.3.1.3 SOLID LIPID NANOPARTICLES (SLN) AND NANOSTRUCTURED LIPID CARRIERS (NLC)

Solid lipids NPs (SLN) were developed after liposomes in the 1990s as an alternative to encapsulate larger amounts of hydrophobic molecules. This system is also highly compatible with cutaneous tissue. However, SLN demonstrated difficulty in maintaining the encapsulated molecules, since with the lipid organization the active was expelled from the system (Desmet et al., 2017; Mukherjee, Ray, and Thakur, 2009).

Nanostructured lipid carriers (NLC) are the second generation of solid lipid NPs and have been developed to improve the loading capacity of molecules since liquid lipids are added to the formulation, with the intention of preventing solid lipid from returning to its organization and expelling the encapsulated active (Shastri, 2017). NLC is known to have high loading capacities of lipophilic molecules, having long shelf life due to its high stability, biocompatibility, and low toxicity (Mukherjee et al., 2009; Shastri, 2017).

The main methods of preparation of both SLN and NLC are through ultrasonication, hot homogenization, high shear homogenization, cold homogenization, high-speed homogenization, solvent evaporation, solvent emulsification, double emulsion, spray drying, and supercritical fluid (Mukherjee et al., 2009; Naseri, Valizadeh, and Zakeri-Milani, 2015).

3.3.2 POLYMERIC-BASED NANOPARTICLES

Polymeric NPs as topical or transdermal delivery systems promote many benefits such as sustained or controlled drug release, chemical, and physical protection of several active agents and compounds, greater drug concentration in solid tumors due to the permeation and retention effect (EPR), targeting specific cells or tissues when conjugated with antibodies, peptides, and others on its surface. Moreover, polymeric NPs have specific characteristics that can confer protection, stability both in formulation and in their release medium, can also reduce dosages over conventional formulations. These characteristics reduce system toxicity and producing sustained release of drugs (Zhang, 2013).

Polymeric NPs can facilitate drug release at specific sites of the skin such as hair follicles or interact with the lipids of skin to mediate drug transport and generate an onsite drug store for sustained release and permeation of the drug onto the skin. In addition, NPs with positive (cationic) charges have advantages such as increased drug permeation because positively charged groups bind to membranes of epithelial cells, depolymerize F-actin and dissolve the skin's junctional proteins which increases permeation of the drug (Wang, 2018; Zhang, 2013).

3.3.2.1 POLYMERIC NANOPARTICLES

The polymeric NPs can be divided into nanocapsules and nanospheres. Nanocapsules are a type of drug reservoir in which a solid material surrounds a liquid or semi-solid core at room temperature, the core being either aqueous or oily for transporting hydrophilic or hydrophobic drugs therein. On the other hand, the nanospheres are already particles whose whole mass is solid, they are usually spheres but may have other shapes and the stabilizer in the form of amphiphile is totally mixed to the polymer and the drug all wrapped in the nanoparticle, there being no differentiated nucleus (Crucho and Teresa, 2017; Vauthier and Bouchemal, 2009)

Several polymers can be used in the composition of the NPs, which may be natural compounds, such as chitosan, alginate, gelatin, albumin, which is indicated to encapsulate hydrophilic drugs, however, these natural polymers vary greatly from batch to batch, hindering product reproducibility. Other synthetic polymers are the most commonly used, have a good purity degree, and have a high reproducibility, they are widely used for hydrophobic drugs, but hydrophilic drugs can also be encapsulated depending on the preparation method chosen. The poly(lactic-co-glycolic acid) (PLGA), poly (ε-caprolactone), glycolic poli acid (PGA), lactate poly acid (PLA) are examples of the most commonly used synthetic polymers (Zhang, 2013).

3.3.2.2 POLYMERIC MICELLES

Polymeric micelles are also classified as polymeric nanocarriers, they have physicochemical characteristics beneficial of drug release in a sustained manner, high stability, biocompatibility, high capacity to solubilize poorly water-soluble agents, promote greater circulation time, greater cell internalization, high transfection efficiency rates, and lower toxicity when compared to conventional formulations. Specific co-polymers are involved in the synthesis of micelles, such as the use of amphiphilic graft, di-block, and tri-block co-polymers with typical core-shell morphology. These three polymers provide important properties to the drug delivery system, such as increased blood circulation time, potential to alter the drug release profile and the possible addition of targeted ligands (Rosa et al., 2018).

3.3.2.3 DENDRIMERS

Dendrimers are classified as polymeric nanocarriers due to their own characteristic architecture with a central molecule as a root and many-branched arms originate symmetrically. Radial controlled symmetry, adjustable size and high encapsulation are structural properties of the dendrimers and surface modifications using binders may also occur (Rosa et al., 2018). Dendrimers easily form suspensions with varying viscosities and are widely used in topical transdermal cosmetic formulations and they have the ability to be used in molecular transport systems and present greater permeability. They can be loaded positively or negatively according to the application and provide benefits such as better solubilization and controlled release of the drug (Hsu et al., 2017).

3.3.3 INORGANICS NANOPARTICLES

The applications of inorganic NPs are related to their advantageous stability when compared with organic NPs. Among these nanocarriers those formed by gold, silver, silica, and metal oxides are the most described in the literature. Thus, these nanocarriers will be presented in this chapter.

3.3.3.1 GOLD NANOPARTICLES

Gold NPs (AuNP) are known to have different features when compared with bulk gold form since in nanoscale gold shows advantageous optical and physical properties. AuNP have the capacity to cause local heat when they are irradiated in a light range of 800 to 1,200 nm. This heat capacity can be used to promote photothermal destruction of tumors cells ref. Other advantages of using AuNP as a delivery system are their low toxicity, stability, biocompatibility, and high capacity to have its surface modified with a wide range of organic molecules. So, their surface can be modified in order to transport drugs molecules, proteins, peptides, DNA, and RNA (Carazo et al., 2018; Labouta and Schneider, 2013). Furthermore, their surface can be modified with some ligands with the aim of improving skin permeation and promote specific molecule targeting (Bessar et al., 2016; Dul et al., 2019). Thus, AuNP has been applied for delivery of several drugs for topical and transdermal applications such as wounds, psoriasis, and cancers (Fratoddi et al., 2019; Mahmoud et al., 2019).

Gold NPs can be synthesized in a wide range of size and shapes (nanorods, nanospheres, nanostars, nanocluster, nanocube, and nanoshell) according to the preparation method used. These different structures can influence AuNP features and skin penetration. In 1951, Turkevich et al. developed the first method of AuNP production, the citrate reduction method, and after it, several modifications and novel methods have been developed for generate specific AuNP sizes and shapes (Vogt et al., 2016; Yeh, Creran, and Rotello, 2012).

3.3.3.2 SILVER NANOPARTICLES

Silver salts are widely known for their broad-spectrum antimicrobial activity, and they are worldwide applied in wound dressing in order to supply a continuous release of silver ions for the treatment of bacterial infections present in wounds, burns, and chronic ulcers. Nanotechnology improved this antimicrobial ability through the great increase of surface area, which promotes greater contact of silver NPs (AgNP) with microorganisms. Nowadays, there are AgNP available on the market due to its antimicrobial activity, like Silverlon® and Acticoat®, which have these NPs incorporated in wounds dressings (Labouta and Schneider, 2013).

Moreover, beyond antimicrobial activity, AgNP has been reported to be used as drug delivery and diagnosis. Some examples of these applications are their use against virus infection and tumors of skin, lung, breast, and hepatoma (Wei et al., 2015).

As described for AuNP, the production method can influence particle size, polydispersity index and shape of AgNP, which can affect the biological activity of these systems. Thus, AgNPs have many routes to be synthesized through chemical, physical, and biological methods. Currently, biological methods using living organisms like bacteria, fungi, plants, and yeasts have been the most applied in order to avoid the use of toxic reducing agents (commonly used in chemical methods) and high energy consumption. Among these biological methods, AgNP produced by plant extract is related to be the most environmentally-friendly (Wei et al., 2015).

3.3.3.3 MESOPOROUS SILICA NANOPARTICLES

Mesoporous silica NPs (MSN) have been applied as drug delivery systems owing to their mesoporous structure and the presence of high surface silanol concentration. These properties are responsible for the great capacity of silica NPs to adsorb and carry molecules. Moreover, these NPs have other advantageous features like biocompatibility, changeable pore size, stability, easy dispersion in aqueous solutions and high surface area (Barua et al., 2019; Labouta and Schneider, 2013).

Currently, many researches have described the large influence of reactions conditions for preparing MSN. Production parameters such as pH values, reaction temperature, silica sources, and type and concentration of surfactant may have a great influence on the type of mesostructures (hexagonal, cubic, lamellar mesophases, and others), particles sizes and morphologies (fibers, tubules, spheres, hollow spheres, etc.). In addition, these productions parameters may also influence the pore volume, which have an important role in the encapsulation of molecules for biological purpose (Wu and Lin, 2013).

3.3.3.4 METAL OXIDES NANOPARTICLES

Titanium dioxide NPs (TiO_2NP) and zinc oxide NPs (ZnONP) have been largely applied on skin due to their photoprotective activity. These NPs are related to increase the retention time in the outermost layers of the skin and improve the ability of TiO_2 and ZnO to absorb UV radiation. Thus, sunscreens formulations containing these kinds of NPs are highly recommended to prevent skin cancer, photo-aging, and sunburn. Besides the protection of these particles against UVA and UVB radiation, both also present antimicrobial activity described in the literature (Carazo et al., 2018; Larese et al., 2015).

3.3.4 HYBRID NANOPARTICLES

The hybrid NPs belong to a new generation of NPs that encompasses attributes of different organic and inorganic materials in the same matrix (Bose et al., 2017; Hood, 2014; Luchini and Vitiello, 2019).

In the development of nanocarriers, polymer lipid hybrids NPs (LPHNPs) have gained significant interest due to their advantages and applications (Dave et al., 2019; Garg et al., 2018). LPHNPs are drug delivery systems composed of a polymer core and a lipid coating. This hybrid morphology offers advantages such as particle size control, sustained or controlled release profile, ease to have your surface functionalized, high incorporation capacity of water-soluble and/or hydrophobic drugs and stability (Dave et al., 2019; Mukherjee et al., 2019).

Structurally, LPHNPs are composed of three distinct functional components, such as (i) internal polymeric matrix (nucleus) containing the drug; (ii) the lipid biocompatible mono-layer favoring drug retention in the polymer core; and (iii) and the PEG-coated outer layer bypassing the immune system and prolonging the systemic circulation time of NPs (Dave et al., 2019; Garg et al., 2018; Mukherjee et al., 2019).

According to the lipid-polymeric arrangement, the LPHNPs can be classified into different groups, polymeric core-lipid shell, and monolithic lipid-polymer hybrid NPs. The first type comprises the simplest form of LPHNPs and consists of a polymeric core coated with a single lipid or bilayer layer (lipid PEG or lipoidal shell). These nanoparticles have an empty inner core, delimited by a lipid coating, containing the polymer-coated core, and surrounded by several layers of lipid coating (lipid PEG). Monolithic lipid-polymer hybrid NPs, also known as mixed LPHNPs, are distributed in a polymer matrix containing the drug (Date et al., 2018; Dave et al., 2019; Garg et al., 2018). Another type also is polymer caged liposomes, which liposomes are coated by polymers. There are also erythrocyte membrane-camouflaged polymeric NPs, which are called biomimetic NPs, and are coated with an erythrocyte cell membrane polymer (Date et al., 2018).

The production methods of LPHNPs include one-step processes and two-step processes. In the conventional two-step method, polymeric NPs and preformed lipid vesicles are mixed and heated, and in the end, LPHNPs are collected by centrifugation. Extrusion or homogenization steps can be further performed to obtain a better-monodispersed form of LPHNPs. In the unconventional two-step method, spray drying and soft lithography particle molding are applied. In a more optimized approach, single-step production, lipids, and polymers are mixed, where they self-aggregate, forming the LPHNPs (nanoprecipitation, emulsification by solvent evaporation, or high-pressure homogenization) (Bose et al., 2017; Dave et al., 2019; Garg et al., 2018; Mukherjee et al., 2019).

Other types of hybrid NPs make up the inorganic NPs with lipid or polymer coatings. Two approaches can be used for the introduction of lipid molecules: supported lipid bilayers (SLB) which the lipids are adsorbed on the NP surface arranged in a lipid bilayer, with internal and external leaflets being composed of the same molecules and hybrid lipid bilayer (HLB) which the inner and outer coating layers are not composed of the same type of molecules. Several examples of inorganic NP are reported in the literature using surface coatings with different lipids, such as NP silica, gold, and iron oxide (Luchini and Vitiello, 2019).

Polymer coating is also highly exploited due to improved physicochemical characteristics, biocompatibility, and stability in biological fluids. One of the most commonly used techniques to produce these hybrid NPs is based on the physical absorption of the polymer on the inorganic surface. In other cases, hybrid materials with covalent interactions between the polymer and the inorganic NPs can be obtained by well-known strategies of polymer grafting such as "grafting-to," "grafting-from" or "grafting-through." Structurally, these hybrid NPs can present core-brush, hybrid nanogel and core-shell architecture. In the first type, the independent polymer chains are arranged to the NPs surface, whereas in the hybrid nanogel, the NPs are incorporated, physically or covalently, inside the crosslinked polymer matrix; and in the core-shell type, the compact polymeric shell is bonded to the surface of INPs (Hood, 2014; Macchione, Biglione, and Strumia, 2018).

3.4 TOPICAL SKIN APPLICATIONS

3.4.1 WOUNDS

Wound healing is a complex process that culminates in a sequence of phases from a coordinated cascade of cellular, molecular, and biochemical events, including hemostasis, inflammation stages, proliferation, and remodeling phase. During the inflammatory phase, the injured endothelium and platelets stimulate the coagulation cascade. Thus, platelets play a key role in healing. The main target is hemostasis, the cascade is initiated and granules are released from the platelets, which contain various growth factors (Kim et al., 2018; Wang et al., 2018). The proliferative phase comprises stages of epithelialization, angiogenesis, granulation tissue formation, and collagen deposition, with intense cellular activity. Finally, in the remodeling stage, a balance exists between the apoptosis of existing cells and the production of new cells. Defects in these phases can lead to excessive scars or chronic wounds (keloids or hypertrophic scar) (Kim et al., 2018; Wang et al., 2018).

For the wound treatments the infection control is an important initial step which can be obtained through topical antiseptic drug treatment, topical, and/or systemic application of anti-bactericidal drugs concomitant to wound debridement techniques for a better environment before receiving other therapies. The gold standard treatment of wounds is based on keeping the wound under controlled moisture conditions from hydrogels, hydrocolloids, alginates, or foams use (Oyarzun-Ampuero et al., 2015).

Nowadays, topical administration of NPs (such as nanoemulsions, nanospheres, micelles, SLN, and liposomes) are applied to control wound infections and promote tissue repairing. Furthermore, the inclusion of these NPs in secondary vehicles (e.g., hydrogels) demonstrates potential effect on wound treatment, mainly the metal NPs. The advantages of using NPs to wounds compared to conventional formulations are related to their controlled drug release, selectivity for the skin membrane and drug deposition into the tissue (Oyarzun-Ampuero et al., 2015).

Silver in its different forms has been widely used as antibiotic therapy for the treatment of open wounds and also several chronic infected wounds. Although some authors discuss their toxicity due to the accumulation of the metal, the formulations containing silver at the nanoscale, proved to be efficient for wound treatment. They are reported to have antimicrobial and anti-inflammatory activity, less toxic, acting on bacterial membrane proteins as well as compounds containing phosphors such as DNA (Chaloupka, Malam, and Seifalian, 2010; Oyarzun-Ampuero et al., 2015).

Gong and colleagues (2013) evaluated a biodegradable hydrogel system containing a natural polymer curcumin encapsulated in micelles. These NPs was thermosensitive and readily converted into a less fluid gel at body temperature. In addition, *in vivo* results showed wound closure enhancement with higher collagen content, better granulation, and higher wound maturity, which confirms its promising application for wound healing (Gong et al., 2013). Moreover, PLGA nanoparticles loaded with peptide LL37 promoted an accelerated wound healing process compared to peptide alone in a full-thickness skin wound animal mouse model. The LL37 belongs to antimicrobial peptides/host defense peptides that are part of the innate immune system (Kumar et al., 2014).

Nanospheres also enhanced large-wound healing by delivering growth factors and cytokines required in the wound healing process from platelet-rich plasma. The platelet-rich plasma-loaded NPs resulted in much faster wound closure as well as dermal and epidermal regeneration in a lapse of 9 days after wounding compared to platelet-rich plasma-loaded fibrin gel in the absence of NPs (La and Yang, 2015).

Polymeric vesicles were successfully evaluated by Geilich et al. (2015) in order to overcome bacterial antibiotic resistance. In this study, silver NPs inhibited the growth of ampicillin-resistant *Escherichia coli* (Geilich et al., 2015). Another antibiotic activity was previously reported by Sugumar and colleagues (2014). The authors showed that eucalyptus oil loaded in nanoemulsion system inhibited the complete growth of *Staphylococcus aureus* within 15 min of interaction. The mechanism of action was associated to the membrane damage. *In vivo* results showed non-skin irritation activity of NPs and higher wound contraction rate when compared to neomycin treatments (Sugumar et al., 2014).

SLN or NLC demonstrated significant improvements in wound closure and re-epithelization degree with consequent inflammatory process restoration degree when used in wounds treatment with no difference between them. Furthermore, administered drug from NPs treatment were plasma levels undetectable compared to free drug treatment suggesting a safety treatment with minimize side effects (Gainza et al., 2014, 2015).

3.4.2 ACNE

Acne is a multifactorial chronic skin disease associated with the pilosebaceous unit and very common throughout the world. Its occurrence can be related to the blockage of the superficial pores of the skin due to excessive production of oil (sebum) and accumulation of impurities, or because of hyperkeratinization due to the presence of bacteria such as *Propionibacterium acnes*, *Acne vulgaris* and in some cases, *Staphylococcus epidermidis* and *Staphylococcus aureus* (Amer et al., 2018; Vyas, Kumar Sonker, and Gidwani, 2014).

The encapsulation of anti-acne agents in nanoparticulate TDDS has been an innovative approach to reduce clinical side effects and preserve their effectiveness (Amer et al., 2018; Vyas et al., 2014). In this scenario, the diversity of nanocarriers drug delivery systems was designed, with emphasis on lipid, vesicular, and polymeric NPs. These carrier systems loaded with anti-acne drugs have demonstrated clinical improvement in signs and symptoms, as well as better support and retention of the active agent in the polysbaceous units, and less skin irritation in preclinical and clinical studies. Some examples are summarized in Table 3.2.

3.4.3 ATOPIC DERMATITIS (AD)

Atopic dermatitis (AD) is a common chronic or recurrent skin inflammatory disease and it affects 2.1 to 4.9% of the adult population and 15 to 20% of children. It is characterized by dry skin, eczema, and intense pruritus, which can be intensified by various agents such as allergens, infections, seasonal, and climatic changes, or psychological stress (Deleanu and Nedelea, 2018; Solomon et al., 2018). Its etiology remains unknown, however, different interactions between the skin barrier, genetic, and environmental factors were associated with the complex pathological pathway (Campana et al., 2017; Deleanu and Nedelea, 2018).

In recent years, it has become clear from several studies that the scientific focus is being driven from conventional therapies to the regimen based on nanocarriers designed for a better therapeutic profile for AD (Shao et al., 2016). In the search for an effective treatment and to reduce the topical adverse effects of corticosteroids commonly used in clinical practice for AD, Siddique et al. (2017, 2019) developed chitosan NPs loaded with hydrocortisone and hydroxytyrosol. These NPs were able to

TABLE 3.2 Examples of Nanoparticles Reported for the Topical Treatment of Acne

Nanoparticles	Drug	Main Results	References
NLC	Adapalene and vitamin C	The multifunctional-NLC improved permeation by the stratum corneum and epidermal retention, in addition to reducing the permeate content *in vitro*. In the animal model, there was a significant reduction in the number of swollen sebaceous glands and histopathological data were similar to healthy animals.	Jain et al. (2016)
	Tretinoin	Tretinoin-NLC showed a lower degree of skin irritation *in vivo* compared to a commercial formulation.	Ghate et al. (2016)
Nanoemulsion	Clindamycin	In an active-controlled, multicentric, phase IV clinical trial, application once daily for 12 weeks showed a significant reduction of total inflammatory and noninflammatory lesions in patients treated with the gel-nanoemulsion compared to control formulation.	Sanmukhani (2014)
	Origanum vulgare L. essential oil	In an animal model, the oil-containing nanoemulsion showed superior healing and antimicrobial effects compared to the reference antibiotic.	Taleb et al. (2018)
Niosome	Benzoyl peroxide and tretinoin	The niosome in gel showed prolonged release of the two encapsulated agents; with low skin permeability and retention in the skin layers. *In vivo*, the treatment was more effective than the control, even with a 4.16-fold lower dose of benzol peroxide.	Gupta et al. (2014)
	Rosmarinic acid	The niosome in gel presented greater permeability and skin retention compared to the simple gel. The niosomes showed a greater capacity of inhibition of the bacterial growth of Staphylococcus aureus and Propionibacterium acnes. *In vivo*, low irritability, decreased inflammation and antibacterial effects were observed in animals treated with the niosome in gel.	Budhiraja and Dhingra (2015)
Polymeric micelles	Retinoic acid	Polymeric micelles have been shown to efficiently improve the *in vitro* delivery of retinoic acid in porcine and human skins compared to the commercial formulation. Delivery efficiency to the polysbaceous units was also higher.	Lapteva et al. (2015)

TABLE 3.2 *(Continued)*

Nanoparticles	Drug	Main Results	References
Cellulose-based nanoparticles	α-mangostin (active component from *Garcinia mangostana* Linn. fruit)	Clinical studies have shown satisfactory results regarding the localized delivery and retention of the nanoparticles in the polysbaceous units and improvement of the clinical signs.	Lueangarun et al. (2019); Pan-In et al. (2015)
		In the double-blinded, split-face, randomized, control study, patients treated with 0.5% topical mangosteen extract in nanoparticle loaded gel (containing alpha-mangostin) for 12 weeks (2 times/day) had a significant reduction of comedones and inflammatory lesions compared with 1% clindamycin gel.	
Nanospheres	Adapalene	The drug was efficiently released from the nanospheres and the hyaluronate gel was shown to improve retention of the nanospheres in the epidermis and dermis. *In vivo*, the formulation showed low irritability.	Sallam and Marín Boscá (2017)
		The formulation showed low irritability and the encapsulation of adapalene in the nanospheres significantly decreases the secretion of proinflammatory cytokines (IL-1α and IL-8). An improvement in the comedolytic properties of the formulation by significantly reducing the size of open utricles in rhino mice compared to commercial treatment.	Ramezanli and Michniak-Kohn (2018)

Abbreviation: NLC: nanostructured lipid carriers; IL: interleukin.

permeate about 2.46 times more than the commercial formulation, as well as having greater affinity at the target site of the skin without spreading to the surrounding tissues, thus it provided substantial safety benefits. In addition, the local and systemic toxicity profile were significantly lower in terms of skin atrophy and hepatic enzyme levels (Siddique et al., 2016). The local and systemic safety and tolerability of these NPs were confirmed in a 28-day randomized clinical trial (Siddique et al., 2019).

Polymeric NPs have also been proposed as a platform to deliver betamethasone valerate, a medium potency topical corticosteroid commonly used in the treatment of AD. Chitosan NPs produced by Md. and colleagues (2019) exhibited optimal physicochemical characteristics and a high drug-binding efficiency, which allowed a better SC penetration for *in vitro* studies (Md et al., 2019). In another study, the targeting and dermal distribution efficiency of betamethasone valerate was significantly improved when these NPs were coated with hyaluronic acid (Pandey et al., 2019). These chitosan NPs coated with hyaluronic acid were also able to entrap a high tacrolimus load and allowed greater skin retention (Zhuo, Abourehab, and Hussain, 2018).

In another approach, nicotinamide, and chitosan NPs loaded with tacrolimus have been shown to significantly improve skin permeation and retention compared to commercial cream. The efficacy on clinical symptoms of the NPs containing a lower dose of tacrolimus (~ 1/3) was superior to commercial formulation (Yu et al., 2018).

3.4.4 INFECTIOUS DISEASES

3.4.4.1 BACTERIAL INFECTIONS

Although human skin acts as a protective barrier against infections, some opportunistic bacterial can colonize and promote skin damage. Among bacterial infections those caused by *Staphylococci* and *Streptococci* are the most common. Some of the most described skin damages by these bacterial are furuncle, impetigo, cellulitis, and carbuncle purulent inflammation of hair follicles (Niska et al., 2018). Besides these bacterial, *Mycobacterium leprae* has also been reported to affect the skin as well as affect peripheral nerves and mucosal surfaces. This bacterial are responsible for a chronic infectious disease named leprosy, also known as Hansen's disease (WHO, 2019).

Since the late 1940s, when penicillin was discovery, to 1970s newer and stronger antibiotics were successfully developed and they promoted a significant improvement in the health status worldwide by reducing mortality and morbidity. However, the inappropriate use and overconsumption of these drugs were responsible for microorganisms antibiotics resistant "superbugs," which caused an antibiotic-resistant era (Huh and Kwon, 2011). Furthermore, in the last 40 years, the discovery of novel antibiotics have been largely decreased, which further reduces the treatment perspectives against resistant "superbugs" (Abed and Couvreur, 2014).

There are several studies in research orientation as well as clinical medical investigation applying both organic and inorganic NPs to improve the treatment against microorganisms antibiotics resistant including AgNP, AuNP, silica NP, TiO_2NP, ZnONP, liposomes, polymeric NPs, dendrimers, and SLNs (Abed and Couvreur, 2014; Jamil, Bokhari, and Imran, 2017). Advantages of inorganic particles over organic ones is that some inorganic NPs (e.g., AgNP, ZnONP, and TiO_2NP) in addition to having ability to encapsulate molecules (e.g., ciprofloxacin, azithromycin, methicillin, and gentamicin) for improving their activity by cellular internalization enhancement and reducing drug-efflux from resistant bacterial, they also present an intrinsic antimicrobial activity, which may be influenced by their composition, particle size, and morphology (Huh and Kwon, 2011; Niska et al., 2018).

3.4.4.2 CANDIDIASIS

Fungal infections in general present high prevalence in diverse populations, besides being responsible for numerous systemic complications. Among fungal infections, topical candidiasis is one of the most common infections and manifests itself in various locations such as the outer layers of the skin, nails, hair, and systemic. About 20 *Candida* species are associated with human or animal infections (*C. albicans, C. tropicalis, C. glabrata, C. parapsilosis, C. krusei, C. guilliermondii*), in which *C. albicans* is present in the most infections (Mathur and Devi, 2017).

Current pharmacotherapy offers effective treatment, however, it has limitations, such as inadequate solubility, poor skin permeation, fungal resistance, and long-term therapeutic effects (Mathur and Devi, 2017). Therefore, strategies have been investigated to allow better results, such as chemical modifications in conventional antifungals, immunotherapy, and the development of nanocarriers (Scorzoni et al., 2017).

In recent years, a variety of nanocarriers have been designed and they demonstrated an improvement on various biopharmaceutical aspects of antifungal agents commonly used in clinical practice, some examples of nanocarriers are micro- and nanoemulsions, micelles, emulgels, liposomes, ethosomes, niosomes, transfersomes, cubosomes, ufasomes, SLN, NLC, nanosponges, inorganic NPs, nanofibers, carbon nanotubes and dendrimers (Mathur and Devi, 2017).

Carbone and colleagues (2019) reported the development of lipid NPs with different essential oils (*Rosmarinus officinalis*, *Lavandula*, *Origanum vulgare* subsp. *hirtum*) for the release of clotrimazole, exploring the possible synergistic effects against *Candida* spp. In this study, the NLCs developed prolonged *in vitro* release and increased the antimicrobial activity against *C. albicans, C. krusei*, and *C. parapsilosis* (Carbone et al., 2019). In another study, clotrimazole-loaded nanoemulsions showed an improved in antifungal activity compared to commercial references. In addition, nanoemulsions allowed prolonged release and *in vivo* tolerability studies confirmed suitability for topical application (Soriano-Ruiz et al., 2019).

3.4.4.3 HERPES

Herpes simplex virus (HSV) infections are highly prevalent throughout the world. HSV type-1 infection causes cutaneous lesions that are usually located at the oral, nasal, and ocular levels. Whereas, the type-2 HSV infections occur most commonly at the skin and mucosal sites. Although the lesions are generally benign, in some cases, it can lead to serious complications such as morbidity, viral encephalitis, lymphocytic meningitis, blindness, and mortality (Suzich and Cliffe, 2018).

The development and application of nanocarriers as a strategy for the treatment of cutaneous infections by HSV has been little explored so far. The research focuses on the design of nanocarriers applied to the mucovaginal and ocular infections therapy due to the recurrence of systemic complications (Agelidis et al., 2019; Orłowski et al., 2018; Yang et al., 2016).

Nowadays, there is on the literature development of lipid-based nanocarriers for the antiviral agent delivery, such as acyclovir and genistein, in order to improve pharmacokinetic aspects and local therapeutic efficacy. Nanoemulsion hydrogel has shown a significantly enhance in bioavailability of acyclovir in *ex-vivo* studies, with higher permeability (2-fold and 1.5-fold) when compared to control gel and marketed cream, respectively (Al-Subaie et al., 2015). Acyclovir-loaded SLNs have also shown an improvement on drug permeation and retention in the dermal layers compared to commercial cream (Jain, Mistry, and Swarnakar, 2011). Further, genistein-loaded nanoemulsions demonstrated promising results in cutaneous delivery and exhibited potential antiherpetic activity *in vitro* against HSV-1 and HSV-2 (Teixeira et al., 2014).

3.4.4.4 LEISHMANIASIS

Neglected tropical diseases (such as leishmaniasis cutaneous, mucocutaneous, and visceral) present unique challenges in developing more effective treatments and prevention methods. Among them, cutaneous leishmaniasis is an infectious disease caused by parasitic protozoa of *the Leishmania* genus transmitted by the insect subfamily of *Phlebotomine* sand flies (Bahrami, Harandi, and Rafati, 2018).

Currently, numerous forms of leishmaniasis are treated with a sparse arsenal of drugs, specifically antimonials, amphotericin B, miltefosine, and paromomycin. Parasitic resistance, clinical side effects, and clinical failures are common occurrences. So, innovations in TTDDS or development of new drugs are strategies that can overcome these barriers (Bahrami et al., 2018; de Souza et al., 2018).

NPs have provided insight against parasitic infections after demonstrating the ability to penetrate target pathogens and minimizing toxic effects through low doses (Bruni et al., 2017). As recently reviewed by De Souza and colleagues (2018), more than 70% of the studies NPs were metallic NPs applied for cutaneous leishmaniasis, while lipid and polymeric NPs were tested in visceral leishmaniasis (> 60%) (de Souza et al., 2018). Table 3.3 summarizes the recent applications of NPs for treatment of cutaneous leishmaniasis reported between 2018 and 2019.

3.4.5 AUTOIMMUNE DISEASES

3.4.5.1 PSORIASIS

Psoriasis is a chronic inflammatory skin disease that affects about 2% of the world's population and it has a multifactorial etiology. It is characterized

TABLE 3.3 Examples of the Recent Applications of Nanoparticles for Treatment of Cutaneous Leishmaniasis Reported between 2018 and 2019

Nanoparticles	Drug	Main Results	References
Mannosylated chitosan NPs	Paromomycin	High drug encapsulation efficiency (EE) and a higher rate of release in acidic media were achieved. The mannose nanoparticles presented a higher rate of cellular uptake in 6 h of treatment, in addition to a greater anti-amastigote activity *in vitro*.	Esfandiari et al. (2019)
Poly butyl-cyanoacrylate NPs	Polymyxin B	The nanoparticles favored a sustained release profile of the drug. The leishmanicidal effects of the drug in the nanoparticles were superior to the control (free), besides showing antimicrobial activity against *Escherichia coli, Pseudomonas, and Klebsiella.*	Souza et al. (2019)
Iron oxide magnetic NPs	–	The magnetic hyperthermia offered by the nanoparticles was shown to be effective in killing infective amastigotes, with approximately 70% reduction in *in-vitro* viability.	Berry et al. (2019)
Pluronic® micelles	–	Micelles of Pluronic® P-123 or F-127 inhibited the survival rate of *L. amazonensis* amastigotes and also induced loss of mitochondrial membrane potential in promastigotes. *In vitro*, the micelles had low cytotoxicity in macrophages.	Oyama et al. (2019)
Poly(lactic-co-glycolic acid) NPs	Amphotericin B	*In vitro* evaluations demonstrated their sustained activity against promastigote macrophages and infected with *L. major* parasites. In an animal model, a single intralesional administration revealed a marked reduction of the lesion compared to the free drug.	Abu Ammar et al. (2019)
Albumin NPs	Meglumine antimonate	Nanoparticles prepared with glutaraldehyde showed higher leishmanicidal activity against *L. major* and lower cytotoxicity against macrophages.	Barazesh et al. (2018)

TABLE 3.3 *(Continued)*

Nanoparticles	Drug	Main Results	References
Deformable liposomes	Sodium stibogluconate	*In vitro* studies deformable liposomes in gel provided greater retention in the deeper layers of the skin. *In vivo*, the safety of the topical application of the formulation was confirmed. A marked leishmanicidal activity in the model of intra-macrophage amastigotes and selectivity were reported. In an animal model, the treatment resulted in efficient wound healing and reduction of the parasitic load.	Dar, Din, and Khan (2018)
Silver NPs	–	Decanethiol functionalized silver nanoparticles showed remarkable and low concentrations of leishmanicidal effects against promastigotes and amastigotes from different strains and *Leishmania* species isolated from different patients (localized cutaneous lesions and diffuse cutaneous lesions). No toxic effects have been reported on macrophages *in vitro*.	Isaac-Márquez et al. (2018)
	Meglumine antimonate	Nanocomposites prepared containing silver nanoparticles, polyvinylpyrrolidone, and different amounts of glucantime (meglumine antimonite) showed potential leishmanicidal activity and the absence of cytotoxicity in macrophages.	Gélvez et al. (2018)

Abbreviation: NP: nanoparticles.

by highly inflamed and scaly erythematous plaques. Histologically the psoriatic phenotype includes epidermal hyperplasia with significant keratinocyte differentiation, increased angiogenesis, and presence of prominent inflammatory infiltrate (Boehncke and Schön, 2015).

Although there is a diversity of therapeutic agents for the treatment of psoriasis, none of them are totally safe and effective for disease treatment without compromising patient compliance. New nanotechnology-based formulations are a promising strategy for overcoming the limitations of conventional formulations, offering a reduction and management of therapeutic doses, a decrease in unwanted effects and an increase in efficacy (Pradhan et al., 2018).

NPs containing anti-psoriatic agents have been successfully developed and they showed promising preclinical results in the topical treatment of psoriasis. In a recent review performed by Pradhan and colleagues (2018), several nanocarriers have been reported as a strategy for targeting the delivery of nucleic acids (small interfering RNA (siRNA) molecules), photosensitizers, calcineurin inhibitors, vitamin D analogs, retinoids, corticosteroids, psoralens, and naturally actives. The main findings show the ability of the nanocarriers to promote greater delivery of the encapsulated drug in the regions of the skin when compared with free drugs, with sustained release profiles and improved therapeutic efficacy (Pradhan et al., 2018).

The most recent studies have reported satisfactory results in the treatment of plaque psoriasis in animal models with the application of methotrexate loaded on different types of NPs (Fratoddi et al., 2019; Tripathi et al., 2018). As reported by Fratoddi and colleagues (2019), topical application of gold NPs functionalized with 3-mercapto-1-propanesulfonate loaded with methotrexate resulted in erythema score decrease and reduction of both epidermal thickness and inflammatory infiltrate, in addition to other findings. Further, no differences were observed in the blood count tests of the treated animals and in the levels of liver enzymes compared to healthy animals (Fratoddi et al., 2019).

In another study, polyphenol-rich extracts (*Cornus mas*) loaded AuNP or AgNP were able to modulate psoriatic inflammation at the molecular and cellular level. These NPs represent an efficient tool for modern psoriasis therapy, with a reductions of immunosuppression-related side effects of biological (Crisan et al., 2018). Berberine oleate, the main alkaloid of the extract of *Mahonia aquidolium*, has also important anti-psoriatic activity.

When it was loaded by liquid-crystalline NPs in hydrogel and topically applied in animals with psoriasis-like it significantly alleviated the symptoms of psoriasis and reduced levels of inflammatory cytokines. In addition, greater retention of the drug was reported through the different layers of the skin, which is paramount for the management of the disease (Freag et al., 2019). In studies performed by Sahu and colleagues (2018), the association of naturally occurring active oils as functional excipients of tacrolimus-containing nanoemulsions exhibited synergistic effects in the treatment of psoriasis-like animals. Topical application has shown a significant reduction of serum cytokines and an improvement in the psoriatic conditions, which indicates the efficacy of the formulation compared to the commercial formulation (Sahu et al., 2018).

3.4.5.2 VITILIGO

Vitiligo is an acquired autoimmune disease characterized by macules and depigmented patches on the skin, which occur secondary to the destruction of melanocytes (Picardo et al., 2015). Epidemiological data show that vitiligo affects 0.5 to 2% of the world's population and several extrinsic and intrinsic factors are associated with its occurrence (Rashighi and Harris, 2017).

Current treatments for vitiligo aim to reduce the oxidative stress of melanocytes, suppress autoimmune targeting, and promote melanocyte regeneration. In this approach, several therapies are suggested according to the type and extent of the lesions, which include the use of topical or systemic immuno-modulators, phototherapy, and, in some cases, surgical options are suggested (Picardo et al., 2015; Rashighi and Harris, 2017). However, there are no available treatment results in effective repigmentation. In addition, in many cases, patients have shown a recurrence of depigmentation within a few months of pharmacotherapy (Boniface et al., 2018).

In recent years nanocarriers design and research of new bioactive molecules have provided great advances as a therapeutic strategy for several autoimmune and inflammatory diseases. However, this technology has been little explored in research and clinical practice in the management of vitiligo. One of the first reports was published by Leeuw et al. (2003) with the development of liposomes containing chelating agents. Clinical studies have shown promising results in repigmentation of areas with refractory and stable vitiligo when topical application of the liposomal formulation was combined with phototherapy (De Leeuw et al., 2003, 2011).

In another study, ultra-deformable liposomes co-loaded with psoralen and resveratrol showed pronounced effects on stimulation of melanocyte proliferation and melanogenesis *in vitro* (Doppalapudi, Mahira, and Khan, 2017). Ethosomes and nanoemulsions have also been shown to improve methoxypsoralene penetration with significant epidermal and dermal accumulation (Barradas et al., 2018; Garg et al., 2016).

Other NPs were used to make feasible the use of plasmids and nucleic acids directed to the target components for vitiligo therapy. Plasmid-loaded niosomes have been shown to maintain plasmid stability, protect against enzymatic degradation and improve cell internalization rates. In addition, effects on melanogenesis, and tyrosinase activity *in vitro* were observed with treatment using pMEL-34 or pAH7/Tyr plasmid genes (Manosroi et al., 2010, 2012). More recently, liquid-crystalline NPs have shown effective internalization of siRNA molecules targeted to the Tyrosinase-related protein 1 (TyRP-1) gene. Gene expression was significantly reduced in 24 h of treatment in Melan-A cells (Tofani et al., 2018).

3.4.6 CANCER

Skin cancer is the most prevalent cancer in the world and its socioeconomic impacts are increasingly driving the emergence of innovative therapies. The use of NPs for topical or transdermal administration to treat this type of malignancy has been widely studied. Topical administration of chemotherapeutic agents through NPs is a promising alternative for reducing adverse effects, reducing doses, enabling localized and targeted therapies (Apalla et al., 2017; Dianzani et al., 2014; Dimitriou et al., 2018; Mofidi et al., 2018). The most used NPs explored for the treatment of skin cancer, aiming at improving the delivery of active agents, are polymeric and lipids-based NPs. The choice of the nanoparticle is related to the therapeutic agent that will be encapsulated, adsorbed, or complexed, and used as single therapy or associated, on multifunctional platforms (Dianzani et al., 2014).

Considering that dacarbazine drug (DZ) is the single US-FDA approved anticancer drug for melanoma chemotherapeutic treatments, with reported non-specific toxic effect on normal cells, Hafeez and Kazmi (2017) reported a newly integrated dacarbazine in NLC (DZNP). They prepare this nanocarrier using the technique of employing the solvent structured method for melanoma treatment. As a result, carried DZ was more effective as a topical formulation when compared to simple drug DZ form (Hafeez and Kazmi, 2017). Another chemotherapeutic drug, paclitaxel was encapsulated in SLN (PTXSLN) and showed an initial burst effect followed by a controlled drug release profile suggesting minimal dose-related side effects (Bharadwaj et al., 2016).

Penetration peptides have been used and the results obtained are promising. They allow NPs to easily overcome biological barriers, not only tissue barriers but also cell membranes, delivering the active agent more efficiently (Gan et al., 2018). The use of immunotherapy has also been closely associated with conventional chemotherapeutic agents in NPs, this association has generated positive results in tumor reduction (Petrilli et al., 2017). For instance, cetuximab, an anti-EGFR antibody, was directly coupled to preformed 5-fluouracil (5-FU)-loaded liposomes. In *in vitro* studies the 5-FU loaded liposomes targeted using skin carcinoma cells, at 72 h, 5-FU liposomes showed a 5-fold lower IC_{50} than 5-FU solution. Additionally, 5-FU immunoliposomes resulted in a 4-fold lower cetuximab IC_{50} than cetuximab solution, demonstrating synergism with a combination index lower than 1 and potential to improve 5-FU and cetuximab cytotoxicity (Petrilli et al., 2017).

Several associations have been done, the use of natural products in NPs has pointed promising results, inclusive when its associated to thermal therapy (Singh et al., 2018) or using photosensitizing molecules to improve the photodynamic therapy (PDT) effect and diagnostic imaging (Jain et al., 2017). The use of RNA was a modern approach and its use to treat skin cancer has been promising. Borgheti-Cardoso and colleagues (2017) reported a liquid crystalline nanodispersion used to carry the siRNA molecule to the cytoplasm of the skin tumoral cells. The developed NPs were essential to transport siRNA directly into the cytoplasm without degradations. As a mean result, the tumor mass was significantly reduced (Borgheti-Cardoso et al., 2017).

3.4.7 LESIONS OF THE AGING

Aging has been defined as the biological, physiological, and social processes of becoming older, resulting in degradation of biophysical organs and tissues properties, in mainly the skin giving rise to lesions of the aging (Limbert et al., 2019). Commonly aging manifestations are wrinkles, lack of skin elasticity, dry skin, hyperpigmentation of the skin and gray hair (Baumann, 2018). Considering that benign or malignant skin disorders are commonly identified in the elderly, the major causes of physiological skin aging was characterized by prolonged sun exposure (Omar, 2006).

Ultraviolet radiation consists mainly of UVB and UVA photons in which the UVA is able to penetrate deeply into the skin, produce free radicals and favor the process of early aging. The UVB does not penetrate as deeply into the skin but usually causes skin burns and consequent damage in superficial skin layers (Yaar and Gilchrest, 2007). In addition, these rays can cause significant DNA damage (Baumann, 2018) with consequent skin cancer onset. It is well-accepted that routine sunscreen and skin care therapies use helps to block the effects of this UV radiation and slow the aging process of the skin (Sunder, 2019; Zouboulis et al., 2018).

In this regard, the use of nanotechnology applied in the anti-aging process has proven to reverse the aging process at the cellular level due to its advantages to the development of cosmetic and skin care formulations enclosing NPs. Among these advantages are reported the more effective skin penetration of the compounds, stability enhancement, efficacy of these formulations and the act like UV filters with improved tolerance (Santos et al., 2019).

Several authors have been published studies evaluating NPs applicability in sunscreen and anti-aging products (Raj et al., 2012) and most of them include lipids-based NPs, polymeric-based NPs, metal-based NPs and other additional nanocarriers (Tables 3.4 and 3.5). For instance, vitamins, tocotrienol, retinoic acid, coenzymes, argireline, and primary or secondary metabolites of plants carried from SLN NPs enhanced dermal action when tested on 20 volunteers (Ajazzuddin, Jeswani, and Jha, 2015). Further, similar SLN coated with polymer resulted in an increased skin penetration and stability of the active agents (European Patent EP2549977 A2, 2013; Ajazzuddin, Jeswani, and Jha, 2015).

Maitra and colleagues (2014), reported reduced facial signs of aging by use of an effective NPs amount in combination with a pigment agent. The formulation formed a film on the skin to create optical effects and increase light transmittance and improve the esthetic appearance of the skin surface (Patent US20140030300, 2014; Ajazzuddin, Jeswani, and Jha, 2015).

TABLE 3.4 Examples of Nanotechnology-based Cosmeceutical as Marketed Products

Nanoparticles	Product Name	Marketed by
Nanosphere	Nanosphere Plus	Dermaswiss
Nanocapsules	Dior Snow Pure®	Dior
	LifePak Nano®	Pharmanex
	Hydra Flash Bronzer®	Lâncome
	Soleil Instant Cooling Sun®	Lancôme
	Soleil Soft-Touch®	Lancôme
	Lancôme soleil	L'Oreal
Liposomes	Dermosome	Microfluidics
	Decorte face cream	Decorte
	Advanced night repair	Estée Lauder
Niosomes	Niosome+ Perfected age treatment	Lancome
	Eusu Niosome Makam pom whitening facial cream	Eusu
SLN	Allure body cream	Chanel
NLC	Phyto NLC active cell repair	Sireh Emas
Nanoemulsion	Precision-solution Déstressante solution nano Émulsion Peaux sensitivity	Chanel
	Nano-Lipobelle H-AECL	Mibelle biochemistry
Gold NPs	Chantecaille nano gold energizing cream	Chantecaille

Abbreviation: NP: nanoparticles; SLN: solid lipid nanoparticles; NLC: nanostructured lipid carriers.

Source: Kaul et al. (2018).

The metal-based NPs as AuNP is used predominantly in consumer products as antibacterial or preservative agent (Arora et al., 2015) while titanium dioxide and zinc oxide are used in sunscreen. NPs containing these minerals form a barrier on the skin able to reflect UVA and UVB with less irritation effect. Furthermore, they are transparent, less greasy, and less smelly forms with increased esthetic appeal when compared to conventional sunscreen that leaves a white chalky layer on the skin (Hong et al., 2015). On the other hand, unwanted penetration of these NPs through the skin reaching systemic circulation was reported by several researchers (Shokri, 2017).

The ZnONPs and TiO_2NPs lower than 200 nm can penetrate the intact skin and promote inadvertent biological damage. Cytotoxicity and neurotoxicity of both NPs on the neural stem cells of mouse has also been reported (Shokri, 2017). So, the health risk of nanocarriers should be assessed during their development (Ajazzuddin et al., 2015).

TABLE 3.5 Examples of Nanotechnology-based Cosmeceutical in the Investigation

Nanoparticles	Main Results	References
Silver NPs	Silver NPs were well tolerated by keratinocytes cells and their pre-treatment protected them from UVB-irradiation-induced apoptosis along with significant reduction in DNA damage.	Arora et al. (2015)
Gold NPs	Increased sun protection factor from 2.43 ± 0.74 to 24.11 ± 0.46% to commercial sunscreens containing latex synthesized gold NPs due to better reflection and scattering of UV radiation compared to non-particulate sunscreen formulation.	Borase et al. (2014)
Polymeric NPs	Polymeric NPs formulation increased dispersion stability of the particles containing a sunscreen themselves and showed efficient ultraviolet absorption capacity.	Patent KR20180024903A (2018)
Bioadhesive NPs	Bioadhesive NPs enhanced UV protection prevents epidermal cellular exposure to UV filters as well as significantly reduced double-stranded DNA damage compared to non-particulate commercial sunscreen formulation.	Deng et al. (2015)
Lipid NPs	Results showed enhanced drug permeation reaches the deepest skin layers and lower fibroblast cell cytotoxicity. In addition, terpene composited lipid NPs were able to drug protection from photodegradation.	Charoenputtakun et al. (2014)
	Liposomes containing several actives agents such as CoQ10, avocado, rosehip, carrot oil, and with vitamins A, B, C, and E, Aloe used for effective Antiwrinkle treatments.	Tripura and Anushree (2017)

Abbreviation: DNA: deoxyribonucleic acid; NP: nanoparticles; TEM: transmission electron microscopy; UV: ultraviolet.

3.5 TRANSDERMAL APPLICATIONS

3.5.1 HORMONAL APPLICATIONS

There are several hormonal diseases that affect a large part of the population which are thyroiditis, diabetes, polycystic ovary syndrome, obesity, some types of tumors (e.g., breast and endometrial cancer), climacteric with menopausal symptoms and uterine fibroids. However, hormonal replacement therapy (HRT) for the vasomotor and genitourinary symptoms associated with a menopausal period is the most effective hormone treatment from transdermal application (Algin-Yapar and Inal, 2014).

The menopausal period is characterized by a decrease in both estrogen and progesterone levels and an increase of follicle-stimulating hormone levels. Thus, HRT from exogenous hormone decreased climacteric symptoms and also had protective effects on osteoporosis and heart disease (Nagai et al., 2018). Since the introduction of the first clinical hormone replacement therapy in 1942 to the present day, several drugs have been evaluated, and the most current therapy involves the skin hormone application associated with nanotechnologies (Sivanandy, Masimasi, and Thacker, 2007).

For instance, TDDS HRT from nanoemulsion containing estradiol and progesterone (0.25 and 10%, respectively) was developed to treat menopausal symptoms. The therapeutic effects were evaluated by several parameters which were laboratory, radiographic, and clinical after a 5-year follow-up. The safety and efficacy of this hormonal NP was successfully confirmed. At baseline, all patients studied had climacteric symptoms that improved significantly with treatment. In addition, treatment did not result in the presence of breast tumors or cardiovascular changes (Botelho et al., 2014).

Nagai and colleagues (2018) investigated the therapeutic effect of transdermal formulation containing raloxifene-loaded NPs on postmenopausal osteoporosis using ovariectomized rats as an animal model study. Raloxifene is classified pharmacologically as a selective estrogen receptor modulator and was confirmed effective for the osteoporosis prevention and treatment. Transdermal nanodispersion was prepared by mixing the raloxifene powder and methylcellulose, the formed dispersion was added in the Carbopol gel (Nagai et al., 2018). These nanodispersion was designed with menthol, whereas, menthol was added to the milled raloxifene dispersion, then stirred and sonicated. The skin penetration of raloxifene NPs was prevented by the SC layer in rat. However, when menthol, a permeation enhancer, was added into the formulation attenuated the SC barrier function, allowing NPs to transit into the underlying EP. Furthermore, the enhanced levels of plasma raloxifene after treatment provided efficient and effective therapy for osteoporosis (Nagai et al., 2018).

Another effective and widely used corticosteroid hormone for transdermal treatment is the triamcinolone acetonide (TAA). Zhenmiao Qin and colleagues (2018), described the preparation of a biocompatible and biodegradable TAA-loaded PLGA NPs. *In vitro* results showed that penetration into scar tissue was 40-fold higher for NPs than for commercial suspensions. In addition, was observed a therapeutic effect with fibroblast proliferation inhibition and absence of intact or damaged skin stimulate (Qin et al., 2019).

3.5.2 INFLAMMATORY DISEASES

Nonsteroidal anti-inflammatory drugs (NSAIDs) are commonly classes of medication for pain and inflammation diseases. However, the NSAIDs as long-term oral administrated to treat inflammation process has been associated with gastric irritation, abdominal pain, ulcers, bleeding disorders and cardiovascular side effects (Figueiredo, 2014; Wongrakpanich et al., 2018). Thus, transdermal application of NSAIDs through nanocarriers represents an interesting strategy and have been offer both targeted therapy and drug safety profile (Stoicea et al., 2017).

Liposomes, niosomes, ethosomes, SLN, NLC, polymeric NPs, nanoemulsion, microemulsion, and dendrimers are NPs successfully explored to delivery of NSAIDs via transdermal routes and were recently reviewed by Kumar and colleagues (2018). For instance, an anti-inflammatory drug called aceclofenac, was SLN carried and formulated as hydrogel. The results showed that after *in vitro* application of aceclofenac-loaded SLN and aceclofenac in gel formulation about 40 and 30% of the drug applied were found to have permeated through the animal skin, respectively (Kumar et al., 2018). Moreover, the aceclofenac-loaded SLN was able to inhibit the inflammation process by peak maximum of 6 hours and also this effect was prolonged, extending the inhibition of carrageenan-induced edema (Raj et al., 2016).

Ketoprofen, other NSAIDs, loaded microemulsion was carried through the skin and its biological effect was evaluated using the rat paw edema model. The ME was prepared by a mixture of ketoprofen, propylene glycol, sorbitan monooleate, polysorbate 80, and canola oil. The *in vivo* biological activity from microemulsion containing 1% of ketoprofen was similar to those obtained using the commercial gel with 2.5% of ketoprofen suggesting the same efficacy with minimized side effects (Aliberti et al., 2017).

Transdermal administration of PAMAM dendrimers carrying more than one NSAIDs, ketoprofen, and diflunisal, in rat model showed a significantly enhanced of accumulative skin permeated amount of both drugs after 24 h of treatment and higher drugs plasma levels in the range of 2.5 and 2.7-fold, respectively, when compared with the drug suspensions without PAMAM dendrimers. These found confirmed the capability of dendrimers as multifunctional transdermal delivery system (Yiyun et al., 2007).

SLN and NLC carrying celecoxib, and NSAIDs with oncological properties, provided prolonged drug release and skin permeation displaying reservoir action compared to a nonparticulate gel formulation containing celecoxib (Üner et al., 2014).

3.5.3 OTHERS

Nicotine replacement therapy (NRT) has been on the market since the early 1990s and its application to the skin provides small nicotine doses that minimize the withdrawal experienced by former smokers. It occurs due to the small and controlled amount of nicotine (Wadgave and Nagesh, 2016). So, this kind of therapy can make stopping smoking easier for patients (Hartmann-Boyce et al., 2018). Recent advancements in nanotechnology for transdermal drug delivery showed an impact on NRT from the use of carbon nanotubes (Girão et al., 2010).

Jeongmin Hong and colleagues (2019) reported a one-step synthesis of nicotine-containing NPs by using a size-controllable nanofiltration technique. Polymeric NPs were prepared by spontaneous emulsification. The results showed that an optimized agitation and emulsification promoted NPs stable for drug delivery as well as the particle size controllability was optimized by nanofiltration from 10 nm to 100 nm. Then, the authors conclude that these findings pave the way to TDDS contain particulate nicotine at the nanoscale range (Hong et al., 2019).

Commonly traditional vaccines have been predominantly administered by the parenteral route, although this route is associated with lack of acceptance by the public, safety issues, local irritation, and other side effects (Kalluri and Banga, 2011). However, immunization through transdermal route have been more effective in achieving therapeutic drug level avoiding the side effect (Hansen and Lehr, 2012).

In the last decades, immune response from transdermal vaccines was obtained from several NPs such as nanoemulsions, liposomes, and polymeric NPs. They were designed to overcome the skin barrier and deliver proteins to the deepest skin layers since they cannot permeate passively across the skin due to their large size and hydrophobicity (Kalluri and Banga, 2011). Pissuwan and colleagues (2011) developed a solid-in-oil dispersion system combined with gold nanorods. This approach provided a higher efficiency delivery of high molecular weight protein and improved the transdermal delivery and skin vaccination (Pissuwan et al., 2011).

Gene therapy from transdermal applications is a promising approach to treat chronic and acquired diseases by repair, replace, add, or silence a gene sequence that is identified as beneficial or detrimental to the proper functioning of the organism (Praça and Bentley, 2018). However, the main challenges of nucleic acid (DNA or RNA) transdermal delivery are related to the effective skin barrier function, as well as its high molecular weight and negative superficial charge under physiological conditions which makes their transport through the negatively charged cell membrane difficult (Rosa et al., 2018). To overcome these advantages, transdermal application of nucleic acids from NPs has been a potential strategy for systemic gene delivery (Yang et al., 2018).

Yang and colleagues (2018) constructed a DNA-loaded liposome with particle size distribution around 90 to 160 nm. In *in vivo* studies were showed that the DNA-loaded liposome applied on the skin was distributed in the epidermal/dermal areas and relieved the AD symptoms in mice while free DNA did not penetrate in deepest layers of the skin but remained on the skin surface (Yang et al., 2018). Previously, Geusens, and colleagues (2010) developed cationic deformable liposomes with the ability to cross the intact human epidermal membrane and high cell transfection efficiencies of siRNA (Geusens et al., 2010).

Promising results were also observed to DNA delivered from functionalized gold nanoparticles conjugated with penetrating peptide (Niu et al., 2017).

3.6 CONCLUSION

As discussed in this chapter, the skin acts as a barrier for passive penetration of several active agents, and because of this, the number of medicines commercially available as TTDDS are limited. In the last few years, these limited potential of TTDDS for passive drug skin permeation has given way to advice of nanotechnology. Drug delivery systems at nanoscale based on NPs showed a high ability to overcome the skin biological barriers and can make more effective the delivery of several drugs classified as hydrophobic, hydrophilic, small, larger molecules, and others.

Despite these advantages, there are few NPs-based medicines approved for clinical use, in large part because the translation of these NPs from research orientation to clinical practice requires efforts focused on transitioning from small-scale preparation to the larger-scale manufacturing process, reliable batch-to-batch quality control, confirmed NPs efficiency and rigorous patient safety from clinical trials.

Singhal and colleagues (2017) reported that TTDDS would be key to the successful exploitation of these NPs since the delivery systems ensure: (i) the payload stability and activity; (ii) determine the interface with the skin and drug partitioning into the skin; (iii) optimizes drug skin absorption or permeation maximizing the fraction of drug content delivered into or across the skin; and (iv) minimizes the drug amount retained within the TTDDS (Singhal, Lapteva, and Kalia, 2017).

KEYWORDS

- **drug delivery system**
- **epidermis**
- **nanoparticles**
- **skin diseases**
- **solid lipid nanoparticles**
- **topical application**
- **transdermal application**

REFERENCES

Abed, N.; Couvreur, P. Nanocarriers for antibiotics: A promising solution to treat intracellular bacterial infections. *Int. J. Antimicrob. Agents*, **2014**, *43*(6), 485–496.

Abu Ammar, A.; Nasereddin, A.; Ereqat, S.; Dan-Goor, M.; Jaffe, C. L.; Zussman, E.; Abdeen, Z. Amphotericin B-loaded nanoparticles for local treatment of cutaneous leishmaniasis. *Drug Deliv. Transl. Res.*, **2019**, *9*(1), 76–84.

Agelidis, A.; Koujah, L.; Suryawanshi, R.; Yadavalli, T.; Mishra, Y. K.; Adelung, R.; Shukla, D. An intra-vaginal zinc oxide tetrapod nanoparticles (ZOTEN) and genital herpesvirus cocktail can provide a novel platform for live virus vaccine. *Front. Immunol.*, **2019**, *10*(March), 500.

Ajazzuddin, M.; Jeswani, G.; Jha, A. Nanocosmetics: Past, present, and future trends. *Recent Patents Nanomed.*, **2015**, *5*(1), 3–11.

Algin-Yapar, E.; Inal, Ö. Transdermal spray in hormone delivery. *Trop. J. Pharm. Res.*, **2014**, *13*(3), 469.

Aliberti, A. L. M.; de Queiroz, A. C.; Praça, F. S. G.; Eloy, J. O.; Bentley, M. V. L. B.; Medina, W. S. G. Ketoprofen microemulsion for improved skin delivery and *in vivo* anti-inflammatory effect. *AAPS PharmSciTech*, **2017**, *18*(7), 2783–2791. 6.

Al-Subaie, M. M.; Hosny, K. M.; El-Say, K. M.; Ahmed, T. A.; Aljaeid, B. M. Utilization of nanotechnology to enhance percutaneous absorption of acyclovir in the treatment of herpes simplex viral infections. *Int. J. Nanomedicine*, **2015**, *10*, 3973–3985.

Amer, S. S.; Nasr, M.; Mamdouh, W.; Sammour, O. Insights on the use of nanocarriers for acne alleviation. *Curr. Drug Deliv.*, **2018**, *16*(1), 18–25.

Amjadi, M.; Mostaghaci, B.; Sitti, M. Recent advances in skin penetration enhancers for transdermal gene and drug delivery. *Curr. Gene Ther.*, **2017**, *17*(2).

Apalla, Z.; Nashan, D.; Weller, R. B.; Castellsagué, X. Skin cancer: Epidemiology, disease burden, pathophysiology, diagnosis, and therapeutic approaches. *Dermatol. Ther. (Heidelb).*, **2017**, *7*(Suppl 1), 5–19.

Arora, S.; Tyagi, N.; Bhardwaj, A.; Rusu, L.; Palanki, R.; Vig, K.; Singh, S. R.; Singh, A. P.; Palanki, S.; Miller, M. E.; Carter, J. E.; Singh, S. Silver nanoparticles protect human keratinocytes against UVB radiation-induced DNA damage and apoptosis: Potential for prevention of skin carcinogenesis. *Nanomedicine Nanotechnology*, B*iol. Med.*, **2015**, *11*(5), 1265–1275.

Ascenso, A.; Salgado, A.; Euletério, C.; Praça, F. G.; Bentley, M. V. L. B.; Marques, H. C.; Oliveira, H.; Santos, C.; Simões, S. *In vitro* and *in vivo* topical delivery studies of tretinoin-loaded ultradeformable vesicles. *Eur. J. Pharm. Biopharm.*, **2014**, *88*(1), 48–55.

Bahrami, F.; Harandi, A. M.; Rafati, S. Biomarkers of cutaneous leishmaniasis. *Front. Cell. Infect. Microbiol.*, **2018**, *8*, 1–8.

Barazesh, A.; Motazedian, M. H.; Sattarahmady, N.; Morowvat, M. H.; Rashidi, S. Preparation of meglumine antimonate loaded albumin nanoparticles and evaluation of its anti-leishmanial activity: An *in vitro* assay. *J. Parasit. Dis.*, **2018**, *42*(3), 416–422.

Barradas, T. N.; Senna, J. P.; Cardoso, S. A.; De Holanda E Silva, K. G.; Elias, M. C. R. formulation characterization and *in vitro* drug release of hydrogel-thickened nanoemulsions for topical delivery of 8-methoxypsoralen. *Mater. Sci. Eng. C*, **2018**, *92*, 245–253.

Barua, S.; Gogoi, S.; Khan, R.; Karak, N. Silicon-Based nanomaterials and their polymer nanocomposites. In *Nanomaterials and Polymer Nanocomposites*; Elsevier, **2019**; pp. 261–305.

Baumann, L. How to use oral and topical cosmeceuticals to prevent and treat skin aging. *Facial Plast. Surg. Clin. North Am.*, **2018**, *26*(4), 407–413.

Berry, S. L.; Walker, K.; Hoskins, C.; Telling, N. D.; Price, H. P. Nanoparticle-mediated magnetic hyperthermia is an effective method for killing the human-infective protozoan parasite leishmania Mexicana *in Vitro. Sci. Rep.*, **2019**, *9*(1), 1–9.

Bessar, H.; Venditti, I.; Benassi, L.; Vaschieri, C.; Azzoni, P.; Pellacani, G.; Magnoni, C.; Botti, E.; Casagrande, V.; Federici, M.; Costanzo, A.; Fontana, L.; Testa, G.; Mostafa, F. F.; Ibrahim, S. A.; Russo, M. V.; Fratoddi, I. Functionalized gold nanoparticles for topical delivery of methotrexate for the possible treatment of psoriasis. *Colloids Surfaces B Biointerfaces*, **2016**, *141*, 141–147.

Bharadwaj, R.; Das, P. J.; Pal, P.; Mazumder, B. Topical delivery of paclitaxel for treatment of skin cancer. *Drug Dev. Ind. Pharm.*, **2016**, *42*(9), 1482–1494.

Boehncke, W. H.; Schön, M. P. Psoriasis. *Lancet*, **2015**, *386*(9997), 983–994.

Bolzinger, M.-A.; Briançon, S.; Pelletier, J.; Chevalier, Y. Penetration of drugs through skin, a complex rate-controlling membrane. *Curr. Opin. Colloid Interface Sci.*, **2012**, *17*(3), 156–165.

Boniface, K.; Seneschal, J.; Picardo, M.; Taïeb, A. Vitiligo: Focus on clinical aspects, immunopathogenesis, and therapy. *Clin. Rev. Allergy Immunol.*, **2018**, *54*(1), 52–67.

Borase, H. P.; Patil, C. D.; Salunkhe, R. B.; Suryawanshi, R. K.; Salunke, B. K.; Patil, S. V. Phytolatex synthesized gold nanoparticles as novel agent to enhance sun protection factor of commercial sunscreens. *Int. J. Cosmet. Sci.*, **2014**, *36*(6), 571–578.

Borgheti-Cardoso, L. N.; Kooijmans, S. A. A.; Fens, M. H. A. M.; Van Der, M. R.; Vicentini, F. T. M. C.; Fantini, M. C. A.; Bentley, M. V. L. B.; Schiffelers, R. M. In situ gelling liquid crystalline system as local SiRNA delivery system. *Ski. Res. Technol.*, **2017**, *24*(2), 165–174.

Bose, R. J. C.; Ravikumar, R.; Karuppagounder, V.; Bennet, D.; Rangasamy, S.; Thandavarayan, R. A. Lipid–polymer hybrid nanoparticle-mediated therapeutics delivery: Advances and challenges. *Drug Discov. Today*, **2017**, *22*(8), 1258–1265.

Botelho, M.; Queiroz, D.; Barros, G.; Guerreiro, S.; Fechine, P.; Umbelino, S.; Lyra, A.; Borges, B.; Freitas, A.; Queiroz, D.; Ruela, R.; Almeida, J.; Quintans Jr, L. Nanostructured transdermal hormone replacement therapy for relieving menopausal symptoms: A confocal Raman spectroscopy study. *Clinics*, **2014**, *69*(2), 75–82.

Bruni, N.; Stella, B.; Giraudo, L.; Della Pepa, C.; Gastaldi, D.; Dosio, F. Nanostructured delivery systems with improved leishmanicidal activity: A critical review. *Int. J. Nanomedicine*, **2017**, *12*, 5289–5311.

Budhiraja, A.; Dhingra, G. Development and characterization of a novel antiacne niosomal gel of rosmarinic acid. *Drug Deliv.*, **2015**, *22*(6), 723–730.

Campana, R.; Dzoro, S.; Mittermann, I.; Fedenko, E.; Elisyutina, O.; Khaitov, M.; Karaulov, A.; Valenta, R. Molecular aspects of allergens in atopic dermatitis. *Curr. Opin. Allergy Clin. Immunol.*, **2017**, *17*(4), 269–277.

Carazo, E.; Borrego-Sánchez, A.; García-Villén, F.; Sánchez-Espejo, R.; Cerezo, P.; Aguzzi, C.; Viseras, C. Advanced inorganic nanosystems for skin drug delivery. *Chem. Rec.*, **2018**, *18*(7, 8), 891–899.

Carbone, C.; Teixeira, M. do C.; Sousa, M. do C.; Martins-Gomes, C.; Silva, A. M.; Souto, E. M. B.; Musumeci, T. Clotrimazole-loaded Mediterranean essential oils NLC: A synergic treatment of candida skin infections. *Pharmaceutics*, **2019**, *11*(5), 231.

Carolina, O. D. S. L.; Spagnol, C. M.; Guillot, A. J.; Melero, A.; Corrêa, M. A. Caffeic acid skin absorption: Delivery of microparticles to hair follicles. *Saudi Pharm. J.*, **2019**, *27*(6), 791–797.

Carter, P.; Narasimhan, B.; Wang, Q. Biocompatible nanoparticles and vesicular systems in transdermal drug delivery for various skin diseases. *Int. J. Pharm.*, **2019**, *555*, 49–62.

Carvalho, A. L. M.; Silva, J. A. Da; Lira, A. A. M.; Conceição, T. M. F.; Nunes, R. De S.; De Albuquerque Junior, R. L. C.; Sarmento, V. H. V.; Leal, L. B.; De Santana, D. P. Evaluation of microemulsion and lamellar liquid crystalline systems for transdermal zidovudine delivery. *J. Pharm. Sci.*, **2016**, *105*(7), 2188–2193.

Chaloupka, K.; Malam, Y.; Seifalian, A. M. Nanosilver as a new generation of nanoproduct in biomedical applications. *Trends Biotechnol.*, **2010**, *28*(11), 580–588.

Charoenputtakun, P.; Pamornpathomkul, B.; Opanasopit, P.; Rojanarata, T.; Ngawhirunpat, T. Terpene composited lipid nanoparticles for enhanced dermal delivery of all-trans-retinoic acids. *Biol. Pharm. Bull.*, **2014**, *37*(7), 1139–1148.

Crisan, D.; Scharffetter-Kochanek, K.; Crisan, M.; Schatz, S.; Hainzl, A.; Olenic, L.; Filip, A.; Schneider, L. A.; Sindrilaru, A. Topical silver and gold nanoparticles complexed with cornus mas suppress inflammation in human psoriasis plaques by inhibiting NF-KB activity. *Exp. Dermatol.*, **2018**, *27*(10), 1166–1169.

Crucho, C. I. C.; Teresa, M. Polymeric nanoparticles : A study on the preparation variables and characterization methods. *Mater. Sci. Eng. C*, **2017**, *80*, 771–784.

Dar, M. J.; Din, F. U.; Khan, G. M. Sodium stibogluconate loaded nano-deformable liposomes for topical treatment of leishmaniasis: Macrophage as a target cell. *Drug Deliv.*, **2018**, *25*(1), 1595–1606.

Date, T.; Nimbalkar, V.; Kamat, J.; Mittal, A.; Mahato, R. I.; Chitkara, D. Lipid-polymer hybrid nanocarriers for delivering cancer therapeutics. *J. Control. Release*, **2018**, *271*, 60–73.

Dave, V.; Tak, K.; Sohgaura, A.; Gupta, A.; Sadhu, V.; Reddy, K. R. Lipid-polymer hybrid nanoparticles: Synthesis strategies and biomedical applications. *J. Microbiol. Methods*, **2019**, *160*(March), 130–142.

De Leeuw, J.; Assen, Y. J.; Van Der Beek, N.; Bjerring, P.; Martino, N. H. A. Treatment of vitiligo with khellin liposomes, ultraviolet light and blister roof transplantation. *J. Eur. Acad. Dermatology Venereol.*, **2011**, *25*(1), 74–81.

De Leeuw, J.; Van Der Beek, N.; Maierhofer, G.; Neugebauer, W. D. A case study to evaluate the treatment of vitiligo with khellin encapsulated in L-phenylalanin stabilized phosphatidylcholine liposomes in combination with ultraviolet light therapy. *Eur. J. Dermatology*, **2003**, *13*(5), 474–477.

De Souza, A.; Marins, D. S. S.; Mathias, S. L.; Monteiro, L. M.; Yukuyama, M. N.; Scarim, C. B.; Löbenberg, R.; Bou-Chacra, N. A. promising nanotherapy in treating leishmaniasis. *Int. J. Pharm.*, **2018**, *547*(1, 2), 421–431.

Deleanu, D.; Nedelea, I. Biological therapies for atopic dermatitis: An update (review). *Exp. Ther. Med.*, **2018**, *4*, 1061–1067.

Deng, Y.; Ediriwickrema, A.; Yang, F.; Lewis, J.; Girardi, M.; Saltzman, W. M. A sunblock based on bioadhesive nanoparticles. *Nat. Mater.*, **2015**, *14*(12), 1278–1285.

Depieri, L. V.; Garcia, P. F. S.; Campos, P. M.; Lopes, B. B. M. V. Advances in the bioanalytical study of drug delivery across the skin. *Ther. Deliv.*, **2015**, *6*(5), 571–594.

Desmet, E.; Van Gele, M.; Lambert, J. Topically applied lipid- and surfactant-based nanoparticles in the treatment of skin disorders. *Expert Opin. Drug Deliv.*, **2017**, *14*(1), 109–122.

Dianzani, C.; Zara, G. P.; Maina, G.; Pettazzoni, P.; Pizzimenti, S.; Rossi, F.; Gigliotti, C. L.; Ciamporcero, E. S.; Daga, M.; Barrera, G. Drug delivery nanoparticles in skin cancers. *Biomed Res. Int.*, **2014**, 1–13.

Dimitriou, F.; Krattinger, R.; Ramelyte, E.; Barysch, M. J.; Micaletto, S.; Dummer, R.; Goldinger, S. M. The world of melanoma: Epidemiologic, genetic, and anatomic differences of melanoma across the globe. *Curr. Oncol. Rep.*, **2018**, *20*(11), 87.

Doppalapudi, S.; Mahira, S.; Khan, W. Development and *in vitro* assessment of psoralen and resveratrol co-loaded ultradeformable liposomes for the treatment of vitiligo. *J. Photochem. Photobiol. B Biol.*, **2017**, *174*, 44–57.

Dul, M.; Nikolic, T.; Stefanidou, M.; McAteer, M. A.; Williams, P.; Mous, J.; Roep, B. O.; Kochba, E.; Levin, Y.; Peakman, M.; Wong, F. S.; Dayan, C. M.; Tatovic, D.; Coulman, S. A.; Birchall, J. C. Conjugation of a peptide autoantigen to gold nanoparticles for intradermally administered antigen specific immunotherapy. *Int. J. Pharm.*, **2019**, *562*, 303–312.

Esfandiari, F.; Motazedian, M. H.; Asgari, G.; Morowvat, M. H.; Molaei, M.; Heli, H. Paromomycin-Loaded mannosylated chitosan nanoparticles: Synthesis, characterization and targeted drug delivery against leishmaniasis. *Acta Trop.*, **2019**, *197*, 105045.

Figueiredo, C. P. R. V. Ibuprofen nanoparticles for oral delivery: Proof of concept. *J. Nanomedine. Biotherapeutic Discov.*, **2014**, *04*(01).

Fratoddi, I.; Benassi, L.; Botti, E.; Vaschieri, C.; Venditti, I.; Bessar, H.; Samir, M. A.; Azzoni, P.; Magnoni, C.; Costanzo, A.; Casagrande, V.; Federici, M.; Bianchi, L.; Pellacani, G. Effects of topical methotrexate loaded gold nanoparticle in cutaneous inflammatory mouse model. *Nanomedicine Nanotechnology, Biol. Med.*, **2019**, *17*, 276–286.

Freag, M. S.; Torky, A. S.; Nasra, M. M.; Abdelmonsif, D. A.; Abdallah, O. Y. Liquid crystalline nanoreservoir releasing a highly skin-penetrating berberine oleate complex for psoriasis management. *Nanomedicine*, **2019**, *14*(8), 931–954.

Gainza, G.; Celdran, D.; Moreno, B.; Javier, J.; Borja, F.; Villullas, S.; Luis, J.; Igartua, M.; Maria, R. The topical administration of RhEGF-loaded nanostructured lipid carriers (RhEGF-NLC) improves healing in a porcine full-thickness excisional wound model. *J. Control. Release*, **2015**, *197*, 41–47.

Gainza, G.; Pastor, M.; Aguirre, J. J.; Villullas, S.; Pedraz, J. L.; Hernandez, R. M.; Igartua, M. A novel strategy for the treatment of chronic wounds based on the topical administration of RhEGF-loaded lipid nanoparticles: *In vitro* bioactivity and *in vivo* effectiveness in healing-impaired Db/Db mice. *J. Control. Release*, **2014**, *185*, 51–61.

Gan, B. K.; Yong, C. Y.; Ho, K. L.; Omar, A. R.; Alitheen, N. B.; Tan, W. S. Targeted delivery of cell penetrating peptide virus-like nanoparticles to skin cancer cells. *Sci. Rep.*, **2018**, *8*(1), 1–13.

Garg, B. J.; Garg, N. K.; Beg, S.; Singh, B.; Katare, O. P. Nanosized ethosomes-based hydrogel formulations of methoxsalen for enhanced topical delivery against vitiligo: Formulation optimization, *in vitro* evaluation and preclinical assessment. *J. Drug Target.*, **2016**, *24*(3), 233–246.

Garg, N. K.; Tandel, N.; Jadon, R. S.; Tyagi, R. K.; Katare, O. P. Lipid–polymer hybrid nanocarrier-mediated cancer therapeutics: Current status and future directions. *Drug Discov. Today*, **2018**, *23*(9), 1610–1621.

Geilich, B. M.; Ven, A. L. Van De; Singleton, G. L.; Sepúlveda, L. J.; Sridhar, S.; Webster, T. J. *Silver Nanoparticle-Embedded Polymer Some Nanocarriers for the Treatment of Antibiotic-Resistant Infections*, **2015**.

Gélvez, A. P. C.; Farias, L. H. S.; Pereira, V. S.; Da Silva, I. C. M.; Costa, A. C.; Dias, C. G. B. T.; Costa, R. M. R.; Da Silva, S. H. M.; Rodrigues, A. P. D. Biosynthesis, characterization and leishmanicidal activity of a biocomposite containing AgNPs-PVP-glucantime. *Nanomedicine*, **2018**, *13*(4), 373–390.

Geusens, B.; Van Gele, M.; Braat, S.; De Smedt, S. C.; Stuart, M. C. A.; Prow, T. W.; Sanchez, W.; Roberts, M. S.; Sanders, N. N.; Lambert, J. Flexible nanosomes (SECosomes) enable efficient sirna delivery in cultured primary skin cells and in the viable epidermis of *ex vivo* human skin. *Adv. Funct. Mater.*, **2010**, *20*(23), 4077–4090.

Ghate, V. M.; Lewis, S. A.; Prabhu, P.; Dubey, A.; Patel, N. Nanostructured lipid carriers for the topical delivery of tretinoin. *Eur. J. Pharm. Biopharm.*, **2016**, *108*, 253–261.

Girão, E. C.; Fagan, S. B.; Zanella, I.; Filho, A. G. S. Nicotine adsorption on single wall carbon nanotubes. *J. Hazard. Mater.*, **2010**, *184*(1–3), 678–683.

Gong, C.; Wu, Q.; Wang, Y.; Zhang, D.; Luo, F.; Zhao, X.; Wei, Y.; Qian, Z. A biodegradable hydrogel system containing curcumin encapsulated in micelles for cutaneous wound healing. *Biomaterials*, **2013**, *34*(27), 6377–6387.

Gupta, A.; Singh, S.; Kotla, N. G.; Webster, T. J. Formulation and Evaluation of a topical niosomal gel containing a combination of benzoyl peroxide and tretinoin for antiacne activity. *Int. J. Nanomedicine*, **2014**, *10*, 171–182.

Hafeez, A.; Kazmi, I. Dacarbazine nanoparticle topical delivery system for the treatment of melanoma. *Sci. Rep.*, **2017**, *7*(1), 16517.

Hansen, S.; Lehr, C. M. Nanoparticles for transcutaneous vaccination. *Microb. Biotechnol.*, **2012**, *5*(2), 156–167.

Hartmann-Boyce, J.; Chepkin, S. C.; Ye, W.; Bullen, C.; Lancaster, T. Nicotine replacement therapy versus control for smoking cessation. *Cochrane Database Syst. Rev.*, **2018**, No. 5.

Hong, J.; Guduru, R.; Khizroev, S.; You, L. One-step fabrication of size-controllable nicotine containing core–shell structures. *Nanoscale Adv.*, **2019**, *1*(4), 1305–1313.

Hong, L.; Salentinig, S.; Hawley, A.; Boyd, B. J. Understanding the mechanism of enzyme-induced formation of lyotropic liquid crystalline nanoparticles. *Langmuir*, **2015**, *31*(24), 6933–6941.

Hood, M. A. Correction to synthetic strategies in the preparation of polymer/inorganic hybrid nanoparticles. doi: 10.3390/Ma7117583. *Materials* (Basel), **2014**, *7*(11), 4057–4087, 7583–7614.

Hsu, H.; Bugno, J.; Lee, S.; Hong, S. Dendrimer based nanocarriers: A versatile platform for drug delivery. *Adv. Rev.*, **2017**, *9*, 1–21.

Hua, S. Lipid-based nano-delivery systems for skin delivery of drugs and bioactives. *Front. Pharmacol.*, **2015**, *6*, **2011–**, **2015**.

Huh, A. J.; Kwon, Y. J. "Nanoantibiotics": A new paradigm for treating infectious diseases using nanomaterials in the antibiotics resistant era. *J. Control. Release*, **2011**, *156*(2), 128–145.

Isaac-Márquez, A. P.; Talamás-Rohana, P.; Galindo-Sevilla, N.; Gaitan-Puch, S. E.; Díaz-Díaz, N. A.; Hernández-Ballina, G. A.; Lezama-Dávila, C. M. Decanethiol functionalized silver nanoparticles are new powerful leishmanicidals *in Vitro. World J. Microbiol. Biotechnol.*, **2018**, *34*(3), 0.

Jain, A.; Garg, N. K.; Jain, A.; Kesharwani, P.; Jain, A. K.; Nirbhavane, P.; Tyagi, R. K. A synergistic approach of adapalene-loaded nanostructured lipid carriers, and Vitamin C Co-administration for treating acne. *Drug Dev. Ind. Pharm.*, **2016**, *42*(6), 897–905.

Jain, S. K.; Verma, A.; Jain, A.; Hurkat, P. Transfollicular drug delivery: Current perspectives. *Res. Reports Transdermal Drug Deliv.*, **2016**, 1.

Jain, S.; Mistry, M. A.; Swarnakar, N. K. Enhanced dermal delivery of acyclovir using solid lipid nanoparticles. *Drug Deliv. Transl. Res.*, **2011**, *1*(5), 395–406.

Jain, S.; Patel, N.; Shah, M. K.; Khatri, P.; Vora, N. Recent advances in lipid-based vesicles and particulate carriers for topical and transdermal application. *J. Pharm. Sci.*, **2017**, *106*(2), 423–445.

Jamil, B.; Bokhari, H.; Imran, M. Mechanism of action: How nano-antimicrobials act? *Curr. Drug Targets*, **2017**, *18*, 363–373.

Kalluri, H.; Banga, A. K. Transdermal delivery of proteins. *AAPS PharmSciTech*, **2011**, *12*(1), 431–441.

Kaur, J.; Kaur, J.; Jaiswal, S.; Gupta, G. Recent advances in topical drug delivery system. *Indo Am. J. Pharm. Res.*, **2016**, *6*(07), 6353–6369.

Kim, H. S.; Sun, X.; Lee, J.; Kim, H.; Fu, X.; Leong, K. W. Advanced drug delivery systems and artificial skin grafts for skin wound healing. *Adv. Drug Deliv. Rev.*, **2018**.

Kiselev, M. A. Conformation of ceramide 6 molecules and chain-flip transitions in the lipid matrix of the outermost layer of mammalian skin, the stratum corneum. *Crystallogr. Reports*, **2007**, *52*(3), 525–528.

Kulkarni, V. S. *Handbook of Non-Invasive Drug Delivery Systems*; Elsevier, **2010**.

Kumar, K.; Her, C.; Comune, M.; Moia, C.; Lopes, A.; Porporato, P. E.; Vanacker, J.; Lam, M. C.; Steinstraesser, L.; Sonveaux, P.; Zhu, H.; Ferreira, L. S.; Vandermeulen, G.; Préat, V. PLGA nanoparticles loaded with host defense peptide LL*37*promote wound healing. *J. Control. Release*, **2014**, *194*, 138–147.

Kumar, L.; Verma, S.; Singh, M.; Chalotra, T.; Utreja, P. Advanced drug delivery systems for transdermal delivery of non-steroidal anti-inflammatory drugs: A review. *Curr. Drug Deliv.*, **2018**, *15*(8), 1087–1099.

La, W.; Yang, H. S. Heparin-conjugated poly(lactic-co-glycolic acid) nanospheres enhance large-wound healing by delivering growth factors in platelet-rich plasma. *Artif. Organs*, **2015**, *39*(4), 388–394.

Labouta, H. I.; Schneider, M. Interaction of inorganic nanoparticles with the skin barrier: Current status and critical review. *Nanomedicine Nanotechnology*, B*iol. Med.*, **2013**, *9*(1), 39–54.

Lademann, J.; Knorr, F.; Richter, H.; Jung, S.; Meinke, M. C.; Rühl, E.; Alexiev, U.; Calderon, M.; Patzelt, A. Hair follicles as a target structure for nanoparticles. *J. Innov. Opt. Health Sci.*, **2015**, *08*(04), 1530004.

Lapteva, M.; Möller, M.; Gurny, R.; Kalia, Y. N. Self-assembled polymeric nanocarriers for the targeted delivery of retinoic acid to the hair follicle. *Nanoscale*, **2015**, *7*(44), 18651–18662.

Larese Filon, F.; Mauro, M.; Adami, G.; Bovenzi, M.; Crosera, M. Nanoparticles skin absorption: New aspects for a safety profile evaluation. *Regul. Toxicol. Pharmacol.*, **2015**, *72*(2), 310–322.

Limbert, G.; Masen, M. A.; Pond, D.; Graham, H. K.; Sherratt, M. J.; Jobanputra, R.; McBride, A. biotribology of the ageing skin—why we should care. *Biotribology*, **2019**, *17*, 75–90.

Lopes, L. Overcoming the cutaneous barrier with microemulsions. *Pharmaceutics*, **2014**, *6*(1), 52–77.

Luchini, A.; Vitiello, G. Understanding the nano-bio interfaces: Lipid-coatings for inorganic nanoparticles as promising strategy for biomedical applications. *Front. Chem.*, **2019**, *7*, 1–16.

Lueangarun, S.; Sriviriyakul, K.; Tempark, T.; Managit, C.; Sithisarn, P. Clinical efficacy of 0.5% topical mangosteen extract in nanoparticle loaded gel in treatment of mild-to-moderate acne vulgaris: A 12-week, split-face, double-blinded, randomized, controlled trial. *J. Cosmet. Dermatol.*, **2019**, *2018*, 1–9.

Macchione, M. A.; Biglione, C.; Strumia, M. Design, synthesis and architectures of hybrid nanomaterials for therapy and diagnosis applications. *Polymers (Basel)*, **2018**, *10*(5), 1–34.

Mahmoud, N. N.; Hikmat, S.; Abu Ghith, D.; Hajeer, M.; Hamadneh, L.; Qattan, D.; Khalil, E. A. Gold nanoparticles loaded into polymeric hydrogel for wound healing in rats: Effect of nanoparticles' shape and surface modification. Int. J. Pharm., **2019**, *565*, 174–186.

Manosroi, J.; Khositsuntiwong, N.; Götz, F.; Werner, R. G.; Manosroi, W.; Manosroi, A. Potent melanin production enhancement of human tyrosinase gene by tat and an entrapment in elastic cationic niosomes: Potential application in vitiligo gene therapy. *Chem. Biol. Drug Des.*, **2012**, *80*(6), 953–960.

Manosroi, J.; Khositsuntiwong, N.; Manosroi, W.; Götz, F.; Werner, R. G.; Manosroi, A. enhancement of transdermal absorption, gene expression and stability of tyrosinase plasmid (PMEL34)-loaded elastic cationic niosomes: Potential application in vitiligo treatment. *J. Pharm. Sci.*, **2010**, *99*(8), 3533–3541.

Mathur, M.; Devi, V. K. Potential of novel drug delivery systems in the management of topical candidiasis. *J. Drug Target.*, **2017**, *25*(8), 685–703.

Matsui, T.; Amagai, M. Dissecting the formation, structure and barrier function of the stratum corneum. *Int. Immunol.*, **2015**, *27*(6), 269–280.

Md, S.; Haque, S.; Madheswaran, T.; Zeeshan, F.; Meka, V. S.; Radhakrishnan, A. K.; Kesharwani, P. Lipid based nanocarriers system for topical delivery of photosensitizers. *Drug Discov. Today*, **2017**, *22*(8), 1274–1283.

Md, S.; Kuldeep, S. J. K. A.; Waqas, M.; Pandey, M.; Choudhury, H.; Habib, H.; Hussain, F.; Hussain, Z. Nanoencapsulation of betamethasone valerate using high pressure homogenization–solvent evaporation technique: Optimization of formulation and process parameters for efficient dermal targeting. *Drug Dev. Ind. Pharm.*, **2019**, *45*(2), 323–332.

Mithun, B. T. S. Challenges facing transdermal drug delivery systems: A conceptual approach. *Res. J. Sci. Technol.*, **2012**, *4*(5), 213–219.

Mofidi, A.; Tompa, E.; Spencer, J.; Kalcevich, C.; Peters, C. E.; Kim, J.; Song, C.; Mortazavi, S. B.; Demers, P. A. The economic burden of occupational non-melanoma skin cancer due to solar radiation. *J. Occup. Environ. Hyg.*, **2018**, *15*(6), 481–491.

Mukherjee, A.; Waters, A. K.; Kalyan, P.; Achrol, A. S.; Kesari, S.; Yenugonda, V. M. Lipid-polymer hybrid nanoparticles as a next generation drug delivery platform: State of the art, emerging technologies, and perspectives. *Int. J. Nanomedicine*, **2019**, *14*, 1937–1952.

Mukherjee, S.; Ray, S.; Thakur, R. Solid lipid nanoparticles: A modern formulation approach in drug delivery system. *Indian J. Pharm. Sci.*, **2009**, *71*(4), 349.

Münch, S.; Wohlrab, J.; Neubert, R. H. H. Dermal and transdermal delivery of pharmaceutically relevant macromolecules. *Eur. J. Pharm. Biopharm.*, **2017**, *119*, 235–242.

Nagai, N.; Ogata, F.; Otake, H.; Nakazawa, Y.; Kawasaki, N. Design of a transdermal formulation containing raloxifene nanoparticles for osteoporosis treatment. *Int. J. Nanomedicine*, **2018**, *13*, 5215–5229.

Naseri, N.; Valizadeh, H.; Zakeri-Milani, P. Solid lipid nanoparticles and nanostructured lipid carriers: Structure, preparation and application. *Adv. Pharm. Bull.*, **2015**, *5*(3), 305–313.

Niska, K.; Zielinska, E.; Radomski, M. W.; Inkielewicz-Stepniak, I. Metal nanoparticles in dermatology and cosmetology: Interactions with human skin cells. *Chem. Biol. Interact.*, **2018**, *295*, 38–51.

Niu, J.; Chu, Y.; Huang, Y. F.; Chong, Y. S.; Jiang, Z. H.; Mao, Z. W.; Peng, L. H.; Gao, J. Q. Transdermal gene delivery by functional peptide-conjugated cationic gold nanoparticle reverses the progression and metastasis of cutaneous melanoma. *ACS Appl. Mater. Interfaces*, **2017**, *9*(11), 9388–9401.

Omar, M. H. Common skin disorders in the elderly. *South African Fam. Pract.*, **2006**, *48*(5), 29–34.

Orłowski, P.; Kowalczyk, A.; Tomaszewska, E.; Ranoszek-Soliwoda, K.; Węgrzyn, A.; Grzesiak, J.; Celichowski, G.; Grobelny, J.; Eriksson, K.; Krzyzowska, M. Antiviral activity of tannic acid modified silver nanoparticles: Potential to activate immune response in herpes genitalis. *Viruses*, **2018**, *10*(10), 524.

Oyama, J.; Lera-Nonose, D. S. S. L.; Ramos-Milaré, Á. C. F. H.; Padilha, F. F. B.; De Freitas, C. F.; Caetano, W.; Hioka, N.; Silveira, T. G. V.; Lonardoni, M. V. C. Potential of pluronics ® P-123 and F-127 as nanocarriers of anti-leishmania chemotherapy. *Acta Trop.*, **2019**, *192*, 11–21.

Oyarzun-Ampuero, F.; Vidal, A.; Concha, M.; Morales, J.; Orellana, S.; Moreno-villoslada, I. *Nanoparticles for the Treatment of Wounds*, **2015**, 4329–4341.

Pandey, M.; Choudhury, H.; Gunasegaran, T. A. P.; Nathan, S. S.; Md, S.; Gorain, B.; Tripathy, M.; Hussain, Z. Hyaluronic acid-modified betamethasone encapsulated polymeric nanoparticles: Fabrication, characterization, *in Vitro* release kinetics, and dermal targeting. *Drug Deliv. Transl. Res.*, **2019**, *9*(2), 520–533.

Pan-In, P.; Wongsomboon, A.; Kokpol, C.; Chaichanawongsaroj, N.; Wanichwecharungruang, S. Depositing α-mangostin nanoparticles to sebaceous gland area for acne treatment. *J. Pharmacol. Sci.*, **2015**, *129*(4), 226–232.

Pastore, M. N.; Kalia, Y. N.; Horstmann, M.; Roberts, M. S. Transdermal patches: History, development and pharmacology. *Br. J. Pharmacol.*, **2015**, *172*(9), 2179–2209.

Patel, D.; Chaudhary, S. A.; Parmar, B.; Bhura, N. *Transdermal Drug Delivery System: A Review.*, **2012**, *1*(4).

Patzelt, A.; Lademann, J. Drug delivery to hair follicles. *Expert Opin. Drug Deliv.*, **2013**, *10*(6), 787–797.

Petrilli, R.; Eloy, J. O.; Praça, F. S. G.; Del Ciampo, J. O.; Fantini, M. A. C.; Fonseca, M. J. V.; Bentley, M. V. L. B. Liquid crystalline nanodispersions functionalized with cell-penetrating peptides for topical delivery of short-interfering RNAs: A proposal for silencing a pro-inflammatory cytokine in cutaneous diseases. *J. Biomed. Nanotechnol.*, **2016**, *12*(5), 1063–1075.

Petrilli, R.; Eloy, J.; Lopez, R.; Lee, R. Cetuximab immunoliposomes enhance delivery of 5-FU to skin squamous carcinoma cells. anticancer. *Agents Med. Chem.*, **2017**, *17*(2), 301–308.

Picardo, M.; Dell'Anna, M. L.; Ezzedine, K.; Hamzavi, I.; Harris, J. E.; Parsad, D.; Taieb, A. Vitiligo. *Nat. Rev. Dis. Prim.*, **2015**, *1*(1), 15011. https://doi.org/10.1038/nrdp., 2015.11.
Pissuwan, D.; Nose, K.; Kurihara, R.; Kaneko, K.; Tahara, Y.; Kamiya, N.; Goto, M.; Katayama, Y.; Niidome, T. A solid-in-oil dispersion of gold nanorods can enhance transdermal protein delivery and skin vaccination. *Small*, **2011**, *7*(2), 215–220.
Praça, F. G.; Bentley, M. V. L. B. Advances in specific gene knockdown of therapeutic targets by RNA interference. *Curr. Pharm. Des.*, **2018**, *24*(23), 2631–2631.
Pradhan, M.; Alexander, A.; Singh, M. R.; Singh, D.; Saraf, S.; Saraf, S.; Ajazuddin. Understanding the prospective of nano-formulations towards the treatment of psoriasis. *Biomed. Pharmacother.*, **2018**, *107*, 447–463.
Qin, Z.; Chen, F.; Chen, D.; Wang, Y.; Tan, Y.; Ban, J. Transdermal permeability of triamcinolone acetonide lipid nanoparticles. *Int. J. Nanomedicine*, **2019**, *14*, 2485–2495.
Radtke, M.; Patzelt, A.; Knorr, F.; Lademann, J.; Netz, R. R. Ratchet effect for nanoparticle transport in hair follicles. *Eur. J. Pharm. Biopharm.*, **2017**, *116*, 125–130.
Raj, R.; Mongia, P.; Ram, A.; Jain, N. K. Enhanced skin delivery of aceclofenac via hydrogel-based solid lipid nanoparticles. *Artif. Cells*, N*anomedicine Biotechnol.*, **2016**, *44*(6), 1434–1439.
Raj, S.; Sumod, U.; Jose, S.; Sabitha, M. Nanotechnology in cosmetics: Opportunities and challenges. *J. Pharm. Bioallied Sci.*, **2012**, *4*(3), 186.
Ramezanli, T.; Michniak-Kohn, B. B. Development and characterization of a topical gel formulation of adapalene-tyrospheres and assessment of its clinical efficacy. *Mol. Pharm.*, **2018**, *15*(9), 3813–3822.
Rashighi, M.; Harris, J. E. Vitiligo pathogenesis and emerging treatments. *Dermatol. Clin.*, **2017**, *35*(2), 257–265. https://doi.org/10.1016/j.det., 2016.11.014.
Roberts, M. Targeted drug delivery to the skin and deeper tissues: Role of physiology, solute structure and disease. *Clin. Exp. Pharmacol. Physiol.*, **1997**, *24*(11), 874–879.
Rosa, J.; Suzuki, I.; Kravicz, M.; Caron, A.; Pupo, A. V.; Praca, F. G.; Bentley, M. V. L. B. Current non-viral SiRNA delivery systems as a promising treatment of skin diseases. *Curr. Pharm. Des.*, **2018**, *24*(23), 2644–2663.
Ruela, A. L. M.; Figueiredo, E. C.; Perissinato, A. G.; Lima, A. C. Z.; Araújo, M. B.; Pereira, G. R. *In vitro* evaluation of transdermal nicotine delivery systems commercially available in Brazil. *Brazilian J. Pharm. Sci.*, **2013**, *49*(3), 579–588.
Sahu, S.; Katiyar, S. S.; Kushwah, V.; Jain, S. Active natural oil-based nanoemulsion containing tacrolimus for synergistic antipsoriatic efficacy. *Nanomedicine*, **2018**, *13*(16), 1985–1998.
Salimi, A.; Mohammad, S. S. Transfollicular drug delivery systems. *Jundishapur J. Nat. Pharm. Prod.*, **2018**, In Press (In Press), 2622–2631.
Sallam, M. A.; Marín Boscá, M. T. Mechanistic analysis of human skin distribution and follicular targeting of adapalene-loaded biodegradable nanospheres with an insight into hydrogel matrix influence, *in vitro* skin irritation, and *In vivo* tolerability. *J. Pharm. Sci.*, **2017**, *106*(10), 3140–3149.
Sanmukhani, J. Clindamycin 1% nano-emulsion gel formulation for the treatment of acne vulgaris: Results of a randomized, active controlled, multicentre, phase IV clinical trial. *J. Clin. Diagnostic Res.*, **2014**, *2011*, 5–9.
Santos, A. C.; Morais, F.; Simões, A.; Pereira, I.; Sequeira, J. A. D.; Pereira-Silva, M.; Veiga, F.; Ribeiro, A. Nanotechnology for the development of new cosmetic formulations. *Expert Opin. Drug Deliv.*, **2019**, *16*(4), 313–330.

Scorzoni, L.; De Paula E Silva, A. C. A.; Marcos, C. M.; Assato, P. A.; De Melo, W. C. M. A.; De Oliveira, H. C.; Costa-Orlandi, C. B.; Mendes-Giannini, M. J. S.; Fusco-Almeida, A. M. Antifungal therapy: New advances in the understanding and treatment of mycosis. *Front. Microbiol.*, **2017**, *8*, 1–23.

Shao, M.; Hussain, Z.; Thu, H. E.; Khan, S.; Katas, H.; Ahmed, T. A.; Tripathy, M.; Leng, J.; Qin, H. L.; Bukhari, S. N. A. Drug nanocarrier, the future of atopic diseases: Advanced drug delivery systems and smart management of disease. *Colloids Surfaces B Biointerfaces*, **2016**, *147*, 475–491.

Shastri, D. H. Effective delivery routes and strategies for solid lipid nanoparticles (SLN) and nanostructured lipid carriers (NLC). *Curr. Pharm. Des.*, **2017**, *23*(43), 6592–6601.

Shokri, J. Nanocosmetics: Benefits and risks. *BioImpacts*, **2017**, *7*(4), 207–208.

Siddique, M. I.; Katas, H.; Amin, M. C. I. M.; Ng, S. F.; Zulfakar, M. H.; Jamil, A. *In-vivo* dermal pharmacokinetics, efficacy, and safety of skin targeting nanoparticles for corticosteroid treatment of atopic dermatitis. *Int. J. Pharm.*, **2016**, *507*(1, 2), 72–82.

Siddique, M. I.; Katas, H.; Jamil, A.; Mohd Amin, M. C. I.; Ng, S. F.; Zulfakar, M. H.; Nadeem, S. M. Potential treatment of atopic dermatitis: Tolerability and safety of cream containing nanoparticles loaded with hydrocortisone and hydroxytyrosol in human subjects. *Drug Deliv. Transl. Res.*, **2019**, *9*(2), 469–481.

Singh, M. C.; Naik, A. S.; Sawant, S. D. Transdermal drug delivery systems with major emphasis on transdermal patcher: A review. *J. Pharm. Research*, **2010**, *3*(7), 1685–1689.

Singh, S. P.; Alvi, S. B.; Pemmaraju, D. B.; Singh, A. D.; Manda, S. V.; Srivastava, R.; Rengan, A. K. NIR triggered liposome gold nanoparticles entrapping curcumin as in situ adjuvant for photothermal treatment of skin cancer. *Int. J. Biol. Macromol.*, **2018**, *110*, 375–382.

Singhal, M.; Lapteva, M.; Kalia, Y. N. Formulation challenges for 21st century topical and transdermal delivery systems. *Expert Opin. Drug Deliv.*, **2017**, *14*(6), 705–708.

Sivanandy, M. S.; Masimasi, N.; Thacker, H. L. Newer hormonal therapies: Lower doses; oral, transdermal, and vaginal formulations. *Cleve. Clin. J. Med.*, **2007**, *74*(5), 369–375.

Solomon, I.; Ilie, M.; Draghici, C.; Voiculescu, V.; Căruntu, C.; Boda, D.; Zurac, S. The impact of lifestyle factors on evolution of atopic dermatitis: An alternative approach (review). *Exp. Ther. Med.*, **2018**, 1078–1084.

Soriano-Ruiz, J. L.; Calpena-Capmany, A. C.; Cañadas-Enrich, C.; Febrer, N. B. De; Suñer-Carbó, J.; Souto, E. B.; Clares-Naveros, B. Biopharmaceutical profile of a clotrimazole nanoemulsion: Evaluation on skin and mucosae as anticandidal agent. *Int. J. Pharm.*, **2019**, *554*, 105–115.

Souza, R. C. J.; Medeiros, M.; Yamashiro-Kanashiro, E. H.; Rocha, M. C.; Cotrim, P. C.; Stephano, M. A.; Lancellotti, M.; Tavares, G. D.; Oliveira-Nascimento, L. Biodegradable nanocarriers coated with polymyxin B: Evaluation of leishmanicidal and antibacterial potential. *PLoS Negl. Trop. Dis.*, **2019**, *13*(5), e0007388.

Stoicea, N.; Fiorda-Diaz, J.; Joseph, N.; Shabsigh, M.; Arias-Morales, C.; Gonzalez-Zacarias, A. A.; Mavarez-Martinez, A.; Marjoribanks, S.; Bergese, S. D. Advanced analgesic drug delivery and nanobiotechnology. *Drugs*, **2017**, *77*(10), 1069–1076.

Sugumar, S.; Ghosh, V.; Nirmala, M. J.; Mukherjee, A.; Chandrasekaran, N. Ultrasonic emulsification of eucalyptus oil nanoemulsion: Antibacterial activity against staphylococcus aureus and wound healing activity in Wistar rats. *Ultrason. Sonochem.*, **2014**, *21*(3), 1044–1049.

Sunder, S. Relevant topical skin care products for prevention and treatment of aging skin. *Facial Plast. Surg. Clin. North Am.*, **2019**.

Suzich, J. B.; Cliffe, A. R. Strength in diversity: Understanding the pathways to herpes simplex virus reactivation. *Virology*, **2018**, *522*(24), 81–91. https://doi.org/10.1016/j.virol., **2018**.07.011.

Taleb, M. H.; Abdeltawab, N. F.; Shamma, R. N.; Abdelgayed, S. S.; Mohamed, S. S.; Farag, M. A.; Ramadan, M. A. Origanum vulgare L. essential oil as a potential anti-acne topical nanoemulsion—*in vitro* and *in vivo* study. *Molecules*, **2018**, *23*(9).

Teixeira, H.; Mattos, C.; Koester, L.; Bassani, V.; Argenta, D.; Misturini, F.; Simoes, C. Factorial design applied to the optimization of lipid composition of topical antiherpetic nanoemulsions containing isoflavone genistein. *Int. J. Nanomedicine*, **2014**, *9*, 4737.

Tofani, L. B.; Depieri, L. V.; Campos, P. M.; Riul, T. B.; Antonietto, K. S.; De Abreu, F. M. C.; Bentley, M. V. L. B. *In vitro* TyRP-1 knockdown based on SiRNA carried by liquid crystalline nanodispersions: An alternative approach for topical treatment of vitiligo. *Pharm. Res.*, **2018**, *35*(5), 104.

Tripathi, P.; Kumar, A.; Jain, P. K.; Patel, J. R. Carbomer gel bearing methotrexate loaded lipid nanocontainers shows improved topical delivery intended for effective management of psoriasis. *Int. J. Biol. Macromol.*, **2018**, *120*, 1322–1334.

Tripura, S. P.; Anushree, H. Novel delivery systems: Current trend in cosmetic industry. *Eur. J. Pharm. Med. Res.*, **2017**, *4*(8), 617–627.

Üner, M.; Yener, G.; Erguven, M.; Karaman, E.; Utku, E. Solid lipid nanoparticles and nanostructured lipid carriers of celecoxib for topical application–preparation, characterization and drug penetration through rat skin. *Curr. Nanosci.*, **2014**, *10*(4), 532–542.

Vauthier, C.; Bouchemal, K. Methods for the preparation and manufacture of polymeric nanoparticles. *Pharm. Res.*, **2009**, *26*(5).

Vogt, A.; Wischke, C.; Neffe, A. T.; Ma, N.; Alexiev, U.; Lendlein, A. Nanocarriers for drug delivery into and through the skin — do existing technologies match clinical challenges? *J. Control. Release*, **2016**, *242*, 3–15.

Vyas, A.; Kumar, S. A.; Gidwani, B. Carrier-based drug delivery system for treatment of acne. *Sci. World J.*, **2014**, *2014*, 1–14.

Wadgave, U.; Nagesh, L. Nicotine replacement therapy: An overview. *Int. J. Heal. Sci.*, **2016**, *10*(3), 425–435.

Wang, P.; Huang, B.; Horng, H.; Yeh, C.; Chen, Y. J. Wound healing. *J. Chinese Med. Assoc.*, **2018**, *81*(2), 94–101.

Wang, Y. Tacrolimus nanoparticles based on chitosan combined with nicotinamide: Enhancing percutaneous delivery and treatment efficacy for atopic dermatitis and reducing dose. *Int. J. Nanomedicine*, **2018**, 129–142.

Wei, L.; Lu, J.; Xu, H.; Patel, A.; Chen, Z. S.; Chen, G. Silver nanoparticles: Synthesis, properties, and therapeutic applications. *Drug Discov. Today*, **2015**, *20*(5), 595–601.

Wickett, R. R.; Visscher, M. O. Structure and function of the epidermal barrier. *Am. J. Infect. Control,* **2006**, *34*(10), S98–S110.

Wongrakpanich, S.; Wongrakpanich, A.; Melhado, K.; Rangaswami, J. A comprehensive review of non-steroidal anti-inflammatory drug use in the elderly. *Aging Dis.*, **2018**, *9*(1), 143.

Wu, S. H.; Lin, H. P. Synthesis of mesoporous silica nanoparticles. *Chem. Soc. Rev.*, **2013**, *42*(9), 3862–3875.

Yaar, M.; Gilchrest, B. A. Photoageing: Mechanism, prevention and therapy. *Br. J. Dermatol.,* **2007**, *157*(5), 874–887.

Yang, G.; Lee, H. E.; Shin, S. W.; Um, S. H.; Lee, J. D.; Kim, K. B.; Kang, H. C.; Cho, Y. Y.; Lee, H. S.; Lee, J. Y. Efficient transdermal delivery of DNA nanostructures alleviates atopic dermatitis symptoms in NC/Nga mice. *Adv. Funct. Mater.*, **2018**, *28*(40).

Yang, X.; Shah, S. J.; Wang, Z.; Agrahari, V.; Pal, D.; Mitra, A. K. Nanoparticle-based topical ophthalmic formulation for sustained release of stereoisomeric dipeptide prodrugs of ganciclovir. *Drug Deliv.*, **2016**, *23*(7), 2399–2409.

Yeh, Y.-C.; Creran, B.; Rotello, V. M. Gold nanoparticles: Preparation, properties, and applications in bionanotechnology. *Nanoscale*, **2012**, *4*(6), 1871–1880.

Yiyun, C.; Na, M.; Tongwen, X.; Rongqiang, F.; Xueyuan, W.; Xiaomin, W.; Longping, W. Transdermal delivery of nonsteroidal anti-inflammatory drugs mediated by polyamidoamine (PAMAM) dendrimers. *J. Pharm. Sci.*, **2007**, *96*(3), 595–602.

Yu, K.; Wang, Y.; Wan, T.; Zhai, Y.; Cao, S.; Ruan, W.; Wu, C.; Xu, Y. Tacrolimus nanoparticles based on chitosan combined with nicotinamide: Enhancing percutaneous delivery and treatment efficacy for atopic dermatitis and reducing dose. *Int. J. Nanomedicine*, **2018**, *13*, 129–142.

Zhang, et al. *Polymeric Nanoparticles-Based Topical Delivery Systems for the Treatment of Dermatological Diseases*. Natl. Heal. Institutes, **2013**, 1–22.

Zhuo, F.; Abourehab, M. A. S.; Hussain, Z. Hyaluronic acid decorated tacrolimus-loaded nanoparticles: Efficient approach to maximize dermal targeting and anti-dermatitis efficacy. *Carbohydr. Polym.*, **2018**, *197*, 478–489.

Zouboulis, C. C.; Elewa, R.; Ottaviani, M.; Fluhr, J.; Picardo, M.; Bernois, A.; Heusèle, C.; Camera, E. Age influences the skin reaction pattern to mechanical stress and its repair level through skin care products. mech. *Ageing Dev.*, **2018**, *170*(2017), 98–105.

PART II
Diverse Transdermal Drug Delivery Systems

CHAPTER 4

Polymeric Micelles in Topical and Transdermal Delivery

Alexsandra C. Apolinário and Luciana B. Lopes

Department of Pharmacology, Institute of Biomedical Sciences, University of São Paulo, Sao Paulo, Brazil,
E-mail: lublopes@usp.br (L. B. Lopes)

ABSTRACT

The use of micelles for delivery of drugs with low solubility in water is most often justified by their ability to dissolve these compounds in their inner core, increasing the apparent solubility and contributing to improve bioavailability. Although micelles formed by low molecular weight surfactants have been more frequently described, they can also be formed by the self-aggregation and organization of amphiphilic block copolymers (ABCs). ABCs are composed of hydrophilic and hydrophobic blocks connected through a covalent bond and can self-aggregate in the presence of solvents, such as the aqueous media. One of the advantages of polymeric micelles over surfactant-based ones is that they are considered more stable. Because the structure-forming polymer influences micelles physical and biological properties, including toxicity, compatibility, and pharmacokinetics (PK) parameters, various properties and multiple functional capabilities can be attributed to the micelles depending on the composition selected. In this chapter, we aim at discussing structural aspects of polymeric micelles, taking into consideration their composition, formation, and main characteristics, as well as their application for topical and transdermal delivery.

Topical and Transdermal Drug Delivery Systems: Applications and Future Prospects. Nayan A. Gujarathi, PhD, Juliana Palma Abriata, PhD, Raj Kumar Keservani, MPharm, Anil K. Sharma, PhD, MPharm (Eds.)

4.1 INTRODUCTION

Micelles are well-known nanocarriers mostly frequently employed for the delivery of compounds with low aqueous solubility because of their ability for dissolution in their inner core, increasing the apparent solubility of drugs. These nanocarriers can be formed by the self-assembly of low molecular weight surfactants or amphiphilic block copolymers (ABCs). The focus of this chapter will be ABCs micelles, which will be referred to as polymeric micelles here for the sake of simplicity.

ABCs are constituted by hydrophilic and hydrophobic blocks connected through a covalent bond. They can self-aggregate when placed in a solvent that is selective for either the hydrophilic or hydrophobic segment, e.g., aqueous media. The hydrophobic portion forms the core, which can encapsulate hydrophobic drugs, while the hydrophilic portion forms the shell, also called corona. As a result, they increase the apparent solubility of poorly soluble drugs and contribute to improve bioavailability (Jones and Leroux, 1999; Kataoka et al., 2012; Owen et al., 2012).

Polymeric micelles have been employed to incorporate drugs and other pharmacologically active compounds, particularly hydrophobic molecules, improving their pharmacokinetic properties, delivering them to the site of action, and reducing the off-target cytotoxicity (Biswas et al., 2016). More recently, the concept of multifunctional micelles has emerged from the idea of incorporating multiple functional capabilities in a single micelle (Movassaghian et al., 2015).

Transdermal and topical drug delivery using micelles is a promising approach to overcome the drawbacks of the skin as an administration route, and improve the cutaneous localization or the transdermal (and thus, systemic) delivery of drugs. The nanometric size of these carriers (especially when the diameter is smaller than the aqueous pore pathways of the SC) as well as features related to molecular structure of surfactants and polymers – such as the ability to denature stratum corneum (SC) proteins and/or insert into this layer and extract intercellular lipids leading to reversible damage to the skin barrier – have been suggested to improve delivery (Morris et al., 2019).

In this chapter, we aim at discussing structural aspects of polymeric micelles, taking into consideration their composition, formation, and main characteristics, as well as the use of these nanocarriers for topical and transdermal delivery. We do not aim to provide a systematic review of the literature, but to present relevant aspects for understanding how these aggregates aid drug penetration into the skin.

4.2 FORMATION OF POLYMERIC MICELLES

At very low concentrations, ABCs only exist as unimers (single chains). As their concentration rises, and the "critical micelle concentration" (CMC) is reached, ABCs chains self-aggregate originating micelles. In these structures, the hydrophobic portion of the copolymer is segregated from the aqueous media where the polymer is dissolved. The term "critical association concentration" (CAC) is sometimes employed for polymeric micelles for differentiation from the "critical micelle concentration" of surfactant-based micelles (Kataoka et al., 2012).

The CMC for polymeric micelles is lower compared to surfactant micelles, as well as the rate of dissociation of the unimers, which lead to the retention of loaded drugs for longer periods of time (Kataoka et al., 2012). Therefore, polymeric micelles are usually more stable. The disassembly of micelles at ABCs concentrations below the CMC has been reported to be rather slow for certain copolymer systems, and this rate of disassembly depends on the physical state of the micelle core. For example, micelles elaborated with copolymers comprising a hydrophobic block with high glass transition temperature (T_g) tend to disassemble more slowly than those with a low T_g (Allen et al., 1999).

The self-assembly of the copolymers into micelles relies not only on CMC values, but also on the hydrophilic/hydrophobic blocks ratio. The relationship between the type of aggregate formed and the structure of the surfactant is mathematically expressed by the "critical packing parameter" (*cpp*), which encompasses the relationship among the head group area (a_0), the hydrophobic tail length (l_c) and its volume (v), as demonstrated in Eqn. (1). Spherical micelles have high curvature and are formed when $cpp \leq ⅓$ (Figure 4.1), cylinders between $⅓ < cpp \leq ½$ and polymersomes when $½ < cpp \leq 1$. It is important to highlight that this condition applies to structures at equilibrium (Israelachvili, 1976).

$$cpp = \frac{v}{a_0 l_c} \quad (1)$$

In this equation, $\boldsymbol{v}$ is the hydrophobic portion volume, $\boldsymbol{a_0}$ is the area of the cross-section of the polar head group, and $\boldsymbol{l_c}$ is the length of the hydrophobic chain(s) molecule of the surfactant.

In addition to micelles, ABCs can also lead to the formation of other nanocarriers like polymersomes, which are polymeric vesicles, and to other out-of-equilibrium structures (Doncom et al., 2017).

The hydrophilic block volume fraction (f) of PEG-based ABCs, which can be obtained from the PEG volume fraction (f_{PEG}) through the conversion of the measured mass fractions using homopolymer melt densities, can also predict the shape of the nanostructure formed (Ahmed and Discher, 2004; Discher and Ahmed, 2006). Spherical micelles are predominantly formed when f of PEG (f_{PEG}) is ~50–70%.

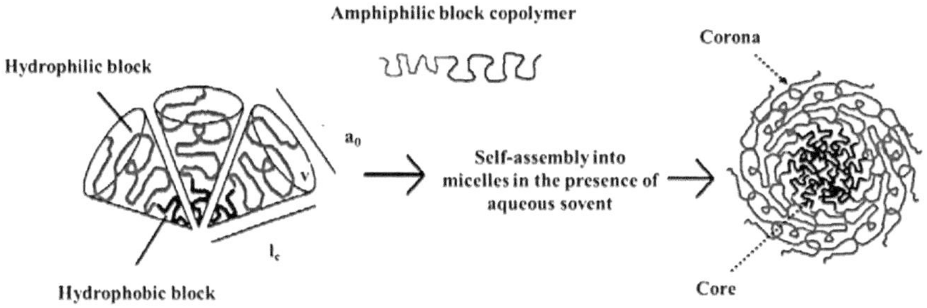

FIGURE 4.1 Representation of amphiphilic copolymer with its hydrophilic (blue segments) and hydrophobic blocks (black segments) and the micelles nanostructure highlighting the core (black area) and corona (blue area) parts.

Source: Adapted (translated and simplified) permission from Apolinário et al., 2017). Open access. https://creativecommons.org/licenses/by/3.0/br/

The result of the balance of forces in the self-assembly process arises respectively from the interaction between water, the hydrophobic and the hydrophilic parts that comprise the amphiphilic molecule. These interactions are strongly dependent on temperature and concentration. Thus, besides *cpp* and f_{PEG} values, other parameters can influence the formation of polymeric micelles, such as the method of preparation, the organic solvent employed and the ratio of the volumes of the organic solvent and water (Dionzou et al., 2016).

The preparation of polymeric micelles encompasses different methods that trigger the self-assembly of the ABCs. The direct dissolution method is the simplest technique, in which ABCs with or without drugs are mixed in an aqueous solution at or above the CMC to self-assemble into micelles. Through the dialysis, the block copolymer is dissolved in a water-miscible organic solvent, and the resultant solution is dialyzed against water so that the organic solvent is replaced by water, which induces self-assembly of block copolymers. Solvent evaporation or the film hydration method requires the dissolution of the polymer in a volatile organic solvent followed by formation of a thin film at the bottom of the flask after the evaporation of the organic solvent under vacuum. This film is then reconstituted in an aqueous phase by vigorous shaking or controlled stirring. For the co-solvent evaporation method, initially, the polymer is dissolved in a volatile water-miscible organic solvent followed by addition of the aqueous phase leading to the formation of micelles (Mandal et al., 2017).

4.3 COMPOSITION AND PROPERTIES OF POLYMERIC MICELLES

Polymeric micelles may have various physical and biological properties according to the ABCs employed, which influence toxicity, pharmacokinetics (PK), pharmacodynamics (PD) and clinical compatibility, among other properties (Deshmukh et al., 2017). The segments of ABCs control self-assembly, loading, and stability of the micelles, in addition to the performance in biological environments (Cabral et al., 2018).

A variety of non-polar drugs can be loaded in the hydrophobic core, leading to the improvement of the drug's apparent solubility and stability in the biological system. This core is shielded by the hydrophilic corona, which protects loaded drugs from interactions with blood components. The small size (generally between 10 and 100 nm) allows polymeric micelles to eventually accumulate in tumors via a passive targeting phenomenon known as the "enhanced permeability and retention effect" (EPR) (Biswas et al., 2016).

PK and PD properties of micelles depend on the functionality of the corona because it stabilizes the structure of the micelles and influences their interactions with proteins and other molecules in the bloodstream. The surface of polymeric micelles can be functionalized to modify their physicochemical and biological properties, which is important for the nanocarrier design and receptor-mediated delivery. Employing end-functionalized block copolymers to build micelles enables reaching this goal (Kataoka et al., 2012).

The shell-forming segments can be formed from several flexible hydrophilic polymers, which assemble into dense chains for steric stabilization. PEG is widely explored since it is water-soluble at room temperature and stabilizes the micelles, may undergo oxidative degradation under biologically relevant conditions, and its molecular weight determines its clearance from the body. For example, PEGs larger than 60 kDa are slowly excreted predominantly through the liver, while PEGs with MW below 60 kDa show faster excretion through the kidneys. Other hydrophilic polymers include poly(vinyl alcohol) (PVA), poly[N-(2-hydroxypropyl) methacrylamide] (PHPMA), poly(glycerol) (PG), poly(N-vinyl-2-pyrrolidone) (PVP), poly(acrylamide) (PAAm), poly(oxazolines) (POxs), poly(acrylic acid), poly(malic acid), zwitterionic polymers, polysaccharides, and poly(amino acids) (PAAs), among others. Features like toxicity, immunogenicity, and nonbiodegradability *in vivo* must be determined (Cabral et al., 2018).

The segregated hydrophobic core serves as a drug reservoir and is formed by the combination of various forces among the block copolymers, which include hydrophobic and electrostatic interactions, metal complexation and hydrogen bonds (Kataoka et al., 2012). The stability of micelles can be affected by various physicochemical aspects of the core-forming blocks, including their crystallinity, stiffness, and chemical cross-linking between polymer segments in the core. Examples of core-forming segments of polymeric micelles are polyethers, polyesters, and polyamino acids for hydrophobic drugs; it has been reported the use of poly(ethylene imine) and poly(amino acid)s for negatively charged biomacromolecules (Cabral et al., 2018).

Overall, ABCs that are sensitive to changes of pH, temperature, ultrasound, enzyme, and light enable a controlled micelle dissociation and drug release triggered by precise stimuli, and can improve the specificity and efficacy of delivery (Movassaghian et al., 2015). A classical example is the copolymer triblock poloxamer (Pluronic®, BASF, Germany), which is constituted by three blocks rather than two blocks: poly(ethylene oxide)–poly(propylene oxide)–poly(ethylene oxide) (PEO–PPO–PEO). Some poloxamers demonstrate a notable temperature-dependent phase behavior when the micelle solution is heated up to 25–37°C; because phase transition leads to gel formation, this property has been explored for controlled drug release after *in situ* gelation at body temperature (Attwood et al., 2007).

4.4 TOPICAL/TRANSDERMAL DELIVERY MEDIATED BY POLYMERIC MICELLES

4.4.1 *SKIN AND ROUTES FOR DRUG DELIVERY*

Drug delivery to the skin encompasses two main routes: cutaneous delivery to promote local accumulation and local therapeutic effect (e.g., for skin cancer, wounds, cutaneous infections, and dermatological conditions), and transdermal delivery, which requires the active compound (such as a drug) to cross various cutaneous layers and reach the blood vessels located in the dermis, and thus, the systemic circulation to elicit systemic effects (e.g., hormonal therapy). For the latter, some advantages have been widely reported: it is considered a non-invasive route, presents reduced fluctuation of bioavailability, enables evasion from the gastrointestinal environment and avoids the first-pass effect. However, cutaneous and transdermal drug delivery are considered a challenge due to skin barrier function (Zhou et al., 2018; Keservani et al., 2020).

This can be explained by both the composition and organization of the tissue. The skin is organized as a multilayer structure, which can be roughly divided in the non-vascularized epidermis (EP) covered by the SC, dermis, and hypodermis (which many authors do not consider part of the skin). The SC is considered the main barrier for penetration; it is composed of corneocytes, non-viable cells filled with keratin filaments, entrenched in a very organized lipid matrix. This "dead" layer prevents the permeation of foreign compounds, including drugs and therefore, is an effective barrier (Pham et al., 2016).

The transepidermal and transappendegeal routes are the two possible pathways described for drug penetration into/across the intact skin. While the first one involves penetration through the intact layers of the skin (either through the transcellular or intercellular routes), the second encompasses entry from the surface of the skin into the subepidermal layers through skin appendages, such as hair follicles and sweat ducts (Prausnitz and Langer, 2008). Depending on the route, features of micelles like size and ABCs properties exert important roles to achieve penetration.

4.4.2 APPLICATION OF POLYMERIC MICELLES FOR TOPICAL/ TRANSDERMAL DELIVERY AND MECHANISMS OF MICELLE-MEDIATED DELIVERY

Many studies have been conducted to demonstrate that polymeric micelles are useful for epidermal targeting and enhancement of penetration of a variety of drugs. Fewer studies have attempted to understand the interaction between the SC and micelles, routes of micelle-mediated skin penetration and mechanisms of penetration enhancement. Table 4.1 exemplifies the use of micelles for topical and transdermal delivery, and depicts

TABLE 4.1 Examples of the Use of Micelles for Topical and Transdermal Delivery*

Copolymer	Purpose	Findings	Routes of Drug Penetration or Topical Approach	References
Methoxypoly(ethylene glycol)-dihexyl substituted polylactide (MPEG-dihexPLA) diblock	Targeted delivery of tacrolimus into the epidermis and upper dermis for the psoriasis therapy	Enhanced deposition in the stratum corneum and penetration into viable layers of the epidermis and upper dermis	The study suggested micelle preferential accumulation in hair follicles	Lapteva et al. (2014)
Glycyrrhizic acid	Delivery of podophyllotoxin aiming at anti-inflammatory properties	Micelles had a higher distribution in epidermis when compared with the clinically used formulation and caused less skin inflammation	The authors discussed that the penetration mechanism need further research	Wang et al. (2016)
Pluronic P123 and Pluronic F127 (mixed micelles)	Dermal terconazol delivery to enhance the treatment of skin fungal infections	Micelles demonstrated a greater extent of terconazol deposition in the rat dorsal skin relative to suspension	The authors hypothesized that the surfactants enter the intracellular matrices of the stratum corneum, interact, and bind to keratin filaments.	Abd-Elsalam et al. (2018)
Hydrophobized hyaluronic acid	Concept proof using a probe as drug model regarding skin delivery	Micelles increased *in vitro* drug deposition in the epidermis and dermis	Enlargement of the stratum corneum further indicated transcellular pathway	Šmejkalová et al. (2017)
Poly (ethylene glycol)-b-poly (δ-valerolactone)	Ketoconazole delivery for improving its skin deposition	Ketoconazole aqueous solubility increases to 86-fold and the micelles showed no skin permeation of the drug enhancing skin deposition	In this case, topical approach was required	Deng et al. (2017b)
Pluronic F68	Delivery of plasmid to enhance its transdermal permeability and promote gene expression	Transdermal permeability of plasmid was enhanced by micelles resulting in longer half-life and distributed transgene expression (for 3 days) in skin besides other organs like stomach, testis, spleen, brain, and spinal cord	The authors reported that micelles could permeate the intact skin not only through the stratum corneum, but through energy-dependent process involving endocytosis	Tong et al. (2012)

TABLE 4.1 (*Continued*)

Copolymer	Purpose	Findings	Routes of Drug Penetration or Topical Approach	References
Y-shaped monomethoxy poly (ethylene glycol)-block-poly(ε-caprolactone)	Improving ketoconazole selective skin deposition and water-solubility	Micelles penetrating skin in depth and decreased ketoconazole's distribution in blood in comparison with a cream	Authors discussed that the micellar drug can adsorb and fuse with the skin altering the flowability of stratum corneum lipids and intercellular lipid lamellae and increasing the permeation into or through the stratum corneum	Deng et al. (2017a)
PEGylated dendron-based copolymers	Enhanced skin permeation and controlled release profile of endoxifen	Enhanced permeation of endoxifen through both mouse (20-fold) and human (4-fold) skin samples compared to ethanol	Authors hypothesized that the permeated dye is not confined to hair follicle regions, which implies that micelles may deliver the encapsulated hydrophobic molecules mainly through and along lipid bilayers	Yang et al. (2014)
Pluronic F127	Achieving stability and penetration of sumatriptan succinate across the skin	Pluronic resulted in improved skin penetration	Authors conclude that micelles penetrate the skin, interact, and disorganize the lipid layers of the stratum corneum	Agrawal et al. (2010)
Methoxy-poly(ethylene glycol)–hexyl substituted polylactide	Delivery of clotrimazole, econazole nitrate, and fluconazole aiming at superficial localization delivering these drugs to the target compartment	Econazole nitrate delivery using micelles resulted in significantly higher drug deposition in both porcine and human skin as compared to liposomal formulation.	Micelles could target delivery to the follicles	Bachhav et al. (2011)

TABLE 4.1 (*Continued*)

Copolymer	Purpose	Findings	Routes of Drug Penetration or Topical Approach	References
poly(Ɛ-caprolactone)-poly(ethylene glycol)-poly(Ɛ-caprolactone)	Improving the bactericidal effect of lauric acid on *Propionibacterium acnes*	The minimum inhibitory concentration and the minimum bactericidal concentration decreased to 10 and 40 g/mL	The authors proposed a skin therapy, but they did not do any assess skin penetration or localization of the drug in the skin	Tran et al. (2016)
Pluronic F127 and Pluronic L121	Dermal delivery of beclomethasone dipropionate	Formulations were enhanced to treat sub-chronic dermatitis (animal model) within a shorter period compared to commercially available cream	The authors cited that the polymeric micelles segregate, leaving the individual polymers to act as chemical penetration enhancers and/or accumulate in the skin appendages	Assem et al. (2019)
α-linolenic acid modified Pluronic F127-block-chitosan	Topical amphotericin delivery	Micelles showed a 310-fold increase in the drug skin deposition and better *in vivo* activity as compared to commercial AMB liposomal gel	The authors cited that the follicular pathway is the main penetration route for amphiphilic micelles and reported that α-linolenic acid weakens the barrier of the packing of the lipid bilayer	Zhou et al. (2019)
Ethoxy-poly(ethylene glycol)-di-hexylsubstituted-poly(lactic acid) (mPEG-hexPLA)	Cutaneous delivery of Vismodegib (a hedgehog pathway inhibitor)	Micelles improved 8,000-fold the aqueous solubility of Vismodegib and allowed the delivery of the drug to the basal epidermis and deeper tissue	Not mentioned	Kandekar et al. (2019)

*The table summarizes the composition, drug incorporated, and possible mechanism of transport and penetration enhancement.

micelle composition, drug employed in the study, main findings related to topical and transdermal delivery and potential mechanisms of penetration. In general, several mechanisms have been demonstrated to contribute to the delivery mediated by nanocarriers, and polymeric micelle-mediated delivery may involve one or more of the mechanisms described below:

1. **Enhancement of Drug Transfer Mediated by the Small Size:** The low interfacial tension associated with the nanometric size of micelles has been suggested to result in a large surface that enables drug transfer to the skin, aiding delivery, and penetration (Kreilgaard, 2002; Kreilgaard et al., 2000; Maghraby, 2008). An example of this mechanism is depicted in the study by Schwarz et al. in which the transport of diazepam, a drug of the benzodiazepine family, was increased when formulations with internal phase diameter of 100–300 nm were employed (Schwarz et al., 1995). However, even though the reduced size may favor penetration, it does not seem sufficient to promote it. As previously demonstrated by Shim et al. (2010), lipid micelles decreased lidocaine hydrochloride delivery to the skin despite displaying diameter in the nanometer range, most likely because they increased the SC rigidity.
2. **Higher Drug Loading Capacity Compared to Simple Solutions:** This property is more relevant for hydrophobic drugs, since they are poorly soluble in aqueous vehicles. The structural organization of the polymers forming the micelles enables solubilization of lipophilic molecules in their core, which increases the loading capacity (Alcolmson et al., 1998; Kreilgaard, 2002; Kreilgaard et al., 2000; Pepe et al., 2013) and creates a higher concentration gradient, the driving force for diffusion.
3. **Penetration-Enhancing Effect of Micelle Components:** ABCs and other penetration enhancers that may be incorporated in the micelles may diffuse out of the system and into the skin. They may lead to a temporary disorganization of the SC barrier and facilitate drug diffusion and/or drug dissolution in the tissue, increasing the transport/translocation of drugs through the tissue (Kreilgaard, 2002; Kreilgaard et al., 2000; Maghraby, 2008; Pepe et al., 2013; Carvalho et al., 2017; Lopes et al., 2015). In addition, based on the mutual penetration enhancement observed among components of another nanocarrier type (microemulsions), it is reasonable to suggest the relevance of a similar effect for polymeric micelles

containing multiple components (Hathout et al., 2010). For microemulsions specifically, it was demonstrated that combining multiple components increased their relative uptake into the SC compared to the pure components, which resulted in a more pronounced penetration of drugs. A similar rationale can be applied for micelles if two or more polymers are combined and/or other components (such as fatty acids) are included.

4. **Enhancement of Skin Hydration:** Considering that water is a penetration enhancer (Williams and Barry, 2012), formulations capable of increasing skin hydration because they either contain large aqueous content or display occlusive effects may contribute to enhance the transport of drugs to the skin. Considering that polymeric micelles are often prepared as dispersion in aqueous vehicles, their composition may contribute to the hydration of the tissue.
5. **Intact Micelle Transfer Across Cutaneous Layers:** Although most studies suggest the inability of nanocarriers to permeate intact skin, the presence of the so-called "edge activators" might enable deformation and transport of selected nanocarriers (Cevc et al., 1998). Additionally, the transfollicular route has been demonstrated to contribute to the transport of nanocarriers. Although most studies have focused on lipid nanoparticles, it is reasonable to expect that the transfollicular transport might also benefit polymeric micelles.

Having described possible enhancement mechanisms, the question that follows is: are there evidences that these mechanisms really play a role on polymeric micelle-mediated drug delivery to/across the skin? As described previously, several characteristics of micelles make them suitable carriers for topical delivery; their small size, usually within 10–100 nm, is an example. When it comes to route, it has been reported that the follicular pathway is the preferred penetration route for amphiphilic micelles (Tong et al., 2012). Confocal microscopy data demonstrated the penetration of Nile Red into hair follicles when loaded into polymeric micelles composed of Pluronic(®) F127. However, penetration of intact 40 nm micelles through energy-dependent processes was also reported, and thus, penetration may occur not only through the SC within 44 nm of intercellular space, but could also involve endocytosis in viable layers (Tong et al., 2012).

Other studies have reported that the penetration-enhancement ability of copolymeric micelles depends on their capability of weakening the barrier function of the SC (Deng et al., 2017a). ABCs are surfactant molecules and seem to act as enhancers for transdermal drug delivery, but their supramolecular aggregates–micelles – have also been investigated as penetration enhancers (Abd-Elsalam et al., 2018; Deng et al., 2017b; Wang et al., 2016). Therefore, to better understand the potential of these structures to promote skin penetration, it is necessary to investigate if the monomer is responsible for the penetration-enhancing effect or if the micelle itself can penetrate and carry drugs into the skin. If the monomer is responsible of the effect, they act on the SC surface or can single copolymer chains pass through the skin and subsequently reform micelles in the receiver compartments?

To shed more light into this matter, Šmejkalová et al. employed FRET to study the penetration of hyaluronic acid polymeric micelles (Šmejkalová et al., 2017). Data suggested that micelle disaggregation occurs in the upper epidermal layers, leading to separate penetration of the vehicle and the loaded compound. Despite the micelle disintegration, the polymeric components enhanced drug penetration through the SC, and confocal images suggest that the predominant penetration pathway for both was not the transfollicular pathway. Their observations were corroborated by Yotsumoto et al. (2018), who studied micelles of poly(ethylene glycol)-poly(amino acid) block copolymers. While the copolymers were located on the skin surface or superficial layers of the SC, the encapsulated compound was gradually transported to deeper layers of the cutaneous tissue. Additionally, the authors also observed a difference in skin hydration when treated with these micelles with encapsulated coenzyme Q10, even though a direct link between treatment with the polymeric micelle and increased hydration was not claimed. The possibility to reduce skin hydration has been suggested for surfactant-based micelles based on the association of this effect and anionic surfactants presence in the SC.

In addition to the effects on the skin itself, polymeric micelles also seem capable of increasing drug solubility, enabling an augment on dose and concentration gradient between the formulation and skin. To exemplify, an optimal micellar system (with 20–30 nm) increased the aqueous solubility of Vismodegib by >8,000-fold (Kandekar et al., 2019). Further support was provided by Yotsumoto et al. (2018), who observed that an improved water solubility resulting from indomethacin and resveratrol encapsulation in polymeric micelles increased the amount of these drugs penetrating into the skin.

4.5 CONCLUSIONS

When it comes to topical and transdermal application, the choice of delivery system, its composition, and properties can impact in a pronounced manner the penetration and distribution of drugs. A large number of studies support the relevance of polymeric micelles to mediate cutaneous penetration and enhance the deposition of drugs in targeted sites of the skin, although the involved mechanisms are still debatable. Although not an easy task, it is necessary to better understand their cutaneous fate to popularize the use of these micelles as topical and transdermal delivery systems.

KEYWORDS

- **amphiphilic block copolymers**
- **apparent solubility**
- **pharmacodynamics**
- **poly(vinyl alcohol)**
- **polymeric micelles**
- **skin penetration**
- **topical**
- **transdermal**

REFERENCES

Abd-Elsalam, W. H.; El-Zahaby, S. A.; Al-Mahallawi, A. M. Formulation and *in vivo* assessment of terconazole-loaded polymeric mixed micelles enriched with Cremophor EL as dual functioning mediator for augmenting physical stability and skin delivery. *Drug Delivery*, **2018**, *25*, 484–492.

Agrawal, V.; Gupta, V.; Ramteke, S.; Trivedi, P. Preparation and evaluation of tubular micelles of Pluronic lecithin organogel for transdermal delivery of sumatriptan. *AAPS PharmSciTech*, **2010**, *11*, 1718–1725.

Ahmed, F.; Discher, D. E. Self-porating polymersomes of PEG – PLA and PEG – PCL: Hydrolysis-triggered controlled release vesicles. *J Controlled Release*, **2004**, *96*, 37–53.

Alcolmson, C. A. M.; Atra, C. H. S.; Antaria, S. H. K.; Idhu, A. M. S.; Awrence, M. J. A. L.; London, C.; Road, M.; Sw, L.; Lilly, E. Effect of oil on the level of solubilization of testosterone propionate into nonionic oil-in-water microemulsions. *Journal of Pharmaceutical Sciences*, **1998**, *87*, 109–116.

Allen, C.; Maysinger, D.; Eisenberg, A. Nano-engineering block copolymer aggregates for drug delivery. *Colloids and Surfaces B: Biointerfaces*, **1999**, *16*, 3–27.

Apolinário, A. C.; Pachioni-vasconcelos, J. D. A.; Pessoa, A.; Rangel-Yagui, C. D. O. Polimerossomos versus lipossomos: A evolução da "bala mágica." *Quimica Nova*, **2017**, *40*, 810–817. http://quimicanova.sbq.org.br/detalhe_artigo.asp?id=6602.

Assem, M.; Khowessah, O. M.; Ghorab, D. Optimization and evaluation of beclomethasone dipropionate micelles incorporated into biocompatible hydrogel using a sub-chronic dermatitis animal model. *AAPS PharmSciTech*, **2019**, *20*, 1–12.

Attwood, D.; Zhou, Z.; Booth, C. Poly(ethylene oxide) based copolymers: Solubilization capacity and gelation. *Expert Opinion on Drug Delivery*, **2007**, *4*, 533–546.

Bachhav, Y. G.; Mondon, K.; Kalia, Y. N.; Gurny, R.; Möller, M. Novel micelle formulations to increase cutaneous bioavailability of azole antifungals. *Journal of Controlled Release*, **2011**, *153*, 126–132.

Biswas, S.; Kumari, P.; Lakhani, P. M.; Ghosh, B. Recent advances in polymeric micelles for anti-cancer drug delivery. *European Journal of Pharmaceutical Sciences*, **2016**, *83*, 184–202.

Cabral, H.; Miyata, K.; Osada, K.; Kataoka, K. Block copolymer micelles in nanomedicine applications. *Chemical Reviews*, **2018**, *118*, 6844–6892.

Carvalho, V. F. M.; Migotto, A.; Giacone, D. V.; De Lemos, D. P.; Zanoni, T. B.; Maria-Engler, S. S.; Costa-Lotufo, L. V.; Lopes, L. B. Co-encapsulation of paclitaxel and C6 ceramide in tributyrin-containing nanocarriers improve co-localization in the skin and potentiate cytotoxic effects in 2D and 3D models. *European Journal of Pharmaceutical Sciences*, **2017**, *109*, 131–143.

Cevc, G.; Gebauer, D.; Stieber, J.; Schatzlein, A.; Blume, G. Ultraflexible vesicles, transfersomes, have an extremely low pore penetration resistance and transport therapeutic amounts of insulin across the intact mammalian skin. *Biochimica et Biophysica Acta*, **1998**, 201–215.

Deng, P.; Teng, F.; Zhou, F.; Song, Z.; Meng, N.; Feng, R. Methoxy poly (ethylene glycol)-b-poly (δ-valerolactone) copolymeric micelles for improved skin delivery of ketoconazole. *Journal of Biomaterials Science, Polymer Edition*, **2017b**, *28*, 63–78.

Deng, P.; Teng, F.; Zhou, F.; Song, Z.; Meng, N.; Liu, N.; Feng, R. (epsilon-caprolactone)-based micelles for skin delivery of ketoconazole : *In vitro* study and *in vivo* evaluation. *Materials Science and Engineering: C*, **2017a**, *78*, 296–304.

Deshmukh, A. S.; Chauhan, P. N.; Noolvi, M. N.; Chaturvedi, K.; Ganguly, K.; Shukla, S. S.; Nadagouda, M. N.; Aminabhavi, T. M. Polymeric micelles: Basic research to clinical practice. *International Journal of Pharmaceutics*, **2017**, *532*, 249–268.

Dionzou, M.; Morère, A.; Roux, C.; Lonetti, B.; Marty, J. D.; Mingotaud, C.; Joseph, P.; Goudounèche, D.; Payré, B.; Léonettie, M.; Mingotaud, A. F. Comparison of methods for the fabrication and the characterization of polymer self-assemblies: What are the important parameters? *Soft Matter*, **2016**, *12*, 2166–2176.

Discher, D. E.; Ahmed, F. Polymersomes. *Annual Review of Biomedical Engineering*, **2006**, *8*, 323–341.

Doncom, K. E. B.; Blackman, L. D.; Wright, D. B.; Gibson, M. I.; Reilly, R. K. O. Dispersity effects in polymer self-assemblies: A matter of hierarchical control. *Chemical Society Reviews*, **2017**, *46*, 4119–4134.

Hathout, R. M.; Mansour, S.; Mortada, N. D.; Geneidi, A. S.; Guy, R. H. Uptake of microemulsion components into the stratum corneum and their molecular effects on skin barrier function. *Molecular Pharmaceutics*, **2010**, *7*, 1266–1273.

Israelachvili, J. L.; Mitchell, D. J.; Ninhan, B. W. *Theory of Self-Assembly of Hydrocarbon Amphiphiles into Micelles and Bilayers*, **1976**, *2*, 1525–1568.

Jones, M.; Leroux, J. Polymeric micelles-a new generation of colloidal drug carriers. *European Journal of Pharmaceutics and Biopharmaceutics,* **1999**, *48*, 101–111.

Kandekar, S. G.; Singhal, M.; Sonaje, K. B.; Kalia, Y. N. Polymeric micelle nanocarriers for targeted epidermal delivery of the hedgehog pathway inhibitor vismodegib: Formulation development and cutaneous biodistribution in human skin. *Expert Opinion on Drug Delivery,* **2019**, *16*, 667–674.

Kataoka, K.; Harada, A.; Nagasaki, Y. Block copolymer micelles for drug delivery : Design, characterization and biological significance. *Advanced Drug Delivery Reviews*, **2012**, *64*, 37–48.

Keservani, R. K.; Bandopadhyay, S.; Bandyopadhyay, N.; Sharma, A. K. Design and fabrication of transdermal/skin drug-delivery system. In *Drug Delivery Systems*; Elsevier, **2020**; pp. 131–178.

Kreilgaard, M. Influence of microemulsions on cutaneous drug delivery. *Advanced Drug Delivery Reviews*, **2002**, *54*, S77–S98.

Kreilgaard, M.; Pedersen, E. J.; Jaroszewski, J. W. NMR characterization and transdermal drug delivery potential of microemulsion systems. *Journal of Controlled Release,* **2000**, *69*, 421–433.

Lapteva, M.; Mondon, K.; Möller, M.; Gurny, R.; Kalia, Y. N. Polymeric micelle nanocarriers for the cutaneous delivery of tacrolimus: A targeted approach for the treatment of psoriasis. *Molecular Pharmaceutics,* **2014**, *11*, 2989–3001.

Lopes, L. B.; Garcia, M. T.; Bentley, M. V. Chemical penetration enhancers. *Therapeutic Delivery*, **2015**, 6(9), 1053–1061.

Maghraby, G. M. E. Transdermal delivery of hydrocortisone from eucalyptus oil microemulsion: Effects of cosurfactants. *International Journal of Pharmaceutics.*, **2008**, *355*, 285–292.

Mandal, A.; Bisht, R.; Rupenthal, I. D.; Mitra, A. K. Polymeric micelles for ocular drug delivery: From structural frameworks to recent preclinical studies. *Journal of Controlled Release*, **2017**, *248*, 96–116.

Morris, S. A. V.; Thompson, R. T.; Glenn, R. W.; Ananthapadmanabhan, K. P.; Kasting, G. B. Mechanisms of anionic surfactant penetration into human skin : Investigating monomer, micelle and submicellar aggregate penetration theories. *International Journal of Cosmetic Science*, **2019**, *41*, 55–66.

Movassaghian, S.; Merkel, O. M.; Torchilin, V. P. Applications of polymer micelles for imaging and drug delivery. *WIREs Nanomedicine and Nanobiotechnology*, **2015**, *7*, 691–707.

Owen, S. C.; Chan, D. P. Y.; Shoichet, M. S. Polymeric micelle stability. *Nano Today,* **2012**, *7*, 53–65.

Pepe, D.; McCall, M.; Zheng, H.; Lopes, L. B. Protein transduction domain-containing microemulsions as cutaneous delivery systems for an anticancer agent. *Journal of Pharmaceutical Sciences,* **2013**, *102*(5), 1476–1487.

Pham, Q. D.; Björklund, S.; Engblom, J.; Topgaard, D.; Sparr, E. Chemical penetration enhancers in stratum corneum-relation between molecular effects and barrier function. *Journal of Controlled Release*, **2016**, *232*, 175–187.

Prausnitz, M. R.; Langer, R. Transdermal drug delivery. *Nature Biotechnology*, **2008**, *26*, 1261–1268.

Schwarz, J. S.; Weisspapir, M. R.; Friedman, D. I. Enhanced transdermal delivery of diazepam by submicron emulsion (SME) creams. *Pharmaceutical Research*, **1995**, *12*, 687–692.

Shim, J.; Jin, M.; Kim, H.; Kim, D.; Geun, S.; Yong, S.; Gyeom, H.; Kim, J. Morphological effect of lipid carriers on permeation of lidocaine hydrochloride through lipid membranes. *International Journal of Pharmaceutics,* **2010**, *388*, 251–256.

Šmejkalová, D.; Muthný, T.;Nešporová, K.; Hermannová, M.; Achbergerová, E.; Huerta-Angeles, G.; Svoboda, M.; Čepa, M.; Machalová, V.; Luptáková, D.; Velebný, V. Hyaluronan polymeric micelles for topical drug delivery. *Carbohydrate Polymers,* **2017**, *156*, 86–96.

Tong, Y.; Yu, T.; Chang, S.; Liaw, J. Nanopolymeric micelle effect on the transdermal permeability, the bioavailability and gene expression of plasmid. *Molecular Pharmaceutics,* **2012**, *9*, 111–120.

Tran, T. Q. M.; Hsieh, M. F.; Chang, K. L.; Pho, Q. H.; Nguyen, V. C.; Cheng, C. Y.; Huang, C. M. Bactericidal effect of lauric acid-loaded PCL-PEG-PCL nano-sized micelles on skin commensal propionibacterium acnes. *Polymers,* **2016**, *8*, 1–19.

Wang, Y.; Zhao, B.; Wang, S.; Liang, Q.; Cai, Y.; Yang, F.; Li, G. Formulation and evaluation of novel glycyrrhizic acid micelles for transdermal delivery of podophyllotoxin. *Drug Delivery,* **2016**, *23*, 1623–1635.

Williams, A. C.; Barry, B. W. Penetration enhancers. *Advanced Drug Delivery Reviews.*, **2012**, *64*, 128–137.

Yang, Y.; Pearson, R. M.; Lee, O.; Lee, C. W.; Chatterton, R. T.; Khan, S. A.; Hong, S. Dendron-based micelles for topical delivery of endoxifen: A potential chemo-preventive medicine for breast cancer. *Advanced Functional Materials*, **2014**, *24*, 2442–2449.

Yotsumoto, K.; Ishii, K.; Kokubo, M.; Yasuoka, S. Improvement of the skin penetration of hydrophobic drugs by polymeric micelles. *International Journal of Pharmaceutics,* **2018**, *553*, 132–140.

Zhou, F.; Xu, H.; Song, Z.; Zhu, L. α-Linolenic acid-modified pluronic 127-CS copolymeric micelles for the skin targeted delivery of amphotericin B. *New Journal of Chemistry,* **2019**, *43*, 444–453.

Zhou, X.; Hao, Y.; Yuan, L.; Pradhan, S.; Shrestha, K.; Pradhan, O.; Liu, H.; Li, W. Nano-formulations for transdermal drug delivery: A review. *Chinese Chemical Letters,* **2018**, *29*, 1713–1724.

CHAPTER 5

Microneedles: An Innovative Strategy for Skin Drug Delivery, Diagnostic, and Cosmetologic Approaches

Marilisa Guimarães Lara

School of Pharmaceutical Sciences of Ribeirão Preto, University of São Paulo, Ribeirão Preto, São Paulo, Brazil, E-mail: mlara@fcfrp.usp.br

ABSTRACT

Microneedles are small devices used to improve the transdermal delivery of drugs. For this purpose, they are inserted into upper layers of skin and consequently open holes for drug transport, improving skin permeability. The application of microneedles is considered both painless and minimally invasive, as well as safe and easy-to-use. These innovative devices have a large application in medicine as therapeutic and diagnostic tools, as well as in cosmetology. The principle of microneedles is discussed in this chapter, along with a review of the many forms of microneedle and its application in drug delivery. Additionally, prerequisites for a suitable performance as well as aspects of the production of microneedles as materials processing are explored. Finally, taking into account the advancements in research and scientific progress in microneedles technology in each of these fields, the uses of microneedles as therapeutic, preventive, diagnostic, and cosmetologic aids are addressed. It is worth to note that microneedles are considered an innovative and promising technology for drug delivery and biomedical applications.

Topical and Transdermal Drug Delivery Systems: Applications and Future Prospects. Nayan A. Gujarathi, PhD, Juliana Palma Abriata, PhD, Raj Kumar Keservani, MPharm, Anil K. Sharma, PhD, MPharm (Eds.)

5.1 THE CONCEPT OF MICRONEEDLES

Microneedles are small devices used to pierce upper layers of skin and consequently to open holes for the improvement of drug delivery into skin (Serrano-Castaneda et al., 2018). These innovative devices have a large application in medicine as therapeutic and diagnostic tools, as well as in cosmetology. They are micron-sized needles with diameters of a few hundreds of microns and lengths of 50 to 9,000 μm. These small devices are inserted into upper layers of skin and therefore do not reach mechanoreceptors and nociceptors. Therefore, their application is considered both painless and minimally invasive. The initial goal of these devices was to increase skin delivery of drugs and optimize transdermal drug delivery (Gupta et al., 2011; van der Maaden et al., 2012; Serrano-Castaneda et al., 2018).

Microneedle technology was first developed in the 1970s for drug delivery purposes, but the advances in microelectronic technology in the 1990s improved their manufacturing techniques and expanded their applications. Novel approaches for microneedles have been proposed such as drug delivery, diagnostic, and prophilatic tools (van der Maaden et al., 2012; Serrano-Castaneda et al., 2018; Duarah et al., 2019). The development of microfabricated tools resulted in the fabrication of microneedle devices of various sizes, geometries, and materials in order to meet specific drug delivery requirements (van der Maaden et al., 2012; Sabri et al., 2019).

The concept of microneedles emerged associated with transdermal drug delivery as a strategy to improve drug transport into skin. In view of that, their application is related to the advantages of transdermal drug delivery such as to avoid first-pass metabolism and gastrointestinal incompatibilities, to present minimum or no side effects, to maintain plasma drug level for a long period of time and to improve drug bioavailability. However, the skin barrier provided mainly by *stratum corneum* (SC) impairs and/or hampers the penetration of a large amount of drugs. Therefore, skin barrier represents a challenge to transdermal drug delivery. Different strategies have been used to increase skin permeability: iontophoresis, sonophoresis, electroporation, use of permeation enhancers, novel drug delivery systems, and others. Among these, the use of microneedles has gained remarkable attention for transdermal drug delivery development since this technology can improve drug transport of different molecules (Prausnitz, 2004; Wu et al., 2008; Gupta et al., 2011; van der Maaden et al., 2012; Babity et al., 2018; Yang et al., 2019).

In addition, microneedles are minimally or non-invasive and painless, resulting in a good patient compliance. They are also safe and easy-to-use devices, allow a self-administration and can be easily removed in case of adverse reactions (van der Maaden et al., 2012; Serrano-Castaneda et al., 2018; Sharma et al., 2019). Another interesting feature of microneedles is that these systems can improve the stability of biomolecules. The formulations of microneedles include viscosity enhancers or sugars that can minimize molecular mobility and dehydration which could reduce the stability of biomolecules (Sabri et al., 2019)

However, there are some limitations and challenges concerning microneedles, such as poor accuracy of dose, possible variations in the delivery of drugs related to skin conditions and external environment, the risk of breaking and bending before insertion, the risk of tip break leaving fragments in the skin and the need of high-cost microelectronic technology to be produced (Serrano-Castaneda et al., 2018; Sharma et al., 2019). Moreover, small dimensions of the microneedles limited the amount of drug available to be delivered. Another aspect that should be considered is the risk of infections, as microneedles affect the barrier function of skin and increase skin permeability. On the other hand, skin permeability can be reversed after the use of microneedles due to skin resealing and therefore, these systems are considered very safe. The effect of repeated application with microneedles on the skin barrier properties was evaluated on a hairless mouse model and no infection or inflammatory markers were found, suggesting that application of microneedles are safe concerning the skin barrier and infection risk (Duarah et al., 2019; Sabri et al., 2019).

The insertion of microneedles into skin creates microchannels that increase drug permeability and thus they facilitate drug delivery. However, skin permeability can be reversed due to skin resealing which is dependent on microneedles geometry and occlusive skin conditions. Occlusion at the site of application retards skin resealing and maintains the generated channels open during the drug delivery period. After the removal of the occlusion, these channels close rapidly, ensuring efficacy and safety of the systems.

In this sense, microneedles are a promising strategy for skin drug delivery and can transport molecules with various physical and chemical properties, including macromolecules such as peptides, proteins, and oligonucleotides. Moreover, the drug delivery from microneedles can be modulated as a controlled release system (Gupta et al., 2011; Serrano-Castaneda et al., 2018; Sabri et al., 2019). The efficiency of microneedles to pierce skin and improve skin delivery of drugs can be affected by several factors. Some properties like mechanical strength, toughness, hardness, and a proper geometry are required to allow a good insertion into skin and disruption of the SC without fractures (Serrano-Castaneda et al., 2018). Another requirement for a good microneedle performance is that they should pierce skin in a reproducible manner to ensure a precise drug delivery. Therefore, a reproducible penetration depth should be achieved. The use of microneedles applicators can be useful to provide a reproducible insertion (van der Maaden et al., 2012). However, during insertion, some deformation at the insertion site may occur due to skin elasticity, which can be prevented by optimizing the microneedles geometry (length and sharpness) and providing a proper pressure by the use of applicators (Wu et al., 2008). The efficiency of microneedles for the improvement of skin permeability is also affected by the insertion area and the insertion time. The increase of insertion time may reduce skin deformation and improve microneedle piercing. Thus, increasing insertion time leads to an increase in the depth of generated holes and therefore, it facilitates drug permeation. However, after a given period of time, the influence of insertion time to skin permeability is reduced (Wu et al., 2008). Biocompatibility and biodegradability are also important requirements for the development of microneedles. Furthermore, the small size in order to ensure painless and non-invasive application as well as drug delivery efficiency should also be considered (Serrano-Castaneda et al., 2018).

5.2 THE FABRICATION OF MICRONEEDLES

Microneedles can be fabricated by several methods based on microelectronic technology and techniques like lithography, and micromolding. Dry etching and wet etching are usually employed. Modern methods have been employed for manufacturing microneedles, such as deep X-ray lithography, dipping, drawing lithography, droplet born air blowing, laser ablation, laser cutting, metal electroplating, microstereolitography, photolithography, pulling pippets, two-photon polymerization and vapor deposition (Sharma et al., 2019).

A wide variety of materials can be used to produce microneedles like silicon, metals, glass, sugars, polymers, and ceramics. The ideal requirements for a material to produce a microneedle are to be compatible, easily available, non-corrosive, and to present high tensile strength and good mechanical strength. Biocompatibility and biodegradability are also very interesting properties of the material for the microneedles` manufacturing process (Duarah et al., 2019; Yang et al., 2019).

Silicon has played an important role in microneedles fabrication, and microneedles of various shapes and sizes may be manufactured with this class of material. Silicon presents good mechanical strength and relatively high hardness, although it is a bit brittle. Metals like alloys, cobalt, nickel, palladium, stainless steel, titanium, and palladium-cobalt arrays are also employed for the microneedle fabrication. Metal microneedles present good mechanical properties like toughness, strength, and hardness, which help to avoid mechanical fractures. Titanium stands out due to its biocompatibility and corrosion resistance, while stainless steel is compatible but corrosive. Glass material is also employed in the manufacture of microneedles. Silica glass is largely used, despite its brittle nature. On the other hand, borosilicate glass is more elastic and can also be employed. Polymers are widely used in microneedles fabrication due to properties like biodegradability, biocompatibility, strength, toughness, low toxicity, and low cost. Examples of polymers that are usually employed for microneedle manufacture are polycarbonate, poly-glycolic acid (PGA), poly-lactic acid (PLA), poly-lactide-co-glycolide acid (PLGA), polyvinyl pyrrolidone (PVP), polyvynilic alcohol (PVA), polyestirene, and chitosan. Carbohydrates like maltose, galactose, mannitol, sucrose, trehalose, and xylitol are good materials for the fabrication of microneedles because they are economic, compatible, and safe. Maltose is largely used by thermal treatment on the manufacturing process. Ceramics is another material for microneedles fabrication with good biocompatibility and strength. Examples of ceramics are alumina, calcium sulfate dehydrate, and calcium phosphate dehydrate (Serrano-Castaneda et al., 2018; Duarah et al., 2019; Sharma et al., 2019; Yang et al., 2019).

Microneedles can be manufactured in several shapes like cylindrical, canonical, pyramid, candle, spike, spear, square, pentagonal, hexagonal, octagonal, rocket, star, etc., with different lengths and widths (van der Maaden et al., 2012; Sharma et al., 2019). The geometry and physical properties of the microneedles affect their insertion into the skin, determining the force required to a proper insertion and the force required for a fracture. The force needed for insertion showed to be dependent on the area of the needle tip while the force for a fracture depends on the wall thickness (Prausnitz, 2004). Therefore, a correct selection of fabricating material, geometry, and manufacturing method are quite important for a good performance of the systems of microneedles (Duarah et al., 2019). Geometry affects the strength of the microneedles and their ability to pierce skin. The sharpness of the microneedles is closely related with skin penetration: the increasing sharpness of the microneedles improves their penetration, although an increased sharpness can reduce the mechanical stiffness. Another geometry parameter that should be considered concerns to the tip of the microneedles, as high tip diameters need higher insertion forces. Geometry also affects drug flux from the microneedles through skin from devices with hollow microneedles. A blunt-tip microneedle compresses more the skin during insertion when compared with a bevel-tip one and this compression leads to a higher risk of clogging. The geometry of a microneedle bore also affects drug fluxes and microneedle strength. Increase in the microneedle bore may increase drug flux but it decreases the microneedle strength. A good combination of these properties is to design a sharp microneedle with the bore off-centered on the side (Wu et al., 2008; van der Maaden et al., 2012).

Microneedles can be applied into the skin manually in the form of a patch or with the use of applicator devices. Applicators are good tools to provide an appropriate pressure to microneedles application which can affect the insertion depth and drug delivery as well as reduce the risk of fractures. These requirements are extremely important because they affect the dose to be delivered, which should be reproducible. The use of applicators should not induce any pain or discomfort. Several applicators have been developed for microneedles administration with varying configurations and based on different strategies: hand-operated design, vibration-based applicators, impact-based applicators, mechanically-driven applicators, or more sophisticated electrically-driven applicators (Wu et al., 2008; van der Maaden et al., 2012; Duarah et al., 2019).

5.3 TYPES OF MICRONEEDLES

Microneedle systems can be designed in several types that follow specific approaches of application.

5.3.1 SOLID MICRONEEDLES

Solid microneedles are able to produce microchannels in the skin through which the drug is permeated. They are usually produced with silicon, stainless steel, or acrylic. These types of microneedles follow the poke and patch approach. The microneedles are applied onto the skin to create pores. Then they are removed from the skin and a patch containing drug formulation is subsequently applied. The transport of the drug is governed by diffusion through the pores created by the microneedle and it is affected by the number of created pores, their size as well as the drug concentration in the formulation. The pores produced by the microneedles should stay open during the drug application period, which can be achieved with occlusion on the site of administration (van der Maaden et al., 2012; Serrano-Castaneda et al., 2018; Sharma et al., 2019). The Poke and Patch approach of solid microneedles is presented in Figure 5.1. In this approach, the microneedle itself does not contain drugs, which will be provided by the patch applied after the remotion of the microneedle. Therefore, it is a two-step process: poke the skin with the microneedle device and the subsequent application of a drug loaded patch (Sharma et al., 2019). This approach can also be combined with iontophoresis technology in order to increase the uptake of charged drugs, although the risk of skin irritation should be considered in this case. Another solid microneedle approach is called the scrape and patch approach. In this case microneedles create micro abrasions when scraped over the skin with the subsequent application of a patch containing drug formulation (van der Maaden et al., 2012; Duarah et al., 2019).

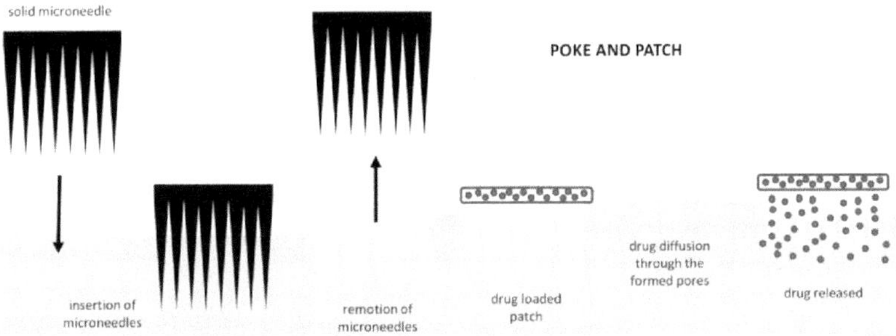

FIGURE 5.1 Solid microneedles and the poke and patch approach.
Source: Based on Serrano-Castañeda et al. (2018); Duarah et al. (2019); Sharma et al. (2019).

5.3.2 COATED MICRONEEDLES

In this case, microneedles are coated with a drug and then inserted into the skin. They are usually produced with stainless steel, titanium or polymers and receive a coating process with drug solution or dispersion. These types of microneedles follows the coat and poke approach (Figure 5.2), where a previously coated microneedle is inserted on the skin followed by a rapid release of drugs. Coated microneedle devices are very versatile because a wide range of drugs can be used in the coating process, including vaccines, DNA, and biomolecules. This approach is a one-step process and therefore it is more easily applied. Coating processes are very important for the coated microneedle development. The thickness of the coating layers and the needle size influence the amount of drug available to be delivered. Therefore, the coating process should be efficient and should not reduce the microneedle sharpness. A layer-by-layer coating improves the efficiency of the process (van der Maaden et al., 2012; Serrano-Castaneda et al., 2018; Sharma et al., 2019; Yang et al., 2019).

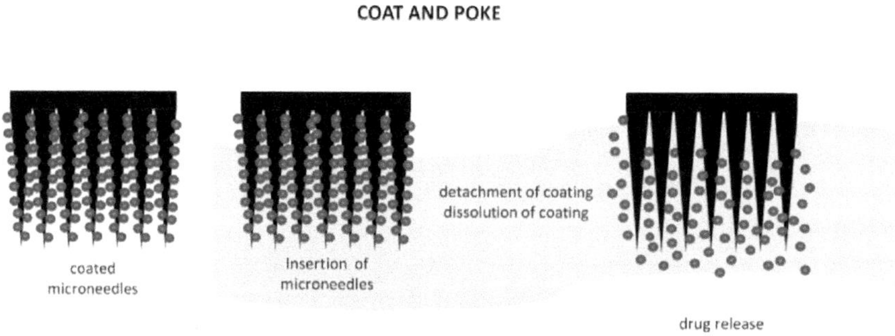

FIGURE 5.2 Coated microneedles and the coat and poke approach.
Source: Based on Serrano-Castañeda et al. (2018); Duarah et al. (2019); Sharma et al. (2019).

5.3.3 DISSOLVING MICRONEEDLES

Dissolving microneedles are fabricated with certain polymers and contain embedded drug. Natural and biodegradable polymers are usually employed. Examples of materials used to fabricate dissolving microneedles are polyvinylpyrrolidone (PVP), carboxymethyl cellulose (CMC), polyvinyl alcohol (PVA), poly-lactide-c0-glycolide (PLGA), chitosan, dextran, and sugars. These microneedles follow the poke and release approach (Figure 5.3) where the microneedles are inserted into the skin and are not removed. Once inserted into the skin, there is a dissolution process, and consequently, the drug is released. It is a one-step process and it is widely accepted. Drug release from dissolving microneedles can be controlled by a correct selection of a polymer and a slow drug release can be achieved. The dissolution of the microneedles is the rate-limiting process. A variation of this approach can be used with porous microneedles in which the drug is released through existing pores in the microneedle, although porous needles are usually more fragile than solid structures and they can break into the skin (van der Maaden et al., 2012; Sharma et al., 2019; Yang et al., 2019).

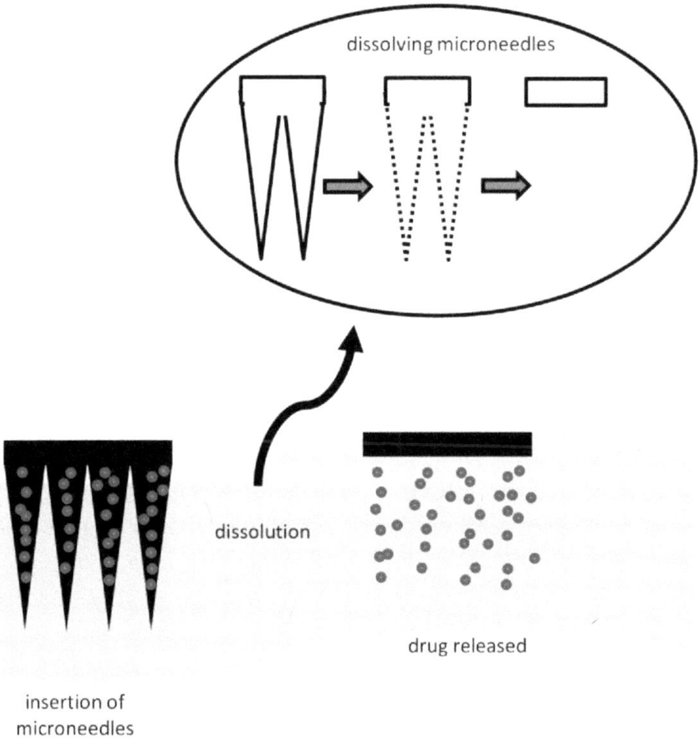

FIGURE 5.3 Dissolving microneedles and the poke and release approach.
Source: Based on Serrano-Castañeda et al. (2018); Duarah et al. (2019); Sharma et al. (2019).

5.3.4 HOLLOW MICRONEEDLES

Hollow microneedles are devices containing small channels that allow the passage of fluids. These types of microneedles are hollow structures filled with drug solution or dispersion, and these liquid formulations can be delivered into the skin like a micro injection. They are usually made with silicon, metal, glass, ceramic, and polymers. Hollow microneedles can deliver high volumes of liquid and doses can be easily controlled. This type of microneedles follows the poke and flow approach (Figure 5.4) where a liquid formulation can be injected and the drug can be delivered into skin. It is a one-step process. The flow of drug solution is the rate-limiting process and a continuous drug delivery through the skin can be achieved by diffusion, pressure, or an electrically driven flow. A limitation of this type of microneedle is the risk of resistance to flow and clogging due to compression of the dermal tissue during the insertion into the skin. This limitation can be avoided considering the correct geometry of the microneedles. The use of hyaluronidases in the infusion solution is suggested with the aim of breaking hyaluronic acid and to facilitate the flow. The drug flow can be modified by several strategies. It is possible to combine microneedle injection applicators with micropumps or pressurized gas or even with iontophoresis technology for charged molecules (van der Maaden et al., 2012; Babity et al., 2018; Serrano-Castaneda et al., 2018; Duarah et al., 2019; Sharma et al., 2019).

5.3.5 HYDROGEL FORMING MICRONEEDLES

These microneedles are made with swelling polymers (like chitosan, PVA: polyvynilic alcohol, PLGA: poly-lactide-co-glycolide acid, poly-methyl-vinyl-ether-co-maleic acid) that are able to absorb a great amount of water and swell *in situ*. Insertion of microneedles into skin provides water to be absorbed and the polymer can swell creating conduits that allow drug release (Figure 5.5) (Sharma et al., 2019).

5.3.6 FUNCTIONAL MICRONEEDLES

The advances on microneedle technology resulted in the development of smart multifunction-responsive microneedles such as multi-round

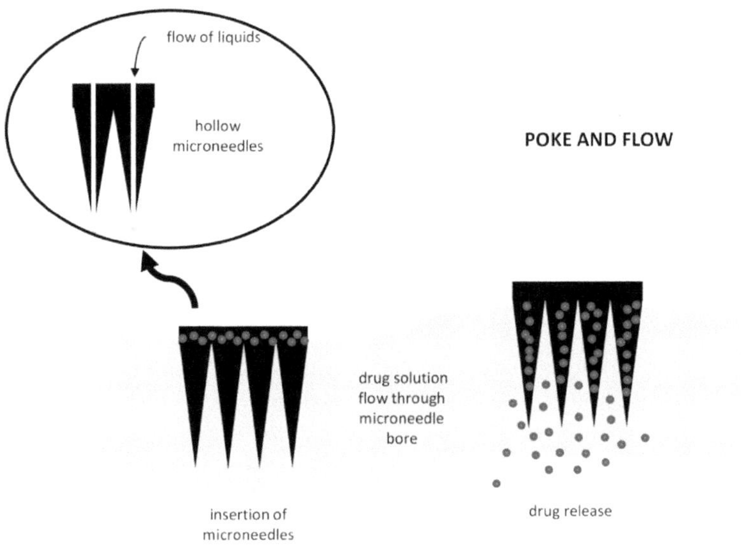

FIGURE 5.4 Hollow microneedles and the poke and flow approach.

Source: Based on Serrano-Castañeda et al. (2018); Duarah et al. (2019); Sharma et al. (2019).

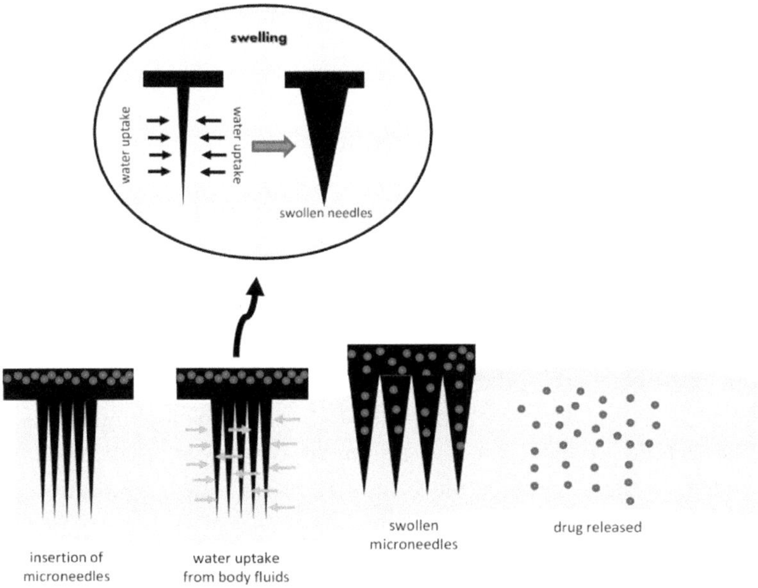

FIGURE 5.5 Hydrogel forming microneedles.

Source: Based on Duarah et al. (2019); Sharma et al. (2019).

microneedles, temperature-responsive microneedles, glucose-responsive microneedles, pH-responsive microneedles and water-soluble microneedles. Smart devices are advantageous compared with conventional microneedles because they comprise an integration of sensors, actuators, and drug formulations which results in an individualized treatment. The sensors can track physiological condition while actuators control the release pattern and the dose delivered to maintain the drug release until the disease indicators improve. This innovative system can afford feedback controlled delivery, i.e., they release the drug according to the real needs of the patient (Duarah et al., 2019; Yang et al., 2019).

There are several studies focusing on advanced smart microneedle devices for diabetes treatment. An example is a glucose-responsive microneedle system to deliver insulin. The enzyme glucose-oxidase reacts with the glucose with oxygen consumption. Hyperglycemic state and hypoxic state are detected by sensors and the dissociation of insulin vesicles is induced to release insulin. Another approach that has been investigated is a pH-responsive microneedle patch to deliver exendin-4. In this case, glucose oxidase converts glucose to $H+$ signals under hyperglycemic conditions. An electrochemical device that is able to monitor pH, temperature, humidity, and glucose level in sweat was investigated to deliver metformin by a thermally activated mechanism (Yang et al., 2019).

5.4 APPLICATIONS OF MICRONEEDLES

5.4.1 THERAPEUTIC APPLICATIONS: DRUG DELIVERY

As mentioned before, microneedles are largely studied for transdermal drug delivery to improve the transport of molecules with various physical and chemical properties. Much attention has been focused on the development of microneedle-based drug delivery systems for the delivery of drugs employed in several diseases. Among them, there can be highlighted research on microneedle systems for the treatment of cancer, diabetes, obesity, Alzheimer's disease, neuropathic pain, and others. Thus, we reinforced the great potential of microneedles to optimize transdermal delivery of several molecules, including peptides and proteins. A lot of research has been done with microneedles as carriers for several drugs such as diclofenac, ibuprofen, ketoprofen, paracetamol, desmopressin, fluorescein sodium, parathyroid hormone, salmon calcitonin, naltrexone, methotrexate, verapamil hydrochloride and amlodipine, nicardipine hydrochloride, heparin, nadroparin sodium, lidocaine, caffeine, doxorubicin, gentamicin, dihydroergotamine mesylate, photosensitizers like 5-aminolevulinic acid (ALA) and derivatives and others. It is worth to mention that microneedle devices can be used associated with other technologies to improve drug delivery like iontophoresis or sonophoresis (Wu et al., 2008; Serrano-Castaneda et al., 2018; Duarah et al., 2019; Sharma et al., 2019; Yang et al., 2019).

Insulin delivery by microneedles have gained remarkable attention and several interesting approaches have been investigated in this field. Clinical trials performed with intradermal microneedles showed efficacy with rapid glucose uptake and fast insulin action, as well as pharmacokinetics (PK) parameters that demonstrate their therapeutic effectiveness. Therefore, microneedles are considered a promising carrier for transdermal delivery of insulin, demonstrating efficacy, high bioavailability, high pharmacological activity and sustained hypoglycemic effect. Due to their painless and easy-to-use application, microneedle devices had a good patient compliance (Serrano-Castaneda et al., 2018; Duarah et al., 2019; Sharma et al., 2019; Yang et al., 2019).

Researches in the insulin delivery also comply the development of glucose-responsive microneedles array patches. In this system, glucose oxidase converted glucose to gluconic acid with the consumption of oxygen. The hyperglycemic state or hypoxic condition could induce dissociation of hyaluronic acid vesicles and therefore insulin is released (Yang et al., 2019).

5.4.2 DIAGNOSTIC APPLICATIONS

Microneedles technology has also been employed for diagnostic and monitoring applications (Babity et al., 2018). In view of that, a lot of research has been done in this field. The following paragraphs will discuss some of the standout areas of these researches.

Microneedles can be used for disease diagnosis in the form of biosensors that provide continuous detection of biomolecules for monitoring health conditions. Hollow microneedles can provide blood sampling in a minimal invasive way. These systems can be used for blood sampling by passive diffusion, capillary forces, and vacuum suction. They can also be combined with other technologies like biosensors and microfluidic chips that enables analysis of blood samples. The use of microneedles to withdraw blood avoids the need to sampling with hypodermic needles or lancets and their complications like patient discomfort and need of high sample volume (Babity et al., 2018; Serrano-Castaneda et al., 2018; Sharma et al., 2019).

A biosensor system based on microneedles to detect nitric oxide have been investigated for cancer diagnosis. A microneedle sensor coupled with endo-microscope with high-resolution imaging combined with electrical detection of nitric oxide was studied as a system for cancer detection. This system is an innovative alternative for traditional procedures that require biopsy sampling of suspicious lesions (Serrano-Castaneda et al., 2018; Sharma et al., 2019).

Microneedles are a good tool for interstitial fluid sampling. The concept of sampling interstitial fluid rather than the blood is an innovative strategy to detect biomarkers. This strategy can be achieved with the use of microneedles for disrupting SC and sampling of interstitial fluid through a painless and minimal invasive technology. Interstitial fluid sampling requires a reduced penetration depth and minimally invasive procedure. Solid microneedles are proposed to puncture skin with subsequently withdraw of interstitial fluid with a cloth or vacuum chamber. Hollow microneedles have also been widely investigated to extract interstitial fluid. The device must present some properties like toughness and flexibility as well as a proper internal diameter to allow the passage of blood cells. The base of the microneedle can be connected to a sensing chamber where it analyzes the withdrawn interstitial fluid that can be instantaneously using a built-in detector. Although this technology was widely investigated for glucose monitoring, the same concept can also be applied for other analytes such as viral cytokines or immunoglobulins for the diagnosis of infections like HIV and Ebola viruses. Swelling microneedles were also proposed for the sampling interstitial fluid followed by subsequent quantification using commercially available tests kits. This strategy has been used for glucose and cholesterol quantification as well as therapeutical drug monitoring such as lithium (Babity et al., 2018; Yang et al., 2019).

Microneedle systems for glucose monitoring have attracted significant attention and a large amount of research has been developed in this field. Microneedles can be functionalized to monitoring glucose by blood or interstitial fluid sampling coupled with glucose detectors, and those systems can provide a continuous glucose monitoring. These approaches are painless and therefore avoid the discomfort of traditional glucose monitoring devices in which the patients pricked their fingers. Glucose sensors microneedles are engineered with immobilized glucose oxidase that reacts with glucose and generates H_2O_2 which is detected by several methods like amperometric, potentiometric, and optical methods. Therefore, the microneedles are functionalized with immobilized oxidative enzymatic detectors and a redox reaction that takes place at the surface of the microneedles. These systems have been studied for glucose and other analytes such as glutamate and lactic acid. Functionalized microneedles can be designed as sensing probes consisting of microneedles coated with conductive materials and immobilized glucose oxidase on the surface of the microneedle. Microneedles can also be integrated with micro-sized electrodes where immobilized enzymes and electrodes are enclosed in hollow microneedles. Another strategy for glucose monitoring is the use of microneedles as a biological fluid sampling as described before and hollow microneedles are developed to monitor glycose by blood or interstitial fluid sampling. Swelling microneedles have also been investigated to extract interstitial fluid and to release drug simultaneously. A very interesting potential application of these swellable microneedle devices is the development of a unique system consisting of a microneedle based enzymatic glucose detector coupled to an insulin –release system that would be able to monitor glucose and to deliver proper amounts of insulin (Babity et al., 2018; Takeuchi and Kim, 2018).

Selective capture of biomarkers is another application of microneedle devices for disease diagnosis because cutaneous protein composition can vary in certain conditions such as inflammatory or immune responses. In this way, antibody-coated microneedles were investigated for selective capture of dermal proteins for diagnosis of systemic diseases. Simultaneous detection of different biomarkers for metabolic acidosis, fatigue, and skin carcinoma designed with microneedles have been also investigated (Babity et al., 2018; Sharma et al., 2019).

An interesting research area that has been emerging is the use of microneedles for biopotential measurements which can be useful for the monitoring of bio signals such as ECGs (electrocardiography), EMG (electromyography) and EEG (electroencephalography) used in monitoring of cardiovascular disease, muscular dysfunction, epilepsy, and others disorders. This concept is based on the measurement of skin impedance and electrical potential generated by electrolytes present in interstitial fluid which is connected to the activity of electrically driven tissues such as the heart and muscles. Electrodes comprised of microneedles are small (requiring a small skin surface area for application), easy to apply, comfortable, and do not require preparation of skin and therefore they are a promising strategy for biopotential measurements (Babity et al., 2018; Serrano-Castaneda et al., 2018).

Finally, microneedles can also contribute to several diseases' diagnosis through the delivery of diagnostic agents in a non-invasive and painless administration. In this sense, delivery of imaging agents like fluorescent dyes through skin can be useful for detecting/monitoring skin disorders and quantification of biomarkers present in the interstitial fluid. Another type of skin delivery that is useful for diagnostic purposes is the delivery of antigens for allergen skin tests or immunotherapy using microneedles containing or coated with a protein allergen solution (Babity et al., 2018; Serrano-Castaneda et al., 2018).

5.4.3 PROPHYLACTIC APPLICATIONS

Microneedles have been used successfully as carrier for vaccine therapy. Intradermal vaccination using microneedles showed to be very effective in human and animal studies. Furthermore, they are painless and safe. Microneedles have been extensively studied as vaccination devices in order to avoid inconveniences and pain of puncture of needle into the dermis. An interesting feature of these systems is that microneedles deliver drugs in the epidermis (EP) and dermis, where there is a rich network of antigen presentatting cells, favoring the immunization process. Although microneedles are considered tools to deliver peptides and proteins, aspects like preservation of protein structure during manufacture, storage, and use should be carefully considered, particularly in the case of vaccine purposes, because they may affect safety and induce unwanted immunogenicity. Different types of microneedles can be used for this purpose, such as dissolving, hollow, solid, and coated microneedles. Several studies have been performed to investigate the potential of MN as a vaccine carrier for different antigens, such as influenza vaccine, BCG (bacillus Calmette-Guérin), antigen ovalbumin, hepatitis B surface antigen, cholera toxin, diphtheria, and tetanus toxoid peptides and DNA vaccines (van der Maaden et al., 2012; Serrano-Castaneda et al., 2018; Sharma et al., 2019; Yang et al., 2019).

5.4.4 COSMETOLOGY APPLICATIONS

Microneedles are used in the cosmetic area for improving skin conditions in several situations. Cosmetic applications of microneedles may be related to promote natural healing of injured skin or to enhance skin permeability of cosmeceutics in a minimally invasive way. For cosmetic purposes, microneedles can be used as a patch for manual application into the skin or using application devices such as high-velocity applicator or Dermaroller® devices. Dermaroller® is a device designed with a cylinder with stainless steel needles and is very common in the cosmetologic use (Serrano-Castaneda et al., 2018; Yang et al., 2019).

Several skin conditions such as seborrheic keratosis, scars, striae, anti-aging, wrinkles, and depigmentation can be improved with the use of microneedles. Microneedles procedures for treatment of burn scars are indicated due to the effect of induced collagen fiber rearrangements in scar tissue, as an alternative strategy for laser treatments. Acne scars can also be treated with a microneedle procedure using repeated Dermaroller® applications for 15 to 20 minutes, which induce dermabrasion. This strategy is based on the ablation concept (injury superficial layers of skin) in order to generate an equilibrate cell proliferation, facilitating skin repair. In addition, microneedles are proposed to deliver active agents for specific skin conditions with the improvement of skin permeability. Some examples of this microneedles application are the delivery of tranexamic acid for the treatment of melasma and retinoic acid microneedles patch for seborrheic keratosis and senile lentigo. Microneedles were also proposed for alopecia areata treatment to improve the delivery of corticosteroids and the bloody supply to the hair follicles providing more nutrients supply. A microneedle patch containing retinyl retinoate and ascorbic acid was proposed for antiwrinkle using dissolving microneedles (Kim et al., 2009; Doddaballapur et al., 2009; Alsantali, 2011; Chandrashekar et al., 2014; Harris, 2015; Machekposhti et al., 2017; Hirobe et al., 2017; Serrano-Castaneda et al., 2018; Yang et al., 2019).

5.5 CONCLUSIONS

The present chapter discussed the concept of microneedle devices and their applicabilities as therapeutic, diagnostic, and cosmetologic tools.

Microneedles are painless, minimally invasive, and easy-to-use and can be employed to deliver several drugs into the skin as well as to provide diagnostic and monitoring advanced devices. Recent advancements in microneedle technology were presented in both therapeutic and diagnostic fields. It is worth mentioning that most of the strategies discussed here are in a scientific development state, and a lot of research is needed to face the challenges concerning technological and clinical aspects. Nevertheless, microneedles have received much attention in the last years and are considered an innovative and promising technology for drug delivery and biomedical applications.

KEYWORDS

- **carboxymethyl cellulose**
- **cosmetic microneedles**
- **diagnostic microneedles**
- **microneedles**
- **polyvinyl alcohol**
- **polyvinylpyrrolidone**
- **skin drug delivery**

REFERENCES

Alsantali, A. Alopecia areata: A new treatment plan. *Clinical Cosmetic Investigational Dermatology*, **2011**, *4*, 107–115.

Babity, S.; Roohnikan, M.; Brambilla, D. Advances in the design of transdermal microneedles for diagnostic and monitoring applications. *Small*, **2018**, *14*(49), e1803186.

Chandrashekar, B.; Yepuri, V.; Mysore, V. Alopecia areata-successful outcome with microneedling and triamcinolone acetonide. *Journal of Cutaneous and Aesthestic Surgery*, **2014**, *7*(1), 63–64.

Doddaballapur, S. Microneedling with dermaroller. *Journal of Cutaneous and Aesthetic Surgery*, **2009**, *2*(2), 110–111.

Duarah, S.; Sharma, M.; Wen, J. Recent advances in microneedle-based drug delivery: Special emphasis on its use in paediatric population. *European Journal of Pharmaceutics and Biopharmaceutics*, **2019**, *136*, 48–69.

Gupta, J.; Gill, H. S.; Andrews, S. N.; Prausnitz, M. R. Kinetics of skin resealing after insertion of microneedles in human subjects. *Journal of Controlled Release*, **2011**, *154*(2), 148–155.

Harris, A. G.; Naidoo, C.; Murrel, D. F. Skin needling as a treatment for acne scarring: An up-to-date review of the literature. *International Journal of Women's Dermatology*, **2015**, *1*(2), 77–81.

Hirobe, S.; Otsuka, R.; Iioka, H.; Quan, Y. S.; Kamiyama, F.; Asada, H.; Okada, N.; Nakagawa, S. Clinical study of a retinoic acid-loaded microneedle patch for seborrheic keratosis or senile lentigo. *Life Sciences*, **2017**, *168*, 24–27.

Kim, S. K.; Park, J. M.; Jang, Y. H.; Son Y. H. Management of hypertrophic scar after burn wound using microneedling procedure. *Burns*, **2009**, *35*, S1, S47.

Machekposhti, S. A.; Soltani, M.; Najafizadeh, P.; Ebrahimi, S. A.; Chen, P. Biocompatible polymer microneedle for transdermal delivery of tranexamic acid. *Journal of Controlled Release*, **2017**, *261*, 87–92.

Prausnitz, M. R. Microneedles for transdermal drug delivery. *Advanced Drug Delivery Re*views, **2004**, *56*(5), 581–587.

Sabri, A. H.; Ogilvie, J.; Abdulhamid, K.; Shpadaruk, V.; McKenna, J.; Segal, J.; Scurr, D. J.; Marlow, M. Expanding the applications of microneedles in dermatology. *European Journal of Pharmaceutics and Biopharmaceutics*, **2019**, *140*, 121–140.

Serrano-Castañeda, P.; Escobar-Chavez, J. J.; Rodriguez-Cruz, I. M.; Melgoza, L. M.; Martinez-Hernandez, J. Microneedles as enhancer of drug absorption through the skin and applications in medicine and cosmetology. *Journal of Pharmacy & Pharmaceutical Sciences*, **2018**, *21*(1), 73–93.

Sharma, S.; Hatware, K.; Bhadane, P.; Sindhikar, S.; Mishra, D. K. Recent advances in microneedle composites for biomedical applications: Advanced drug delivery technologies. *Materials Science & Engineering: C, Materials for Biological Applications*, **2019**, *103*, 109717.

Takeuchi, K.; Kim, B. Functionalized microneedles for continuous glucose monitoring. *Nano Convergence*, **2018**, *5*(1), 28.

Van Der Maaden, K.; Jiskoot, W.; Bouwstra, J. Microneedle technologies for (trans)dermal drug and vaccine delivery. *Journal of Controlled Release*, **2012**, *161*(2), 645–655.

Yang, J.; Liu, X.; Fu, Y.; Song, Y. Recent advances of microneedles for biomedical applications: Drug delivery and beyond. *Acta Pharmaceutica Sinica B*, **2019**, *9*(3), 469–483.

CHAPTER 6

Emulsion, Nanoemulsion, and Microemulsion in TDDS

Patrícia Mazureki Campos, Guilherme Dos Anjos Camargo, Ana Beatriz Klosowski, and Priscileila Colerato Ferrari

Pharmaceutical Sciences Department, State University of Ponta Grossa, Ponta Grossa, Paraná, Brazil,
E-mail: patriciamazurekicampos@yahoo.com.br (P. M. Campos)

ABSTRACT

Colloidal carriers such as emulsions and submicron emulsions (nano-emulsions and microemulsions) have gained considerable interest in the dermal and transdermal delivery of therapeutic agents over the last few decades, and their fundamentals and applications are described in this chapter. These systems are composed of oily and aqueous phases stabilized by one or more surfactants. They differ according to size range droplet, thermodynamic, and kinetic stability, appearance, surfactant content, and spontaneity of formation. Emulsions are turbid systems, which are not spontaneously formed and are thermodynamically unstable, presenting the highest droplet size range. Despite the advantages of lipid-based systems for TDDS, emulsion applications have been shown limited or insufficient for transdermal deliveries, mainly due to a strong dependency between the observed pharmacological effects and the droplet size, affecting the drug permeation into the skin. Nanometric-based systems, such as nano-emulsions and microemulsions, can promote higher skin accumulation and penetration, due to their small droplet size. Nanoemulsions possess the ability to increase the solubility of poorly soluble drugs, as well as the

Topical and Transdermal Drug Delivery Systems: Applications and Future Prospects. Nayan A. Gujarathi, PhD, Juliana Palma Abriata, PhD, Raj Kumar Keservani, MPharm, Anil K. Sharma, PhD, MPharm (Eds.)

bioavailability of transdermal systems and to show better effects when compared to oral formulations. Microemulsions are able to interact with the stratum corneum (SC) promoting and enhancing skin permeation of drugs for transdermal application. Some examples concerning dermal and transdermal drug delivery by these colloidal carriers are presented and discussed in this chapter.

6.1 INTRODUCTION

The transdermal route can be a way to transport molecules up to the systemic circulation, without causing pain and discomfort to the patient, usually related to the parenteral route, and preventing stomach degradation related to the oral route. The transdermal route is carried out with the application of the formulation on the healthy skin without lesions. The drug initially penetrates through the stratum corneum (SC), which is the outermost layer, then crosses the deepest epidermis (EP) and dermis without accumulation of substances in the dermal layer. When this drug reaches the dermis, it becomes available for systemic absorption through the microcirculation present in the dermis (Alkilani et al., 2015). The advantages and disadvantages of the transdermal route of administration are shown in Table 6.1.

In order to allow better permeation and consequent transdermal drug delivery of molecules, some factors need to be taken into account, such as (i) permeation can be improved by the use of chemical permeation enhancers which partially fluidize the SC lipids such as ethanol, terpenes, glycol, and azones; (ii) use of physical methods such as sonophoresis, iontophoresis, electroporation, and microneedles; (iii) use of nanotechnology, with particle size ranging from 1 to 1,000 nm, by means of nanoparticles, nanoemulsion, liposomes, ethosomes, among others. So, one or more than one of them can be used to achieve transdermal delivery (Iqbal et al., 2018).

Furthermore, to have an appropriate drug delivery, the substances must present suitable physicochemical properties, such as molecular weight lower than 500 Daltons to cross the skin layers. The chemical structure should be amphiphilic and/or a surfactant may be added to modify the skin phospholipids. The log P value should be between 1 and 3 to ensure enough lipophilic solubility for skin permeation and should also have some aqueous solubility of at least 1 mg/mL due to the presence of incoming body fluids in contact with this substance. The melting point must be below 200°C and the recommended daily dose of dermal drug should be 10–40 mg, depending on the drug (Iqbal et al., 2018).

TABLE 6.1 Advantages and Disadvantages of Transdermal Drug Delivery

Advantages	**Disadvantages**
✓ Prolonged duration of action	× Possibility of irritation or allergic reaction at the site of application
✓ Reduction in the dosing frequency	× Impossibility of administering ionic drugs
✓ More uniform plasma levels	× It cannot achieve high drug levels in blood
✓ Useful for potent drugs	× Limited development for large molecules
✓ Improvement of bioavailability	
✓ Reduction of adverse effects	× It cannot deliver drugs in a pulsatile fashion
✓ Drug flexible administration by remotion of the transdermal delivery system from the skin	× Long time adherence is difficult
✓ Avoidance of first-pass metabolism, gastrointestinal absorption, and/or enzymatic degradation	× The pharmacokinetic and pharmacodynamic characteristics of the drug should be suitable for slow entry and sustained transdermal delivery
✓ Noninvasive administration compared to the parenteral route	

Source: Escobar-Chávez et al. (2012); Reddy and Satyanandam (2014); Tanwar and Sachdeva (2016); Vinod et al. (2015).

The major difficulty for a good permeation of the drug through the skin is the physical barrier, which is the outermost layer of the EP, called SC. To overcome this barrier and to achieve proper drug transportation, it is often necessary to employ active or passive methods. In the active method, an external driving force that allows the drug permeation is used by means of physical methods such as iontophoresis, electroporation, abrasion/suction, sonophoresis, and radiofrequency. Through the passive methods, substances or carriers that increase skin permeability are used, such as systems formed by lipids, surfactants, and polymers and/or substances that enhance permeation, such as fatty acids, organic solvents, alcohols, and esters. In addition, it is of utmost importance that the assets are incorporated into formulations suitable for such dermal delivery, such as creams, gels, and nanocarriers – nanoparticles, liposomes, niosomes, dendrimers, or carbon nanotubes (Manikkath et al., 2018).

The emulsified systems are attractive because they can deliver small and large molecules due to their heterogeneity features. These systems allow dermal and transdermal drug deliveries, and they can be classified as macroemulsions, microemulsions, and nanoemulsions with the particularities illustrated in Table 6.2. In addition, the stability parameter is one of the main factors guiding the choice of preparation, such as microemulsion, compared with others.

TABLE 6.2 Comparison of Macroemulsions, Nanoemulsions, and Microemulsions with Respect to Size, Shape, Stability, Method of Preparation, and Polydispersity

	Macroemulsions	**Nanoemulsions**	**Microemulsions**
Size	1–100 μm	20–500 nm	10–100 nm
Shape	Spherical	Spherical	Spherical, lamellar
Stability	Thermodynamically unstable, weakly kinetically stable	Thermodynamically unstable, kinetically stable	Thermodynamically stable
Method of Preparation	High and low energy methods	High and low energy methods	Low energy method
Polydispersity Index	Often high (>40%)	Typically low (<10–20%)	Typically low (<10%)

Source: Adapted from: Gupta et al. (2016).

6.2 EMULSION

6.2.1 BASIC CONSIDERATIONS

Emulsions can be defined as heterogeneous systems consisting of two immiscible phases, oily and aqueous phases, being one phase dispersed as droplets in another continuous phase. Because they are thermodynamically unstable preparations, it is necessary to add a third component as the surfactant or emulsifying agent to increase the stability of the formulation. Emulsions may be given different names depending on the characteristics of the dispersed and continuous phases; for example, if the oil phase is dispersed in the aqueous phase, it is "oil in water" (O/W) emulsion. To analyze the type of emulsion, it is necessary to evaluate the emulsifier or surfactant used, which is characterized by the hydrophilic-lipophilic balance (HLB). HLB values usually range from 1 to 20, and the higher this value is, the greater the hydrophilic character of the surfactant is.

According to the Bancroft rule, it was believed that the phase in which the surfactant would best solubilize would constitute the continuous phase of the emulsion, but based on further studies, an amendment to the Bancroft rule was proposed, determining the continuous phase by micelle formation, regardless of the surfactant solubility in both phases. Emulsions can be classified as O/W or "water in oil" (W/O), and can also be multiple, in this case being formed by more than two phases, for example, oil droplets dispersed in oily continuous phase water droplets (O/W/O) or water droplets dispersed in aqueous continuous phase oil droplets (W/O/W). In addition, emulsions may vary in viscosity or consistency determining their form as liquid, in lotion preparations, or even semi-solid, in the production of creams (Albert et al., 2019; Gershkovich et al., 2008; Khan et al., 2011; Lu and Gao, 2010; Menaa, 2014; Otto et al., 2008; Hu et al., 2016).

One of the advantages presented by emulsions is their wide application in products, mainly in the cosmetic and pharmaceutical areas, due to their ability to convey lipophilic actives in the oil phase and also hydrophilic in the aqueous phase, mask unpleasant taste and odor of substances, reduce irritation or toxicity, improve skin absorption and penetration, positively influencing the bioavailability of the compound of interest, and promoting controlled and/or targeted release of assets to specific sites. In the case of topical emulsions, possible degradation of the active in the gastrointestinal tract or first-pass hepatic metabolism is avoided, allowing the action of the substance to be prolonged at the application site. However, emulsions also present disadvantages and one of the main ones would be the instability of these systems, presenting short shelf life, since over time the separation of the phases that constitute the emulsion may occur, by different mechanisms addressed in the Stability topic, requiring the addition of surfactants to increase the period through which the original emulsion characteristics can be maintained (Lu and Gao, 2010; Otto et al., 2009).

6.2.2 FORMULATION DEVELOPMENT

To obtain emulsions, there are different preparation methods, as follows:

1. **Continental and Dry Gum Method:** The emulsifying agent, usually acacia gum, is mixed with the oil phase and the aqueous phase is mixed. The difference from continental and dry gum method would be in the proportion of the components (Kale and Deore, 2017).

2. **Wet Gum Method:** This method uses the same proportion of components as the dry gum method, but the preparation method is different. Firstly, the emulsifier mucilage (acacia) is formed and later the oil is added drop wise to the continuous grinding mucilage (Kale and Deore, 2017).
3. ***In Situ* Soap Method:** For emulsion formation, oil, and sodium hydroxide solution are added and need to be mixed with stirring (Kale and Deore, 2017);
4. **Mechanical Method:** The emulsion forming components, which would be oil, water, and emulsifier, are mixed and shaken with mechanical force using proper equipment (Kale and Deore, 2017).

   In addition to the proposed methods for emulsion development, mechanisms, or theories to explain the emulsification process are proposed, which are described below:
5. **Surface Tension Theory:** In the formation of the emulsion, when the two immiscible liquids are in contact, the surface tension causes resistance, so the addition of the surfactant decreases this force and enables the emulsification by forming large to small droplets of one liquid in the other liquid (Kale and Deore, 2017; Khan et al., 2011).
6. **Oriented Wedge Theory:** Monomolecular layers of the emulsifying agent are formed and directed around the droplets constituting the internal phase of the emulsion. In these formed monomolecular layers, the apolar end of the surfactant faces the oil phase, while the polar end is directed to the aqueous phase. If the apolar end is smaller, there is an oil-in-water (O/W) emulsion, while if the polar end is smaller, the emulsion will be water-in-oil (W/O) (Kale and Deore, 2017; Khan et al., 2011).
7. **Plastic or Interfacial Film Theory:** The emulsifying agent is present, as thin-film, at the interface between the immiscible liquids of the emulsion surrounding the inner phase droplet. This film prevents the instability signs due to its strength and flexibility (Kale and Deore, 2017; Khan et al., 2011). The film of emulsifying agent can be monomolecular (monolayer), multimolecular (multilayers of hydrophilic colloids) or formed by solid particles (clays, silicates of magnesium or aluminum, magnesium hydroxide) (Khan et al., 2011).

Due to the wide variety of components used in emulsion composition and process limitations, it is necessary to make an association between different techniques to have a complete characterization of the physicochemical properties of macroemulsions. In general, phase diagram constructions are done, considering variables such as surfactant amount, composition, temperature, and structure organization with the evaluation of the behavior phases. Rheology is used to know the system viscosity, which is dependent on some factors, such as the nature or feature of the oil and water phases, proportion of constituents and particle size distribution. The droplet size, which affects directly the macroemulsion stability, can be determined by dynamic light scattering (DLS) or photon correlation spectroscopy (PCS) techniques. Zeta potential, which is influenced by pH, ionic strength, concentration, and temperature, consequently affecting the system stability. And, the measurement of interfacial tension by rotating devices (Lu and Gao, 2010).

6.2.3 STABILITY ISSUES

In the process of developing emulsions, it is necessary to verify that these preparations are stable, to maintain or to preserve the initial characteristics. Macroemulsions are subjected to instability signs due to their intrinsic features. Therefore, coalescence, flocculation, creaming/sedimentation, and Ostwald ripening are often associated with this type of emulsified system (Khan et al., 2011). There are several steps and details in the composition and production of the macroemulsions that should be taken into account in order to have tolerable stability, such as surfactant amount, phase viscosity, enough energy in the production process to generate small and homogeneous droplets, suitable temperature, time of production and the solubility of the dispersed/continuous phases (Khan et al., 2011; Hu et al., 2016).

The main evaluation methods for macroemulsion are the: oil/water ratio with oil proportion until approximately 0.7 to avoid major instability signs; macroscopical analysis (physical features as aspect and color); particle size characterization determined by optical microscope, Andreasen apparatus and Coulter counter apparatus; viscosity and electrophoretic features as particle electric charge/zeta potential (Khan et al., 2011).

6.2.4 EMULSIONS APPLIED IN DERMAL AND TRANSDERMAL DRUG DELIVERY

Fat-soluble substances may permeate the skin through the SC. In the early 1900s, there were many reports of poisoning following topical application of aniline dyes to dye clothes and shoes. At the beginning of the 20th century, several *in vivo* studies were carried out, in which systemic absorption of topically applied substances was observed by measuring the compounds in blood and urine. At the same time, the first transdermal patch was developed to maintain active blood levels for a period of time, and one of the most well-known examples of transdermal patch drug would be nicotine, which is used by smokers to treat cigarette addiction. Then came several transdermal drugs to treat different pathologies, such as nitroglycerin used in angina pectoris, estradiol for female hormone replacement, fentanyl in pain, and scopolamine for the treatment of menstrual cramps (Szunerits and Boukherroub, 2018).

Percutaneous absorption involves the transfer of drug from the skin into the microcirculation and can be considered as a passage through a series of diffusional barriers. Factors affecting drug absorption are numerous: individual variations, skin hydration and temperature, age, site of application, disease, vehicles, and formulations, penetration enhancers, and surfactants used (Marti-Mestres and Nielloud, 2002). The three main factors to determine the transdermal efficacy of drugs are mobility of drug in the vehicle, release of drug from the vehicle, and permeation of drug into the skin. These factors affect either the thermodynamic activity that delivers the drug into the skin or the permeability of the drug in the skin, particularly the SC (Peltola et al., 2003). Numerous authors have observed a direct influence of particle size on skin penetration (Marti-Mestres; Nielloud, 2002), i.e., the small particle size ensures close contact to the SC thus increasing the amount of encapsulated drug penetrating into the viable skin (Jenning et al., 2000) and, also, a strong dependency between the observed pharmacological effects and the oil droplet size (Friedman et al., 1995). Progress made in the last years in this field are concentrated on the various aspects of the drug release and the influence of droplet size. In the last 20 years, evaluations have focused on transdermal penetration and promising pharmaceutical vehicles with negligible skin barrier perturbations and low irritancy. The skin penetration of a broad variety of drugs was investigated after application of emulsions for local and systemic activity (Schwarz et al., 1995; Lehmann et al., 2001; Kreilgaard, 2001; Kotyla et al., 2008). A pronounced penetration enhancement was achieved when drugs were formulated in submicron vehicles rather than in standard emulsions. The drug applied topically in emulsion creams was strongly dependent on oil droplet size. Furthermore, submicron emulsions increased transdermal drug delivery and bioavailability, prolonging the activity. The authors suggested that the smaller particles penetrated to a significantly greater extent and provided a larger surface area for absorption than the larger particles, thereby providing greater efficacy as a delivery system.

In this way, macroemulsion applications have been shown limited and/or insufficient for transdermal deliveries. The superior transdermal flux from submicron emulsions has been shown to be mainly due to their high solubilization potential for lipophilic and hydrophilic drugs. This generates an increased thermodynamic activity towards the skin (Kreilgaard et al., 2000; Alvarez-Figueroa; Blanco-Méndez, 2001). The drug thermodynamic activity in the submicron emulsion is a significant driving force for the release and permeation of the drug into the skin because it reflects the relative activities of the drug in different phases since the drug can be released from the internal phase to the external phase and then from the external phase to the skin; the relative activities may determine the skin permeation capability (Delgado-Charro et al., 1997; Chao et al., 2012). Due to the small droplet size and large amount of inner oil phase in submicron emulsions, the density of droplets and their surface area are assumed to be high. Therefore, droplets settle down to close contact with the skin providing a high concentration gradient and improved drug permeation.

Thus, several advances have been achieved with the reduction in the particle size of emulsified systems reaching micro and nanometric scales to promote higher skin penetration, less adverse effects, and more effective for transdermal and dermal uses with the named microemulsion and nanoemulsion systems as discussed in the next sections.

6.3 NANOEMULSION

6.3.1 *BASIC CONSIDERATIONS*

Nanoemulsion is a thermodynamically unstable system composed of two immiscible liquids as oil-in-water (O/W) or water-in-oil (W/O) stabilized by the presence of an interfacial layer of surfactant and co-surfactant; it can even be a multiple nanoemulsion (W/O/W). The surfactant solubility favors the emulsification, for example, the aqueous solubility of a surfactant will produce O/W nanoemulsion. Nanoemulsion is a colloidal dispersion with diameter droplets less than 500 nm and it is an isotropic system with clear or cloudy appearance (McClements, 2012; Singh et al., 2017; Rai et al., 2018).

Both the oily and aqueous phases contain soluble ingredients. When the oil components are chosen, their molecular weight and molecular structure should be large enough to prevent Ostwald ripening, as well as the concentration and surfactant type, which influence the stability to avoid coalescence (Mason et al., 2006). The type and concentration of the phase components determine the main properties of the nanoemulsion (Salvia-Trujillo et al., 2017).

Usually, O/W nanoemulsion contains oil/lipid (5–20%). The oil phase composition is set according to the loaded molecule, specifically, long, medium, and short-chain of triglycerides, and also, unsaturated oils are commonly used, and this composition dictates the molecule bioavailability (Singh et al., 2017).

The surfactants used in nanoemulsion are lecithin, sodium deoxycholate, tween (polyoxyethylene-660-hydroxystearate) 20, 40, 60, and 80, as well as span (sorbitan monolaurate) 20, 40, 60, and 80. Likewise, amphiphilic proteins, sodium dodecyl sulfate, poloxamer family, polysaccharides, and PEG containing block polymers. The co-surfactants used are polyethylene glycol, propylene glycol, ethylene glycol, ethanol, propanol, and glycerin (Singh et al., 2017; Rai et al., 2018). The choice of surfactant blend dominates the droplet size, toxicity, pharmacological properties, and stability (Ali et al., 2014; Li et al., 2015). In addition, a stable nanoemulsion could be established under the phase diagram construction with the right proportion of oil and aqueous phases plus surfactant/co-surfactant (Rai et al., 2018). The nanoemulsion and microemulsion compositions are quite similar, what changes is the higher surfactant-to-oil ratio for the microemulsion, plus a wider range of surfactants used for nanoemulsion (McClements, 2012).

The advantages of nanoemulsion are high stability during the storage, high surface area due to the small droplets and good bioavailability. In addition to the high solubilizing ability by the chemical composition for hydro and lipophilic loaded molecules, it can protect them from oxidation and hydrolysis. It can be administered through different routes as oral, intravenous, and topical, enabling efficient penetration of the molecules, thus, with several pharmaceutical/cosmetic applications. On the other hand, the disadvantages are the temperature sensitivity and pH changes, Ostwald ripening instability, the use of high amount of surfactant/co-surfactant to stabilize the nanodroplets, the inability to solubilize high-melting molecules and expensive production methods (Lovelyn and Attama, 2011; Rai et al., 2018).

Regarding skin permeation, nanoemulsion can increase molecule solubility due to the loading molecule into the micelles formed by the surfactant/co-surfactant blend. Thereafter, it may accumulate on the regions of the SC and hair follicle or extract lipids from the SC disrupting it, which increases the local concentration and promotes passive diffusion of the loaded molecule, finally, providing better skin bioavailability by the enhancement permeation mechanism (Shakeel et al., 2008). Nanoemulsions have been studied to deliver small and macromolecules in order to improve the skin penetration. They can overcome the molecule limitations in crossing the skin and supply enough amounts of molecules to exert their therapeutic effects. Thus, they have been applied for dermatological, cosmetic, and transdermal purposes (Nastiti et al., 2017).

The superior characteristics of nanoemulsion when compared with macroemulsion applied for skin delivery are outlined in Figure 6.1.

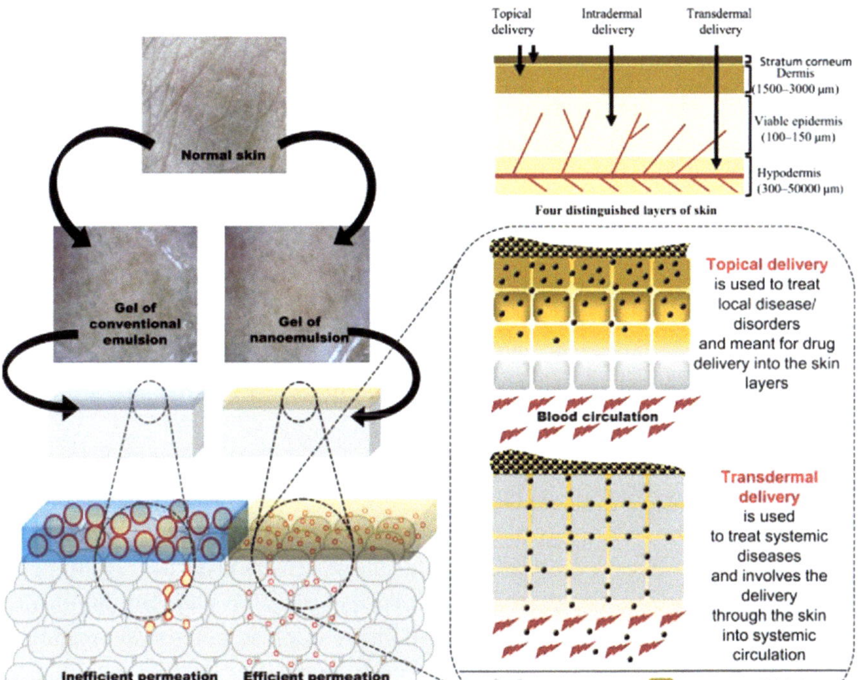

FIGURE 6.1 Representation of nanoemulsion delivery applied on the skin versus conventional formulation.

Source: Reprinted with permission from Rai et al. (2018). © Elsevier.

When considering droplet size of nanoemulsions, the most suitable is in the range of 150 to 350 nm, being more favored below 60 nm for transdermal application and on the tops of 1,000 nm for dermal application (Rai et al., 2018).

6.3.2 FORMULATION DEVELOPMENT

The nanoemulsion formation theory is based on the multiphase colloidal dispersion characterized by stability and clarity aspects. In addition, the application of high energy throughout the droplets formation is essential to the nano size, which is one of the differences from the microemulsion that uses thermodynamic self-assembly. Nanoemulsion requires extreme share to reach above Laplace pressure, which means enough stress to provide nanodroplets (Devarajan; Ravichandran, 2011).

The nanoemulsion preparation methods are classified as high-energy and low-energy methods. High-energy methods, through mechanical devices, apply powerful disruptive forces to break down the droplets and disperse them in the external phase. Usually, ultrasonicators, microfluidizer, and high pressure homogenizers are the pieces of equipment used at small and large scale. They are flexible to the different oil phases, despite the cost and high temperature produced. Low-energy methods, in turn, are possible after the extensive consideration of rational aspects from the oil and aqueous components, surfactant/co-surfactant blend, HLB and production temperature. The methods included are spontaneous emulsification and phase inversion. The challenge of the low-energy methods is the utilization of the stored energy into the system to provide the nanodroplets (Singh et al., 2017).

The factors that affect the droplet dynamics are the shear rate on the droplet of immiscible liquid into a dispersed phase, the role of surfactants to low interfacial tension to downsize and stabilize the nanodroplets and the relative viscosity of oil and water phases (Mason et al., 2006). For instance, during the homogenization process of O/W nanoemulsion, the low viscosity and the decrease in the oil phase interfacial tension promoted by the adequate surfactant blend will produce smaller droplet size due to the facilitated droplet disruption (Salvia-Trujillo et al., 2017).

The key aspect to prepare a nanoemulsion is to elect an adequate method embracing the optimum stirring speed/shear stress associated with the selection and orderly addition of proper components (Rai et al., 2018).

To optimize the nanoemulsion production, the pursued aspects are droplet size, zeta potential, equilibrium solubility, dissolution efficiency and monomodal distribution (Pund et al., 2014). The selection of an experimental design that includes stirring speed, pressure applied, ultrasonic amplitude, temperature, concentration of surfactants and oil phase, as well the ternary phase diagram are ways of improving it (Arora et al., 2014; Rai et al., 2018).

Several techniques are used to characterize the physicochemical profile of the nanoemulsion such as DLS to measure size and polydispersity index; electrophoresis mobility for zeta potential; transmission electronic microscopy to evaluate the morphology; drug content through sensitive techniques related to the drug; viscosity measurement by a rotatory viscometer at different temperature and rate; and also, simpler techniques as organoleptic evaluation, electrical conductivity, pH measurement, refractive index, dye solubilization, dilution test, phase analysis, and interfacial tension (Thakur et al., 2012; Jaiswal et al., 2015; Jain and Thareja, 2019).

6.3.3 STABILITY ISSUES

Nanoemulsions are considered kinetically stable systems (without any apparent flocculation or coalescence in the long-term storage) with extremely small droplet size and uniform size distribution (McClements, 2012; Rai et al., 2018). During the preparation process, the volume ratios of the phases define the stability and droplet number. In general, the greater volume represents the external phase (Singh et al., 2017).

The Brownian motion associated with a surface charge of the droplets impairs the sedimentation, flocculation, and coalescence (Rai et al., 2018). The energy magnitude of the surfactant film around the droplet content defines the kinetic stability. In other words, it is the proximity hindrance by the presence of repulsive interactions acting among the droplets to the stability maintenance (steric and electrostatic) promoted by stabilizers as polymers, charged/nonionic surfactants, and polymeric surfactants (McClements, 2012). Further, low solubility between the phases is desirable to prevent the Ostwald ripening, because this is the main mechanism of instability for the nanoemulsion (Sharma et al., 2010).

The nanoemulsion stability can be identified during long-term storage (period of six months) by pursuing droplet size, polydispersity index and zeta potential due to the aggregation and the instability signs (coalescence, flocculation, Ostwald ripening and/or phase separation), as well as drug content, encapsulation efficiency (EE), and viscosity. In addition, it can be submitted to temperature modifications as cooling and heating cycles plus centrifugation test for short-term storage (Lovelyn and Attama, 2011; McClements, 2012; Rai et al., 2018).

6.3.4 NANOEMULSIONS APPLIED IN DERMAL DRUG DELIVERY

In comparison with conventional forms and due to the small droplet size, nanoemulsion accumulates on the skin. It gathers the modification of lipid bilayers and cellular arrangement and the molecule deposition on reservoirs, which enhance the molecule penetration into the skin (Mou et al., 2008). One example of a small molecule is the caffeine carried in the aqueous phase of a W/O nanoemulsion with oil phase composed of Transcutol HP, isopropanol, and Lauroglycol 90. The nanoemulsion was developed to treat skin cancer lesions and supplied a steady-state flux of caffeine through the skin (Shakeel and Ramadan, 2010). For macromolecules, nanoemulsion increased their entry and holding in the follicular opening to act on deeper skin lesions by the similar oily composition from nanoemulsion and skin sebum (Wu et al., 2001).

Two types of skin premalignant lesions encompassing the lentigo maligna and the actinic keratosis have undergone clinical trials in phases IV and I, respectively. Nanoemulsions associated with photodynamic therapy (PDT) were used aiming to increase the 5-aminolevulinic acid (ALA) entry into the skin layers, where the lesions were located, to prevent their development. Further, with the same molecule loaded into nanoemulsion and PDT, it was used to treat basal cell carcinoma and clear the lesions (Singh et al., 2017). For melanoma treatment, the acai oil was used as a photosensitizer in nanoemulsion mediated by PDT. The necrotic area in the tumor represented the high concentration reached by the acai oil and the effectiveness in the treatment of this lethal skin cancer (Monge-Fuentes et al., 2017). Furthermore, dacarbazine was topically applied from nanoemulsion, sized around 130 nm, on a region with melanoma tumor, which had its size reduced when compared to a conventional suspension measuring approximately 5 μm (Tagne et al., 2008).

Using dermal nanoemulsion, tacrolimus, an immunosuppressive agent, was topically administered to treat uveitis with improved safety and efficacy (Garg et al., 2013). The increased concentration of a microbicidal agent promoted by the nanoemulsion was efficient to reduce methicillin-resistant *Staphylococcus aureus* in infected wounds as well as inflammation (Cao et al., 2017). For the vitiligo management, flavonoids, and psoralen were loaded into nanoemulsion to posterior photoactivation by ultraviolet rays (Shivasaraun et al., 2019). The increased skin penetration was also observed for triptolide plus the skin hydration with participation of hair follicles and SC to the nanoemulsion. This delivery system improved the symptoms of eczema/dermatitis inflammation when compared to oral formulation with fewer gastrointestinal side effects (Yang et al., 2017).

In the search for more suitable and effective treatments for psoriasis, nanoemulsions have been developed. A synergic approach was proposed with tacrolimus and Kalonji oil (functional excipient) vehiculated into a nanoemulsion sized around 93 nm. The nanoemulsion produced 4.33-fold higher dermal bioavailability with a significant reduction in serum cytokines linked to psoriasis when compared to marketed formulations (Sahu et al., 2018). Another synergy with clobetasol propionate and calcipotriol had their concentrations maximized in the skin when delivered by the nanoemulsion (Kaur et al., 2017). Likewise, good anti-inflammatory effect was achieved with pioglitazone from a nanoemulsion due to the decrease in inflammatory cytokines such as IL-6, IL-1β and TNF-α. Nanoemulsion promoted high pioglitazone retention in the skin and ensured prolonged effect with good tolerability (Espinoza et al., 2019).

Regarding other skin inflammatory conditions, nanoemulsion with olive oil and capsaicin showed better tolerance to pain and less edema, which means greater analgesic and anti-inflammatory effects than those obtained with a commercial capsaicin cream (Ghiasi et al., 2019). Another condition reported is the oxidative stress with a great generation of reactive oxygen species (ROS). So, nanoemulsion can increase the level of protection against oxidative stress, as demonstrated with *Punica granatum* seed oil. This oil is rich in polyphenol, which when entrapped into the nanoemulsion did not cause hemolysis or phototoxicity, being a good option to treat skin after sunlight exposure (Baccarin et al., 2015). For dry skin and atopic dermatitis (AD), a positively charged nanoemulsion containing ceramide 3, cholesterol, and palmitic acid was produced. It was shown to result in higher skin hydration and elasticity as well as less erythema for several women with different ages after regular application of this nanoemulsion (Yilmaz and Borchet, 2006).

6.3.5 NANOEMULSIONS APPLIED IN TRANSDERMAL DRUG DELIVERY

Nanoemulsions are suitable to increase the solubility of poorly soluble drugs, and they can also be used to increase the bioavailability of transdermal systems, which expands their use for different pathologies. Nanoemulsion gel enhanced the skin permeation through the skin layers of fluvastatin and demonstrated new bone formation in the trabecular region of osteoporotic rat femurs (Kaur and Ajitha, 2019). The steady-state flux of aceclofenac and significant skin permeation was reached with a nanoemulsion composed of Labrafil, Triacetin, Transcutol P and Tween 80 (Shakeel et al., 2007).

Through the skin penetration routes, ketoprofen, an anti-inflammatory drug with gastrointestinal side effects, was delivered with a transdermal nanoemulsion, which had quick and increased penetration (Sakeena et al., 2010) and the same was achieved with meloxicam (Khurana et al., 2013). In a clinical trial, the osteoarthritis of the knee was treated with 3% diclofenac nanoemulsion to increase the analgesic activity and to decrease the daily pain with efficacy and tolerability (Singh et al., 2017).

The beta-adrenergic propranolol has been utilized to treat infantile hemangiomas. The antiproliferative effect of this drug is attributed to vasoconstriction, angiogenesis inhibition and cellular apoptosis induction, but its oral administration can bring systemic side effects. For this reason, a nanoemulsion with 26 nm droplet size provided propranolol retention on the skin, specifically in the dermis (site of action), and low biodistribution in the bloodstream. Thus, the propranolol vehiculated into the nanoemulsion was considered an effective transdermal delivery system (Marques et al., 2018).

The nanoemulsion remarkable properties were used for transcutaneous immunization as a different strategy to produce durable immune response. A solid nanoemulsion containing imiquimod was developed to induce high quality response of T cells against virus infections and to activate and to mature the skin-derived dendritic cells. This was achieved across the skin with the nanoemulsion and constituted an alternative vaccine due to the enhanced cutaneous permeation (Lopez et al., 2017).

Hypertension treatments have been managed with different drugs to decrease blood pressure. However, these drugs may present low bioavailability and high hepatic first-pass metabolism when administered by oral route. One example to overcome these limitations was the carvedilol delivered from nanoemulsion prepared by the spontaneous emulsification method. This nanoemulsion enhanced the carvedilol bioavailability through transdermal administration with higher flux value when compared to the oral marketed form (Pratap et al., 2012).

Another chronic disease treated by the transdermal delivery system is the diabetes. Glibenclamide, an oral hypoglycemic drug, was entrapped into a nanoemulsion in order to prevent hyperglycemia and other diabetes complications. The drug reached expressive flux through the skin due to high thermodynamic activity by the concentration of surfactant/co-surfactant from the nanoemulsion. The *in vivo* evaluation showed pronounced glibenclamide bioavailability from this transdermal nanoemulsion when compared to the oral suspension, thus, with better control of hyperglycemia and prevention of complications associated with diabetes (Wais et al., 2013).

6.4 MICROEMULSION

6.4.1 BASIC CONSIDERATIONS

Microemulsion can be described as a thermodynamically stable system, because the phases are not prone to separation, remaining stable for long periods. The composition consists of at least three compounds: two immiscible liquids – oil and water – with surfactants or co-surfactants to make an interfacial film that results in the formulation stability. However, water is not mandatory by having non-aqueous ME made of other immiscible liquids in water (Rakshit et al., 2019; Pankajkumar-Patel et al., 2019; Tung et al., 2019; Shukla et al., 2018; Callender et al., 2017; Nastiti et al., 2017).

In general, the dispersed phase is in minor amount and creates the microemulsion particles, which indicate the microemulsions classification according to Winsor, as follows:

- **Type I:** Particles are formed by oil dispersed in aqueous media. The system is biphasic with an excess of oil phase and the surfactant is solubilized in the aqueous phase. Microemulsion is classified as O/W;
- **Type II:** Particles are dispersed in oil including surfactants, in a reverse phase of water, constituting the W/O biphasic system with an excess of W/O emulsion and aqueous phase lower excess;

- **Type III:** Surfactants are in a middle phase between the oil and aqueous phases, in which both lack surfactants, forming a three-phase system with an excess of oil phase, a middle phase with bicontinuous microemulsion and aqueous phase lower excess. Both aqueous and oil phases are surfactant-deficient;
- **Type IV:** This microemulsion is an extension of III, where there is higher surfactant concentration in the middle phase, resulting in a monophasic system.

Depending on the surfactant choice, the type of microemulsion can be easily modified (Shiri et al., 2019; Goswami et al., 2019; Callender et al., 2017).

Microemulsion represents a useful drug carrier system for several purposes encapsulating different actives and drugs, including hydrophilic and lipophilic compounds. The microemulsion advantages and disadvantages are found in Figure 6.2.

6.4.2 FORMULATION DEVELOPMENT

Despite the name, microemulsions are spontaneously formed with particle size between 10 and 100 nm (Goswami et al., 2019; Callender et al., 2017; Nastiti et al., 2017; Muzaffar et al., 2013), which is much smaller than macro and nanoemulsions. This happens because of the surfactant density on the film interface, in which the harder the film is, the higher the packing density will be and, consequently, the smaller the particle formed (Callender et al., 2017; Muzaffar et al., 2013). The spontaneous formation occurs because of the optimal proportions of compounds and temperature, being the temperature range specific to the chemical composition. Also, in practice, low energy is necessary to form the microemulsion, for example, gentle stirring or heat (Nastiti et al., 2017). Figure 6.3 show the microemulsion formation.

The choice of components and their proportions are critical to form a stable system with adequate particle sizes. Usually, the oil phase includes fatty acids, alcohols such as isopropyl myristate, isopropyl palmitate; medium-chain triglycerides, terpenes, and other permeation enhancers. These components can be used in isolation or combined. The aqueous phase may include sodium chloride and buffers, preservatives, and, also, permeation enhancers. Viscosity agents such as Carbopol ®, Aerosil ® and gelatin can be incorporated to reduce the flow and to generate a consistent final product (Nastiti et al., 2017).

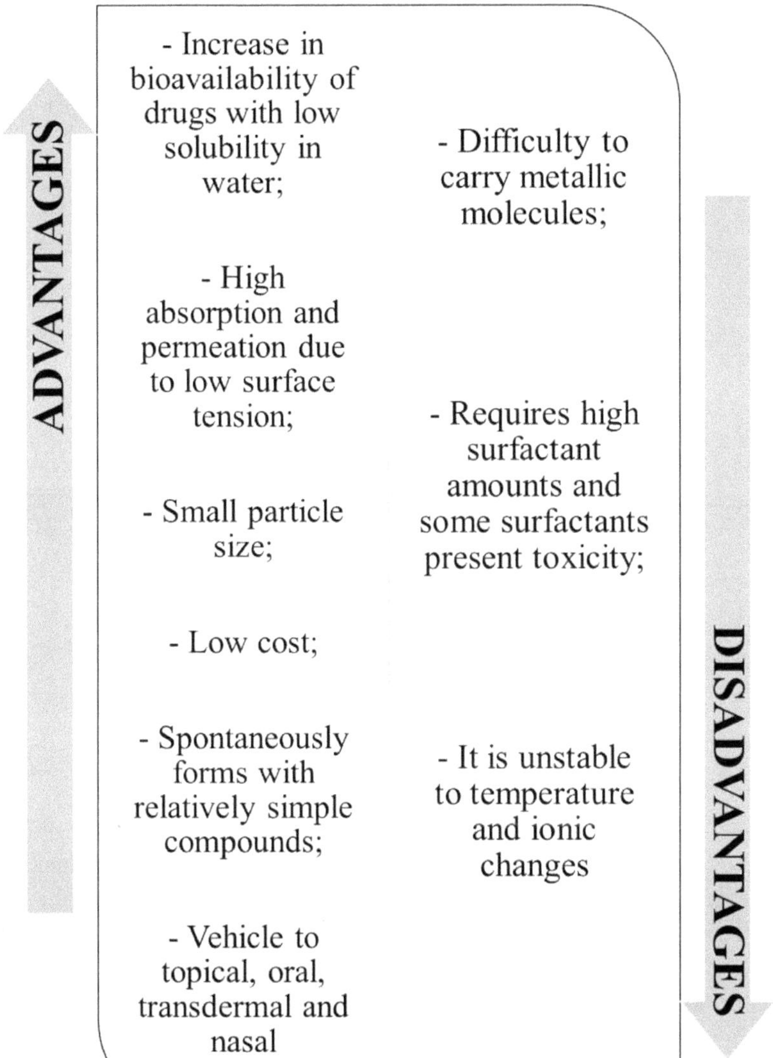

FIGURE 6.2 Advantages and disadvantages of microemulsions.

Source: Goswami et al. (2019); Shukla et al. (2018); Callender et al. (2017); and Nastiti et al. (2017).

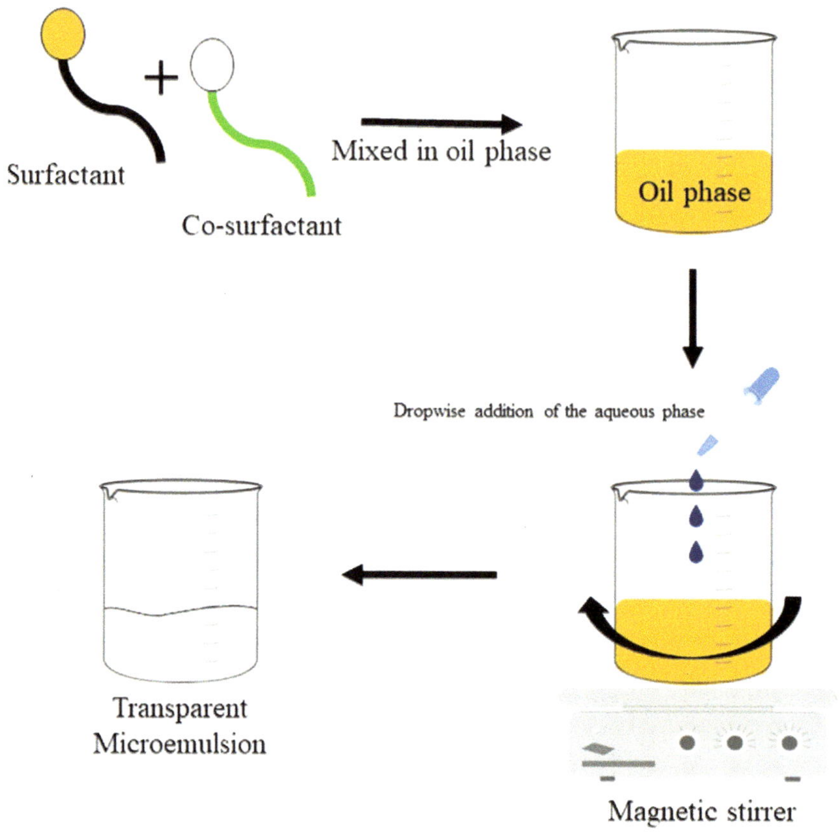

FIGURE 6.3 Microemulsion preparation method.
Source: Adapted from Goswami et al. (2019).

A wide range of surfactants and co-surfactants are used, considering that an adequate combination is required to reduce effectively the interfacial tension and to produce a stable system with appropriate particle sizes and few skin irritations. The surfactants most used are Tween ® (polysorbates), Cremophor ® and Transcutol ®. Lecithin, an amphiphilic compound, has been investigated as an ideal surfactant because it is a natural compound with low irritation degree. Co-surfactants, usually, are short and medium chain alcohols and polyglyceryl derivatives, including ethanol, isopropanol, isopropyl myristate and propylene glycol (Nastiti, 2017). For skin formulations, the nonionic are the preferred surfactants due to minor skin irritation (Nastiti et al., 2017; Callender et al., 2017).

Although microemulsion is easy to produce, there are some factors that affect its formation such as:

1. **Surfactant Properties, Oil Phase and Temperature:** The concentration of surfactants is responsible for the phases behavior. An increase in concentration will decrease the degree of polar group dissociation, resulting in a type II microemulsion. However, the system dilution with water can increase the dissociation of the polar groups, which results in a type I microemulsion. Similarly, the oil phase has the penetration ability that can influence curvature, providing the swelling of the surfactant tail. Regarding temperature, its increase can result in a W/O system and the decrease in O/W system, therefore, a medium temperature should be kept, but, with oil or water excess a bicontinuous structure will be formed.
2. **Chain Size and Kind of Cosurfactant:** Small chain of cosurfactant can result in positive curvature and create type I microemulsion, and the addition of long-chain surfactants can create a type II microemulsion.
3. **Nature of Oil and Surfactant Hydrophobic Chain Length:** The nature of oil as the aromatic ring can increase the transition phase from O/W to W/O, while a surfactant with hydrophobic long chain can result in O/W–W/O ME as bicontinuous phase (Goswami et al., 2019).

There are significant differences in microemulsion properties when compared with nanoemulsions and emulsions. Microemulsion presents thermodynamic stability, which prevents phase separation. Other emulsions are cloudy, while microemulsions are isotropic, transparent, and limpid, constituted by one phase due to the small particle sizes. Microemulsion does not reflect the light and it is not visible in optical microscopy. Moreover, a difference related with preparation methods exists, because emulsions usually require high energy to be prepared, while microemulsions do not (Subongkot, 2019; Tung et al., 2019; Ferreira et al., 2019; Shiri et al., 2019; Pankajkumar-Patel et al., 2019; Shukla et al., 2018; Callender et al., 2017; Nastiti et al., 2017; Muzaffar et al., 2013).

In addition, microemulsions can be formulated by trial and error or ternary diagram seeking different techniques to characterize them such as: interfacial tension, particle size, zeta potential, conductivity, viscosity, structure, drug release and stability (Tung et al., 2019; Callender et al., 2017; Nastiti et al., 2017).

Interfacial tension is used to evaluate the surfactant's ability to emulsify and reduce the tension between the oil and aqueous phases, detecting which factors should be modified to reduce this tension. Particle size can be determined by DLS or PCS that is used to analyze the fluctuations of incident laser intensity through the Brownian motion of particles. This technique will guide the kind of emulsion formed such as macro, nano or microemulsion. Zeta potential gives the superficial charge and indicates the behavior of electrostatic interaction between the microemulsion particles. Conductivity is used to confirm the kind of microemulsion, since water is a great ions conductor, high conductivity will show the water amount in the system. Viscosity can indicate microemulsion stability, based on diffusion constant and particle-particle interactions (Tung et al., 2019; Callender et al., 2017; Nastiti et al., 2017).

Furthermore, ternary phase pseudodiagrams are built to indicate the regions of microemulsion as a function of composition in water, oil, and surfactants (Nastiti et al., 2017). This enables the identification of more suitable compounds and their relative proportions to obtain physically stable systems. An ideal formulation of microemulsion corresponds to a system with the desired physicochemical properties such as viscosity, stability, and pH, as well as a specific area of the dispersed phase. The construction of these diagrams requires the delineation of a large number of samples, resulting in a lengthy task, which makes it necessary to search for alternative methods that circumvent related issued problems while being as efficient as the traditional method (da Silva et al., 2015).

The formulation release test can be performed using diffusion modification in Franz cells, where the receptor compartment consists of a buffer solution and the donor compartment is fixed with a cellophane membrane containing the microemulsion. At specific time intervals, a sample is taken from the receptor medium and the same amount is replenished to maintain the sink conditions and the molecule release is analyzed with specific wavelength UV spectroscopy or other suitable methods (Goswami et al., 2019).

6.4.3 STABILITY ISSUES

For the microemulsion stability tests, accelerated stability and long-term stability tests may be performed. For long-term stability, the formulation is maintained under three different conditions – room temperature, refrigeration (4–8°C), and higher temperature (50 ± 2°C) – and afterwards evaluated. Accelerated stability study can be performed by centrifugation, freeze/thaw cycle, and heating/cooling for thermodynamic stability of the formulation (Goswami et al., 2019).

For example, Mouri et al. (2016) used the centrifugation technique to verify the stability of a microemulsion system. Centrifugation was carried out at 8,000 rpm for 15, 30, and 45 minutes to test the stability system against the high gravitational force to produce phase separation. Regarding temperature, the microemulsion was kept at room temperature and 40°C for 3 months with periodic checking of the appearance and those authors concluded that there was no phase separation, turbidity, or precipitation of the drug even after centrifugation for 45 minutes at 8,000 rpm, indicating an optimum system stability.

Noelia et al. (2017) used DLS to evaluate particle size and polydispersity index before and after microemulsion undergoing stress conditions: changes in pH, ionic strength, centrifugation, and temperature. All microemulsions were evaluated in duplicate with 10 measurements. The microemulsions studied were highly stable and showed no significant differences in droplet size or distribution when pH, ionic force and centrifugation varied. In all cases, the average particle size was less than 15 nm. In addition, MEs showed good thermal stability even during cold storage without change in size, despite the thermal dependence on the emulsifier rheological properties. During freezing, a slight increase in size (less than 21 nm) was observed while the optical transparent appearance was maintained, the microemulsions were just destabilized during heating at 150°C.

Ferreira et al. (2019) verified the stability of the clotrimazole and itraconazole incorporated into microemulsion formulations by placing them separately in an amber-capped vial and stored at 40°C ± 2°C and 5 to 8 ±3°C for 3 months, with sampling at 0, 30, 60, and 90 days. Samples were evaluated for drug content using the developed and validated HPLC method. The accelerated stability results showed that a clotrimazole derivative was formed, as well as a possible interaction between itraconazole and benzyl alcohol when the microemulsion was conditioned at elevated temperatures (40°C). On the other hand, studies performed at 5°C showed that the microemulsion was stable for at least 3 months, since there was no degradation and no peaks were observed in the HPLC analysis, which allows inferring that it was possible to guarantee the stability of the formulation under cooling.

6.4.4 MICROEMULSIONS APPLIED TO DERMAL DRUG DELIVERY

The skin has an intense barrier function, allowing only a few substances to penetrate under certain conditions. Considering the potential of microemulsions to promote skin permeation of drugs, there are two possible mechanisms that guide the optimization of drug permeation:

- Due to the large amount of surfactants in the formulation, microemulsions can interact with the SC temporarily disrupting its lipid bilayer. As a result, lipids, and keratinocytes go from the ordered crystalline form to the disordered liquid form and the skin permeation is increased at a specific rate, facilitating penetration of the molecules due to pore formation or some paths to cross (Shukla et al., 2018; Damasceno et al., 2011; Oliveira et al., 2004).
- The physicochemical properties of the constituents used and the ratio of their components result in another mechanism. Since the polar and nonpolar components contribute to the increase in the microemulsion thermodynamic activity by reducing the interfacial tension of the skin surface and the drug solubilized in the vehicle due to close contact with the epithelium lipids, resulting in faster permeation through the deeper layers of the skin (Shukla et al., 2018; Resende, 2013).

The interaction with the SC and increased penetration of the molecules carried by the microemulsion can be explained by the lipid-protein-partitioning theory. According to that theory, permeation through the skin depends on the interaction of the drug with the intercellular compartment of the lipid bilayer; interaction with keratinocytes and interaction of the excess of co-solvent or penetration enhancer with the lipid bilayer of SC. This can result in temporary disruption of the skin barrier and promotion of drug transfer through the skin. Once the barrier is weakened, drug diffusion can occur in three ways: I) Intercellular: passage between cells, ideal for hydrophilic substances; II) intracellular: drug passes through cells, ideal for lipophilic substances; and iii) follicular: drug is transferred via hair follicle or other cutaneous attachments (Shukla et al., 2018).

Microemulsions can be developed for the delivery of many skin-acting compounds for either dermatological, cosmetic, or transdermal effects. The choice of a microemulsion rather than conventional pharmaceutical forms is due to the increased action of the components on the skin, the phase structure and particle size. Since the presence of sebum in the hair follicles is similar to the oil phase, the surfactants and alcohol used in the formulation facilitate the transfollicular transport for both hydro and lipophilic compounds (Nastiti et al., 2017).

Melanoma, basal cell carcinoma and squamous cell carcinoma are the three types of skin cancer. Melanoma is the most aggressive and most resistant to chemotherapy. In addition, the substances used so far have limitations, which affect their therapeutic efficacy, however, microemulsions can overcome these issues. Nonionic surfactants are often used in microemulsions to deliver the chemotherapeutics to be well tolerated by the skin. An example is microemulsions carrying temozolomide acid hexyl ester, a prodrug, with low solubility and instability, used in anticancer therapy. The literature indicates that drug-loaded into topical microemulsions in skin cancer treatment can deliver high dosage of drugs in a sustained manner with better effectiveness, protection of the loaded actives against degradation and reduced skin irritation when compared to conventional formulations. In the same way, phytochemical compounds incorporated into microemulsions may improve chemoprevention strategies in the melanoma treatment, which makes the dermal therapy more useful and acceptable (Ravikumar and Tatke, 2019).

Another example was achieved with the Imiquimod, a topical immunostimulant, used for the treatment of actinic keratosis and basal cell carcinoma. This drug also presents low solubility in most pharmaceutical excipients and very low skin penetration property. The microemulsion system developed showed greater skin deposition than the commercial formulation, despite the lower drug content carried into this system. The study also showed the importance of the cosurfactant in the microemulsion structure and the increased performance of this system, as it facilitated drug solubilization in the microemulsion and improved drug absorption capacity through the skin, as well as the formation of a clear and viscous microemulsion. The microemulsion was shown to be a promising vehicle for the imiquimod delivery into the skin and its composition with isostaric acid could benefit the biological activity, while the flexibility conferred by cosurfactant was required for efficient skin drug deposition (Pescina et al., 2018).

The role of immune responses, specifically, dendritic skin cells from the dermis has aroused great interest in research. Dermal administration of allergens could target these cells to create immunotolerance and many adverse effects could be prevented. Thus, Kiselmann et al. (2018) created a topical microemulsion to deliver proteogenic allergens in the EP, which were able to reach the dendritic cells of the dermis. They evaluated the physicochemical properties as well as the influence of different preservatives on the microemulsion stability. *In vitro* and *in vivo* skin penetration assays indicated the protective efficacy of the dermal allergen-specific immunotherapy. The microemulsion developed had low surfactant concentration and was great for topical administration of proteogenic allergens. *In vivo* trials proved the skin penetration studies and indicated the immunomodulatory and allergen protective effects of the microemulsion investigated.

Another work verified the dermal application for peptides with therapeutic and cosmetic interest. Skin penetrations of these peptides are challenging because of their high molecular weight, hydrophobicity, polarity, and susceptible to enzymatic degradation. A microemulsion based on W/O system was developed for carrying GEKG tetrapeptide plus skin penetration enhancers. The study found that the standard cream allowed low penetration of GEKG molecules into the skin due to limited physicochemical properties of the peptide for dermal penetration. However, the microemulsion formulations with enhancers led to an increased penetration and the system rapidly transported the peptide into the skin, particularly, in the dermis, the target compartment as well as through the skin. These results indicated that the microemulsion was excellent to carry hydrophobic peptides and suitable for cosmetic and pharmaceutical therapies (Sommer et al., 2018).

Zhang and Michniak-Kohn (2018) investigated microemulsion and microemulsion-gel formulations for dermal delivery of clotrimazole, an antifungal agent. The microemulsion showed different microstructures by containing isopropyl myristate as the oil phase, Labrasol® and Cremophor EL® as surfactant and cosurfactant and water. Dermal delivery of the drug was evaluated at different concentrations in the dermal tissue compared with control, which evidenced the significant effect of the microemulsion to improve the dermal retention. The highest concentration was observed for microemulsion based on O/W, suggesting that the microemulsion microstructure is an important feature to improve the efficiency of delivery. Microemulsion gel formulations prepared by incorporating 1.0% (w/w) Carbopol 980® showed higher dermal concentration of clotrimazole. Microemulsions were shown to produce up to 2.4 times better results than the commercial cream used. Their results provide more insight for the understanding of the microemulsion permeation enhancement mechanism and, also, a significant guide for the formulations applied to dermal drug delivery.

Several topical fungal infections affect countless people in tropical areas, especially those caused by *Tricophyton*, *Microsporum*, and *Epidermophyton*, which cause skin lesions on different parts of the body. Treatments are used to combat these fungi, but the effectiveness of the treatment is limited due to the poor penetration, restricted to the SC, resulting in a longer and more complicated treatment for the patient. Thus, a thickening agent can be used to form a microemulsion gel to increase skin adhesion and to carry antifungal actives through the SC by the skin penetration improved ability by the microemulsion. For example, Clotrimazole in microemulsion had an effective action without causing any signs of irritation. Another example was the Fluconazole-loaded into microemulsion system that improved permeation into the skin when compared to a saturated solution and had a greater antifungal effect against *C. albicans* (Boonme et al., 2016).

Microemulsions can carry many natural plant-derived actives for both medicinal and cosmetic applications. Microemulsions have been used to carry several topical phytochemicals for the purpose of increasing skin permeation and reducing skin irritation, such as *Triptolide*, a derived substance from Chinese medicine, which is used for various benefits such as immunosuppressive and anticarcinogenic treatment, but, with great toxicity. The substance delivery through the microemulsion caused a more controlled and sustained release when compared to the free compound solution and this led to a reduction in topical irritation (Boonme et al., 2016).

Savic et al. (2017) developed a system to improve the penetration of tacrolimus into the skin and searched for the use of lecithin-based substances as potentially non-irritating agents. The microemulsion proposed had in its composition ethanol, isopropanol, and/or propylene glycol as cosurfactants, it had varied caprylic/capric triglycerides and propylene glycol monocaprylate as oil phase. The influence of excipients on the size of the microemulsion region in pseudo-ternary phase diagrams and their ability to form different types of microemulsions was evaluated. The microemulsion structure and tacrolimus location were investigated using electronic paramagnetic resonance spectroscopy. In addition, stability studies showed no change in tacrolimus content in a period of one year of storage at room temperature. Further, *ex vivo* penetration studies with porcine ear skin identified the superiority of two reference-related microemulsions, leading to the development of microemulsions as promising carriers for tacrolimus dermal delivery.

6.4.5 MICROEMULSIONS APPLIED TO TRANSDERMAL DRUG DELIVERY

Kupper et al. (2017) developed a transdermal microemulsion W/O hydrogel system for two natural polymers: collagen and hyaluronic acid, with low and high molecular weight. The systems were evaluated for conductivity, phase inversion studies, particle size, polydispersity index and rheological properties. The results showed that the aqueous phase structure in the microemulsion is controlled by the ratio and amount of the surfactant mixture (sorbitan ester derivatives). Also, the results showed that depending on the weight of the collagen molecule and its surface activity, there was an influence on the particle size of the microemulsions. While, for the hyaluronic acid with high molecular weight, due to its water-binding capacity, the hydrogel formation altered the microemulsion rheological properties, thus providing the formulation with a viscous consistency.

Lin et al. (2018) developed a new microemulsion W/O for a natural Wen-Luo-Tong (WLT) extract containing mainly icariine, yellow hydroxysaflor A (HSYA) and gallic acid to be applied to the skin as a potential treatment for peripheral neuropathy. The optimized formulation consisting of 10% (w/w) water extract solution, 60% isopropyl myristate, 30% (w/w) propylene glycol, significantly, increased the permeation with cumulative amounts of HSYA, icariin, and gallic acid compared to the controls (water extract solution). This new formulation also increased the number of components that penetrated through the skin of rats. Around 10 components were detected in the Franz cell receptor solution using a UPLC-Q-TOF/MS system after application of the microemulsion formulation for 24 h on the skin *in vitro*. However, only one component was detected after the control was applied.

Mishra et al. (2016) prepared and evaluated the microemulsion and microemulsion-based gel formulations of Zaltoprofen, a non-steroidal anti-inflammatory, for transdermal administration. Pseudo-ternary phase diagrams were used to prepare the formulations. Permeation studies were performed using excised pigskin. The composition of the optimized formulation was 1% (w/w) zaltoprofen, 20% (w/w) Capryol 90, 50% (w/w) (2:1, Cremophor RH 40 and Transcutol P). The optimized formulation showed a 22.11 nm particle size, 0.251 polydispersity index and –11.4 mV zeta potential. The microemulsion gel was considered safe in the study of skin irritation. The microemulsion gel activity was observed in the hot plate and rat paw edema tests. Thus, the results suggested that microemulsion could be a promising formulation for transdermal administration of Zaltoprofen.

Another study investigated a microemulsion for transdermal administration of calcium phenoprofen to eliminate its oral gastrointestinal adverse effects. The microemulsion was prepared using the water titration method with oleic acid as oil phase, Tween 80 as surfactant and propylene glycol as a cosurfactant. Oleic acid was selected as an oil phase due to its good solubilization capacity. The selected microemulsion formulations were characterized by optical birefringence, transmission electron microscopy, pH, percentage transmittance, electron conductivity, drug content, droplet size, rheological properties, and stability evaluation. *In vitro* study of calcium phenoprofen release from microemulsions through membrane synthesis and hairless rat skin was developed. The optimized formula presented the highest transdermal permeation and highest skin permeation rate. The rat paw edema of the optimized formula was highly significant ($P < 0.001$) when compared to calcium phenoprofen single gel. The microemulsion was observed to be a promising system for transdermal administration of calcium phenoprofen (Farghaly et al., 2017).

6.5 CONCLUSION

Emulsions, nanoemulsions, and microemulsions have been applied as carrier systems for the transdermal and dermal delivery of therapeutic agents. These lipid-based systems mainly vary in droplet size range, composition, stability, and appearance, resulting in the possibility of application as drug carriers with different physicochemical properties and therapeutic classes. A vast number of excipients are available for the formulation of these colloidal systems, with desired characteristics for TDDS. The *in vitro* and *in vivo* potentials of these systems influence the performance of the therapeutic agent administered by dermal and transdermal routes and have been established by numerous investigations, according to the studies presented.

KEYWORDS

- **colloidal systems**
- **hydrophilic-lipophilic balance**
- **nanotechnology**
- **skin pathologies**
- **systemic absorption**
- **topical application**

REFERENCES

Albert, C.; Beladjine, M.; Tsapis, N.; Fattal, E.; Agnely, F.; Huang, N. Pickering emulsions: Preparation processes, key parameters governing their properties and potential for pharmaceutical applications. *Journal of Controlled Release,* **2019**, *309*, 302–332.

Ali, M. S.; Alam, M. S.; Alam, N.; Siddiqui, M. R. Preparation, characterization and stability study of dutasteride loaded nanoemulsion for treatment of benign prostatic hypertrophy. *Iranian Journal of Pharmaceutical Research*, **2014**, *13*(4), 1125–1140.

Alkaline, A. Z.; McCrudden, M. T. C.; Donnelly, R. F. Transdermal drug delivery: Innovative pharmaceutical developments based in disruption of the barrier properties of the stratum corneum. *Pharmaceutics*, **2015**, *7*, 438–470.

Alvarez-Figueroa, M. J.; Blanco-Méndez, J. Transdermal delivery of methotrexate: Iontophoretic delivery from hydrogels and passive delivery from microemulsions. *International Journal of Pharmaceutics*, **2001**, *215*, 57–65.

Arora, R.; Aggarwal, G.; Harikumar, S. L.; Kaur, K. Nanoemulsion based hydrogel for enhanced transdermal delivery of ketoprofen. *Advances in Pharmaceutics*, **2014**, *2014*, 1–12.

Baccarin, T.; Mitjans, M.; Lemos-Senna, E.; Vinardell, M. P. Protection against oxidative damage in human erythrocytes and preliminary photo safety assessment of Punica granatum seed oil nanoemulsions entrapping polyphenol-rich ethyl acetate fraction. *Toxicology In Vitro,* **2015**, *30*, 421–428.

Boonme, P.; Kaewbanjong, J.; Amnuaikit, T.; Andreani, T.; Silva, A. M.; Souto, E. B. Microemulsion and microemulsion-based gels for topical antifungical therapy with phytochemicals. *Current Pharmaceutical Design*, **2016**, *22*, 4257–4263.

Callender, S. P.; Mathews, J. A.; Kobernyk, K.; Wettig, S. D. Microemulsion utility in pharmaceuticals: Implications for multi-drug delivery. *International Journal of Pharmaceutics*, **2017**, *526*, 425–442.

Cao, Z.; Spilker, T.; Fan, Y.; Kalikin, L. M.; Ciotti, S.; LiPuma, J. J.; Makidon, P. E.; Wilkinson, J. E.; Baker, J. R. Jr.; Wang, S. H. Nanoemulsion is an effective antimicrobial for methicillin-resistant Staphylococcus aureus in infected wounds. *Nanomedicine (Lond.)*, **2017**, *12*, 1177–1185.

Chao, Y.; Huang, C.; Fu, L.; Huang, Y.; Tsai, Y.; Wu, P. The effect of submicron emulsion systems on transdermal delivery of kaempferol. *Chemical and Pharmaceutical Bulletin,* **2012**, *60*, 1171–1175.

Da Silva, J. D. F.; Da Silva, Y. P.; Piatnicki, C. M. S.; Bockel, W. J.; Mendonça, C. R. B. Microemulsões: Componentes, características, potencialidades em química de alimentos e outras aplicações. *Quimica Nova*, **2015**, *38*, 1196–1206.

Damasceno, B. P. G. L.; Silva, J. A.; Oliveira, E. E.; Silveira, W. L. L.; Araújo, I. B.; Oliveira, A. G.; Egito, E. S. T. Microemulsão: Um promissor carreador para moléculas insolúveis. *Revista de Ciências Farmacêuticas Básica e Aplicada [Microemulsion: a promising carrier for insoluble molecules. Journal of Basic and Applied Pharmaceutical Sciences]*, **2011**, *32*, 9–18.

Delgado-Charro, M. B.; Iglesias-Vilas, G.; Blanco-Mendez, J.; Lopez-Quintela, M. A.; Marty J. P.; Guy, R. H. Delivery of a hydrophilic solute through the skin from novel microemulsion systems. *European Journal of Pharmaceutics and Biopharmaceutics*, **1997**, *43*, 37–42.

Devarajan, V.; Ravichandran, V. Nanoemulsion: As modified drug delivery tool. *International Journal of Comprehensive Pharmacy*, **2011**, *4*(1), 1–6.

Escobar-Chávez, J. J.; Díaz-Torres, R.; Rodríguez-Cruz, I. M.; Domínguez-Delgado, C. L.; Morales, R. S.; Ángeles-Anguiano, E.; Melgoza-Contreras, L. M. Nanocarriers for transdermal drug delivery. *Research and Reports in Transdermal Drug Delivery*, **2012**, *1*, 3–17.

Espinoza, L. C.; Silva-Abreu, M.; Calpena, A. C.; Rodríguez-Lagunas, M. J.; Fábrega, M. J.; Garduño-Ramírez, M. L.; Clares, B. Nanoemulsion strategy of pioglitazone for the treatment of skin inflammatory diseases. *Nanomedicine*, **2019**, *19*, 115–125.

Farghaly, D. A.; Aboelwaka, A. A.; Hamza, M. Y.; Mohamed, M. I. Microemulsion for topical delivery of fenoprofen calcium: *In-vitro* and *in-vivo* evaluation. *Journal of Liposome Research.*, **2017**, 126–136.

Ferreira, P. G.; Lima, C. G. S.; Noronha, L. L.; Moraes, M. C.; Silva, F. C.; Viçosa, A. L.; Futuro, D. O.; Ferreira, V. F. Development of a method for the quantification of clotrimazole and itraconazole and study of their stability in a new microemulsion for the treatment of aporotrichosis. *Molecules*, **2019**, *24*, 1–15.

Garg, V.; Jain, G. K.; Nirmal, J.; Kohli, K. Topical tacrolimus nanoemulsion, a promising therapeutic approach for uveitis. *Medical Hypotheses*, **2013**, *81*, 910–904.

Gershkovich, P.; Wasan, K. M.; Barta, C. A. A review of the application of lipid-based systems in systemic, dermal/transdermal, and ocular drug delivery. *Critical Reviews in Therapeutic Drug Carrier Systems,* **2008**, *26*, 545–584.

Ghiasi, Z.; Esmaeli, F.; Aghajani, M.; Ghazi-Khansari, M.; Faramarzi, M. A.; Amani, A. Enhancing analgesic and anti-inflammatory effects of capsaicin when loaded into olive oil nanoemulsion: An *in vivo* study. *International Journal of Pharmaceutics*, **2019**, *559*, 341–347.

Goswami, P.; Choudhury, A.; Dey, B. K. Microemulsion–A potential carrier for improved bioavailability. *International Journal of Pharmaceutical & Biological Archives*, **2019**, *10*, 69–77.

Gupta, A.; Eral, H. B.; Hatton, T. A.; Doyle, P. S. Nanoemulsions: Formation, properties and applications. *Soft Matter,* **2016**, *12*(11), 2826–2841.

Hu, Y. T.; Ting, Y.; Hu, J. Y.; Hsieh, S. C. Techniques and methods to study functional characteristics of emulsion systems. *Journal of Food and Drug Analysis,* **2016**, *25*, 16–26.
Iqbal, B.; Ali, J.; Baboota, S. Recent advances and development in epidermal and dermal drug deposition enhancement technology. *International Journal of Dermatology,* **2018**, *57*, 646–660.
Jain, A. K.; Thareja, S. *In vitro* and *in vivo* characterization of pharmaceutical nanocarriers used for drug delivery. *Artificial Cells, Nanomedicine and Biotechnology*, **2019**, *47*, 524–539.
Jaiswal, M.; Dudhe, R.; Sharma, P. K. Nanoemulsion: An advanced mode of drug delivery system. *3Biotech*, **2015**, *5*, 123–127.
Jenning, V.; Gysler, A.; Schafer-Korting, M.; Gohla, S. H. Vitamin A loaded solid lipid nanoparticles for topical use: Occlusive properties and drug targeting to the upper skin. *European Journal of Pharmaceutics and Biopharmaceutics*, **2000**, *49*, 211–218.
Kale, S. N.; Deore, S. L. Emulsion microemulsion and nanoemulsion: A review. *Systematic Reviews in Pharmacy,* **2017**, 8, 39–47.
Kaur, A.; Katiyar, S. S.; Kushwah, V.; Jain, S. Nanoemulsion loaded gel for topical co-delivery of clobitasol propionate and calcipotriol in psoriasis. *Nanomedicine,* **2017**, *13*, 1473–1482.
Kaur, R.; Ajitha, M. Transdermal delivery of fluvastatin loaded nanoemulsion gel: Preparation, characterization and *in vivo* anti-osteoporosis activity. *European Journal of Pharmaceutical Sciences*, **2019**, *136*, 104956.
Khan, B. A.; Akhtar, N.; Khan, H. M. S.; Waseem, K.; Mahmood, T.; Rasul, A.; Iqbal, M.; Khan, H. Basics of pharmaceutical emulsions: A review. *African Journal of Pharmacy and Pharmacology,* **2011**, *5*, 2715–2725.
Khurana, S.; Jain, N. K.; Bedi, P. M. Nanoemulsion based gel for transdermal delivery of meloxicam: Physic-chemical, mechanistic investigation. *Life Sciences*, **2013**, *92*, 383–392.
Kiselmann, C.; Dobler, D.; Schmidts, T.; Eicher, A. C.; Möbs, C.; Pfützner, W.; Runkel, F. Development of a skin-friendly microemulsion for dermal allergen-specific immunotherapy. *International Journal of Pharmaceutics*, **2018**, *550*, 463–469.
Kotyla, T.; Kuob, F.; Moolchandani, V.; Wilson, T.; Nicolosi, R. Increased bioavailability of a transdermal application of a nano-sized emulsion preparation. *International Journal of Pharmaceutics*, **2008**, *347*, 144–148.
Kreilgaard, M. Dermal pharmacokinetics of microemulsion formulations determined by *in vivo* microdialysis. *Pharmaceutical Research*, **2001**, *18*, 367–373.
Kreilgaard, M.; Pedersen, E. J.; Jaroszewski, J. W. NMR characterization and transdermal drug delivery potential of microemulsion systems. *Journal of Controlled Release*, **2000**, *69*, 421–433.
Kupper, S.; Klosowska-Chomiczewska, I.; Szumala, P. Collagen and hyaluronic acid hydrogel in water-in-oil microemulsion delivery systems. *Carbohydrate Polymers*, **2017**, *175*, 347–354.
Lehmann, L.; Keipert, S.; Gloor, M. Effects of microemulsions on the stratum corneum and hydrocortisone penetration. *European Journal of Pharmaceutics and Biopharmaceutics*, **2001**, *52*, 129–136.
Li, W.; Chen, H.; He, Z.; Han, C.; Liu, S.; Li, Y. Influence of surfactant and oil composition on the stability and antibacterial activity of eugenol nanoemulsion. *LWT Food Science and Technology,* **2015**, *62*, 39–47.

Lin, H.; Michniak-Kohn, B.; Xia, Z.; Xu, L.; Kang, Q.; Chen, C.; Ma, S.; Wu, Q. Transdermal delivery of compounds with different lipophilicity and molecular weight from W/O microemulsions analyzed by UPLC-Q-TOF/MS and LC-MS/MS. *Current Drug Delivery*, **2018**, *15*, 1009–1019.

Lopez, P. A.; Denny, M.; Hartmann, A. K.; Alflen, A.; Probst, H. C.; Von, S. E.; Tenzer, S.; Schild, H.; Stassen, M.; Langguth, P.; Radsak, M. P. Transcutaneous immunization with a novel imiquimod nanoemulsion induces superior T cell responses and virus protection. *Journal of Dermatological Science*, **2017**, *87*, 252–259.

Lovelyn, C.; Attama, A. A. Current state of nanoemulsions in drug delivery. *Journal of Biomaterials and Nanobiotechnology*, **2011**, *2*, 626–639.

Lu, G. W.; Gao, P. Emulsions and microemulsions for topical and transdermal drug delivery. *Chapter 3 in Handbook of Non-Invasive Drug Delivery Systems*, **2010**.

Manikkath, J.; Sumathy, T. K.; Manikkath, A.; Mutalik, S. Delving deeper into dermal and transdermal drug delivery: Factors and mechanisms associated with nanocarrier-mediated strategies. *Current Pharmaceutical Design,* **2018**, *24*, 3210–3222.

Marques, T. Z. S.; Santos-Oliveira, R.; Siqueira, L. B. O.; Cardoso, V. S.; Freitas, Z. M. F.; Barros, R. C. S. A.; Villa, A. L. V.; Monteiro, M. S. S. B.; Santos, E. P.; Ricci-Junior, E. Development and characterization of a nanoemulsion containing propranolol for topical delivery. *International Journal of Nanomedicine*, **2018**, *13*, 2827–2837.

Marti-Mestres, G.; Nielloud, F. Emulsions in health care applications–An overview. *Journal of Dispersion Science and Technology*, **2002**, *23*(1–3), 419–439.

Mason, T. G.; Wilking, J. N.; Meleson, K.; Chang, C. B.; Graves, S. M. Nanoemulsions: Formation, structure, and physical properties. *Journal of Physics: Condensed Matter,* **2006**, *18*, R635–R666.

McClements, D. J. Nanoemulsion versus microemulsion: Terminology, differences, and similarities. *Soft Matter*, **2012**, *8*, 1719–1729.

Menaa, F. Emulsions systems for skin care: From macro to nano-formulations. *Journal of Pharmaceutical Care & Health Systems,* **2014**, *1*, 1, 2.

Mishra, R.; Prabhavalkar, K. S.; Bhatt, L. K. Preparation, optimization, and evaluation of Zaltoprofen-loaded microemulsion and micro emulsion-based gel for transdermal delivery. *Journal of Liposome Research,* **2016**, *26*, 297–306.

Mojeiko, G.; Brito, M. De; Salata, G. C.; Lopes, L. B. Combination of microneedles and microemulsions to increase celecoxib topical delivery for potential application in chemoprevention of breast cancer. *International Journal of Pharmaceutics*, **2019**, *560*, 365–376.

Monge-Fuentes, V.; Muehlmann, L. A.; Longo, J. P.; Silva, J. R.; Fascinelli, M. L.; De Souza, P.; Faria, F.; Degterev, I. A.; Rodriguez, A.; Carneiro, F. P.; Lucci, C. M.; Escobar, P.; Amorim, R. F.; Azevedo, R. B. Photodynamic therapy mediated by acai oil (*Euterpe oleracea* Martius) in nanoemulsion: A potential treatment for melanoma. *Journal of Photochemistry and Photobiology B, Biology,* **2017**, *166*, 301–310.

Mou, D.; Chen, H.; Du, D.; Mao, C.; Wan, J.; Xu, H.; Yang, X. Hydrogel-thickened nanoemulsion system for topical delivery of lipophilic drugs. *International Journal of Pharmaceutics*, **2008**, *353*, 270–276.

Mouri, A.; Legrand, P.; Ghzaoui, A. E.; Dorandeu, C.; Maurel, J. C.; Devoisselle, J. M. Formulation, physicochemical characterization and stability study of lithium-loaded microemulsion system. *International Journal of Pharmaceutics*, **2016**, *502*, 117–124.

Muzaffar, F.; Singh, U. K.; Chaunan, L. Review on microemulsion as futuristic drug delivery. *International Journal of Pharmacy and Pharmaceutical Sciences*, **2013**, *5*, 39–53.

Nastiti, C. M. R. R.; Ponto, T.; Abd, E.; Grice, J. E.; Benson, H. A. E.; Roberts, M. S. Topical nano and microemulsions for skin delivery. *Pharmaceutics*, **2017**, *9*, 1–25.

Noelia, M. C.; Alicia, N. C.; Gabriel, L. Physical and chemical stability under environmental stress of microemulsions formulated with fish oil. *Food Research International*, **2019**, *119*, 283–290.

Oliveira, A. G.; Scarpa, M. V.; Correa, M. A.; Cera, L. F. R.; Formariz, T. P. Microemulsões: Estrutura e aplicações como sistema de liberação de fármacos. *Química Nova*, **2004**, *27*, 131–138.

Otto, A.; Du Plessius, J.; Wiechers, J. W. Formulation effects of topical emulsions on transdermal and dermal delivery. *International Journal of Cosmetic Science*, **2009**, *31*, 1–19.

Pajic, N. B.; Ilic, T.; Nikolic, I.; Dobricic, V.; Pantelic, I.; Savic S. Alkyl polycglucoside -based adapalede-loaded microemulsions for targeted dermal delivery: Structure, stability and comparative biopharmaceutical characterization with a conventional dosage form. *Journal of Drug Delivery Science and Technology*, **2019**, *54*, 1–12.

Pankajkumar-Patel, N.; Peris-García, E.; Ruiz-Angel, M. J.; Carda-Broch, S. Modulation of retention and selectivity in oil-in-water microemulsion liquid chromatography: A review. *Journal of Chromatography A*, **2019**, *1592*, 31–100.

Peltola, S.; Saarinen-Savolainen, P.; Kiesvaara, J.; Suhonen, T. M.; Urtti, A. Microemulsions for topical delivery of estradiol. *International Journal of Pharmaceutics*, **2003**, *254*, 99–107.

Pescina, S.; Garrastazu, G.; Favero, E. D.; Rondelli, V.; Cantù, L.; Padula, C.; Santi, P.; Nicoli, S. Microemulsion based on TPGS and isostearic acid for imiquimod formulation and skin delivery. *European Journal of Pharmaceutical Sciences*, **2018**, *125*, 223–231.

Pratap, S. B.; Brajesh, K.; Jain, S. K.; Kausar, S. Development and characterization of a nanoemulsion gel formulation for transdermal delivery of carvedilol. *International Journal of Drug Development and Research*, **2012**, *4*, 151–161.

Pund, S.; Shete, Y.; Jagadele, S. Multivariate analysis of physicochemical characteristics of lipid-based nanoemulsifying cilostazol-quality by design. *Colloids and Surfaces B Biointerfaces*, **2014**, *115*, 29–36.

Rai, V. K.; Mishra, N.; Yadav, K. S.; Yadav, N. P.; Nanoemulsion as pharmaceutical carrier for dermal and transdermal drug delivery: Formulation development, stability issues, basic considerations and application. *Journal of Controlled Release*, **2018**, *270*, 203–225.

Rakshit, A. K.; Naskar, B.; Moulik, S. P. Commemorating 75 years of microemulsion. *Current Science*, **2019**, *116*, 898–912.

Ravikumar, P.; Tatke, P. Advances in encapsulated dermal formulations in chemoprevention of melanoma: An overview. *Journal of Cosmetic Dermatology*, **2019**, *18*, 1606–1612.

Reddy, B. V.; Satyanandam, S. Transdermal drug delivery system: A review. *International Journal of Chemistry and Pharmaceutical Sciences*, **2014**, *2*, 870–877.

Resende, Q. G. M. *Estudo da Formação de Microemulsões Contendo Óleo Essencial de Citrus Sinensis (L.) Osbeck: Efeito Dos co-Tensoativos e Avaliação da Interação das Microemulsões com Modelo de Estrato Córneo [Study of the formation of microemulsions containing essential oil of Citrus sinensis (L.) Osbeck: Effect of co-tensoactives and evaluation of the interaction of microemulsions with stratum corneum model.].* Dissertation, Universidade Federal de Sergipe, **2013**.

Sahu, S.; Katiyar, S. S.; Kushwah, V.; Jain, S. Active natural oil-based nanoemulsion containing tacrolimus for synergistic antipsoriatic efficacy. *Nanomedicine (Lond.)*, **2018**, *13*, 1985–1998.

Sakeena, M. H.; Yam, M. F.; Elrashid, S. M.; Munavvar, A. S.; Azmin, M. N. Anti-inflammatory and analgesic effects of ketoprofen in palm oil esters nanoemulsion. *Journal of Oleo Science*, **2010**, *59*, 667–671.

Salvia-Trujillo, L.; Soliva-Fortuny, S.; Rojas-Graü, M. A.; McClements, D. J.; Martín-Belloso, O. Edible nanoemulsions as carriers of active ingredients: A review. *Annual Review of Food Science and Technology*, **2017**, *8*, 20.1–20.28.

Savic, V.; Todosijevic, M.; Ilic, T.; Lukic, M.; Mitsou, E.; Papadimitriou, V.; Avramiotis, S.; Markovic, B.; Cekic, N.; Savic, S. Tacrolimus loaded biocompatible lecithin-based microemulsions with improved skin penetration: Structure characterization and *in vitro/in vivo* performances. *International Journal of Pharmaceutics*, **2017**, *529*, 491–505.

Schwarz, J. S.; Weisspapir, M. R.; Friedman, D. I. Enhanced transdermal delivery of diazepam by submicron emulsion (SME) creams. *Pharmaceutical Research*, **1995**, *12*,687–692.

Shakeel, F.; Baboota, S.; Ahuja, A.; Ali, J.; Aqil, M.; Shafiq, S. Nanoemulsions as vehicles for transdermal delivery of aceclofenac. *AAPS PharmSciTech*, **2007**, *8*, E104.

Shakeel, F.; Baboota, S.; Ahuja, A.; Ali, J.; Shafiq, S. Celecoxib nanoemulsion: Skin permeation mechanism and bioavailability assessment. *Journal of Drug Targeting*, **2008**, *16*, 733–740.

Shakeel, F.; Ramadan, W. Transdermal delivery of anticancer drug caffeine from water-in-oil nanoemulsions. *Colloids and Surfaces B: Biointerfaces*, **2010**, *75*, 356–362.

Sharma, N.; Bansal, M.; Visht, S.; Sharma, P. K.; Kulkarni, G. T. Nanoemulsion: A new concept of delivery system. *Chronicles of Young Scientists*, **2010**, *1*, 2–6.

Shiri, M. S. Z.; Henderson, W.; Mucalo, M. R. A review of the lesser-studied microemulsion-based synthesis methodologies used for preparing nanoparticle system of noble metals, Os, Re, Ir and Rh. *Materials*, **2019**, *12*, 1–26.

Shivasaraun, U. V.; Sureshkumar, R.; Karthika, S.; Nethravathi, P. Flavonoids as adjuvant in psoralen-based phytochemotherapy in the management of vitiligo/leukoderma. *Asian Journal of Pharmaceutics.*, **2019**, *13*, 85–92.

Shukla, T.; Upmanyu, N.; Agrawal, M.; Saraf, S.; Saraf, S.; Alexander, A. Biomedical applications of microemulsions through dermal and transdermal route. *Biomedicine & Pharmacotherapy*, **2018**, *108*, 1477–1494.

Singh, Y.; Meher, J. G.; Raval, K.; Khan, F. A.; Chaurasia, M.; Jain, N. K.; Chourasia, M. K. Nanoemulsion: Concepts, development and applications in drug delivery. *Journal of Controlled Release*, **2017**, *252*, 28–49.

Sommer, E.; Neubert, R. H. H.; Mentel, M.; Tuchscherer, B.; Mrestani, Y.; Wohlrab, J. Dermal peptide delivery using enhancer molecules and colloidal carrier systems, Part III: Tetrapeptide GEKG. *European Journal of Pharmaceutical Sciences*, **2018**, *124*, 137–144.

Subongkot, T. Development and mechanistic study of a microemulsion containing vitamin E TPGS for the enhancement of oral absorption of celecoxib. *International Journal of Nanomedicine*, **2019**, *14*, 3087–3102.

Szunerits, S.; Boukherroub, R. Heat: A highly efficient skin enhancer for transdermal drug delivery. *Frontiers in Bioengineering and Biotechnology*, **2018**, *6*, 1–13.

Tagne, J. B.; Kakumanu, S.; Nicolosi, R. J. Nanoemulsion preparations of the anticancer drug dacarbazine significantly increase its efficacy in a xenograft mouse melanoma model. *Molecular Pharmaceutics*, **2008**, *5*, 1055–1063.

Tanwar, H.; Sachedeva, R. Transdermal drug delivery system: A review. *International Journal of Pharmaceutical Sciences and Research*, **2016**, *7*, 2274–2290.

Thakur, N.; Garg, G.; Sharma, P. K.; Kumar, N. Nanoemulsion: A review on various pharmaceutical application. *Global Journal of Pharmacology*, **2012**, *6*, 222–225.

Tung, N. T.; Vu, V. D.; Nguyen, P. L. DoE-based development, physicochemical characterization, and pharmacological evaluation of a topical hydrogel containing betamethasone dipropionate microemulsion. *Colloids and Surfaces B: Biointerfaces*, **2019**, *181*, 480–488.

Vinod, V.; Vijaya, V. C. H. S. Overview on transdermal drug delivery by semisolid systems: Emulgel. *International Journal of Pharmacy and Analytical Research*, **2015**, *4*, 256–263.

Wais, M.; Samad, A.; Nazish, I.; Khale, A.; Aqil, M.; Khan, M. Formulation development ex-vivo and in-vivo evaluation of nanoemulsion for transdermal delivery of glibenclamide. *International Journal of Pharmacy and Pharmaceutical Sciences*, **2013**, *5*, 747–754.

Wu, H.; Ramachandran, C.; Bielinska, A. U.; Kingzett, K.; Sun, R.; Weiner, N. D.; Roessler, B. J. Topical transfection using plasmid DNA in a water-in-oil nanoemulsion. *International Journal of Pharmaceutics*, **2001**, *221*, 23–34.

Yang, M.; Gu, Y.; Yang, D.; Tang, X.; Liu, J. Development of triptolide-nanoemulsion gels for percutaneous administration: Physicochemical, transport, pharmacokinetics and pharmacodynamics characteristics. *Journal of Nanobiotechnology*, **2017**, *15*, 1–15.

Yilmaz, E.; Borchet, H. H. Effect of lipid-containing, positively charged nanoemulsion on skin hydration, elasticity and erythema – an *in vivo* study. *International Journal of Pharmaceutics*, **2006**, *307*, 232–238.

Zhang, J.; Michniak-Khon, B. B. Investigation of microemulsion and microemulsion gel formulations for dermal delivery of clotrimazole. *International Journal of Pharmaceutics*, **2018**, *526*, 345–352.

CHAPTER 7

Pulmonary Drug Delivery System

Nayan A. Gujarathi,[1] Akshada A. Bakliwal,[2] Bhushan R. Rane,[3] Raj Kumar Keservani,[4] Tulshidas S. Patil,[1] Meenendra Tripathi,[4] and Gulam Mohammed Husain[5]

[1]Shri Vile Parle Kelavani Mandal's Institute of Pharmacy, Dhule–424001, Maharashtra, India, E-mail: nayang5678@gmail.com (N. A. Gujarathi)

[2]Sandip Institute of Pharmaceutical Sciences, Nashik–422213, Maharashtra, India

[3]Shri D.D. Vispute College of Pharmacy and Research Center, New Panvel–410206, Maharashtra, India

[4]Faculty of B. Pharmacy, CSM Group of Institutions, Prayagraj–212111, Uttar Pradesh, India

[5]National Research Institute of Unani Medicine for Skin Disorders, Ministry of Ayush, Hyderabad–500038, Telangana, India

ABSTRACT

In recent years, the pulmonary drug delivery has incredible scientific and bioscience dynamism efficaciously utilized in the management for lung ailments, by targeting at desired site to augment therapeutic effect and reducing side effects using minute drug dosages. In the 21st century, there is a vast shift in inhaled therapy for systemic effect. Despite that, the pulmonary tract will be in general considered as most favorable and

Topical and Transdermal Drug Delivery Systems: Applications and Future Prospects. Nayan A. Gujarathi, PhD, Juliana Palma Abriata, PhD, Raj Kumar Keservani, MPharm, Anil K. Sharma, PhD, MPharm (Eds.)

enthralling administration route for delivering the active pharmaceutical ingredient to treat pulmonary tract diseases in particular acute bronchitis, chronic respiratory distress, bacteriological contaminations as well as systemic infections. The pulmonary medicine contributes variegated advantages above the other routes, and several techniques are created for fabricating particulate matter for lung therapy. Advancement in the characterization of pulmonic dosage form includes *in-vitro, in-vivo, and ex-vivo* evaluation parameters. This chapter also focuses on recent research work and the usage of lung drug delivery systems.

7.1 BACKGROUND

The pulmonary drug delivery system has incredible scientific and biomedical enthusiasm for current years and has advanced treatment for lung infections, by virtue of upgraded region focusing on and decreased fundamental reactions accompanied by delivering microparticulate drug dosages (Salama et al., 2008). Besides, the large effective surface and high penetrability of the respiratory system, the 21st century has glimpsed the change in perspective to inhaled treatment for systemic use. However, the respiratory tract will in general be considered as promising and reliable administration route for drug moiety anticipated to treat pulmonary tract (e.g., bronchopneumonia, chronic obstructive bronchitis, microbial contaminations) and also systemic diseases (e.g., diabetes). A wide assortment of specialists has been regulated up to lung through the extravascular therapy for the management of different disease states (Jaspart et al., 2007). The most incessant utilization of pulmonic medications is for the management of chronic obstructive pulmonary disease (COPD) and bronchial asthma more specifically some drugs can be incorporated such as anti-asthmatic, corticosteroids, cholinergic even some antidiabetic drugs can be administered. Nonetheless, the respiratory course has been accepting expanded consideration since the mid-1990s as an option in contrast to parenteral drug delivery, most patiently for the delivery of protein and peptide therapeutics. Furthermore, the acid susceptible drugs are also administered via nasal route directly to lungs. Around the world, more than 300 million individuals experience the ill effects of COPD and asthma; it is only because of air pollution.

7.2 INTRODUCTION

7.2.1 ANATOMY OF PULMONARY DRUG DELIVERY SYSTEM

The human lung comprises of 5 lobules and 10 broncho pulmonary, lung is highly perfused organ (Figure 7.1). Lung lobules consist of 3–5 terminal bronchioles which are arranged adjacent to each other. Every bronchiole in lung contains acinus which is the smallest and basic part of the lung, it consist of alveolar ducts, sacs, and alveoli. The alveolar blood boundary in its least complex structure comprises a solitary epithelial and endothelial cell. While this morphologic organization promptly encourages the trade, it can even now speak to a noteworthy boundary to enormous particles (Patton and Byron, 2007; Salama et al., 2008).

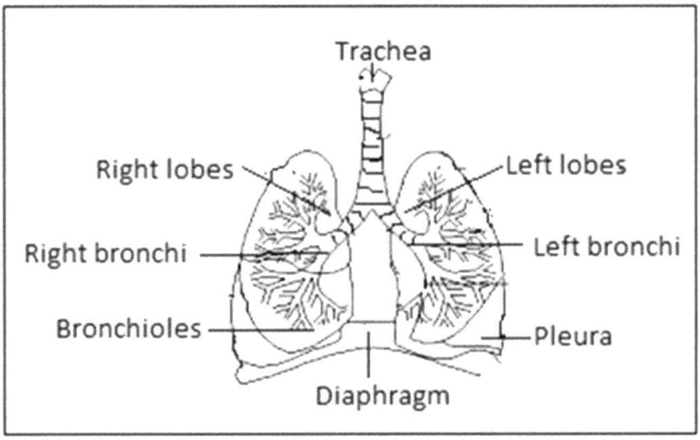

FIGURE 7.1 Anatomy of pulmonary drug delivery system.

Prior to entering the fundamental flow, solutes must navigate a thin layer of liquid, the epithelial covering liquid. This layer will in general gather at the edges of the alveoli and is secured by a weakened layer of surfactant. In contrast to the bigger airway routes, the alveolar region is fixed with a surface-active layer comprising of phospholipids (mainly phosphatidyl choline and phosphatidyl glycerol) and a few key apoproteins. The surfactant lining liquid assumes a significant role in keeping up alveolar liquid homeostasis and penetrability and takes an interest in different protection systems. Late investigations recommend that the surfactant may hinder distribution out of the alveoli. The pulmonary routes, from the upper airway routes to the terminal bronchioles, are fixed with a viscoelastic, gel-like mucous layer with thickness of 0.5–5.0 mm (Ugwoke et al., 2007).

The secretion lining comprises of two layers: a liquid layer of low consistency, which encompasses the cilia (periciliary liquid layer) and mucous at top which is more viscous layer. Mucus acts as a protective layer, comprises a complex mixture of glycoproteins which is secreted by the goblet cells and local glands. The lung tissue is exceptionally vascularized, which focus on particularly lipophilic and lower subatomic drug particulate matter (Liang et al., 2015; Zhou et al., 2015).

In any case, late advancement inside biotechnology has produced a novel protein and peptide medications to which organization to the respiratory tract, to get initial delivery appears to be worthwhile contrasted with for example parenteral or gastrointestinal organization (tablets, containers, and so on.). For instance, the low metabolic movement in the lungs permits initial delivery without liver entry. Nonetheless, the respiratory system itself confines the passage of particulate issue by different methods: for example, geometry of the aviation routes and clearance mechanism of the lungs. Subsequently, inward breath particles must be efficiently enhanced to achieve ingestion locales in the alveolar epithelium (Kaur et al., 2014).

7.2.2 ADVANTAGES AND DISADVANTAGES OF PULMONARY DRUG DELIVERY SYSTEM

This route can offer considerable benefits and drawbacks in comparison to other drug delivery systems which are as listed below:

- **Advantages:**
  - It offers neighborhood activity inside the respiratory tract;
  - Provides fast medication activity;
  - Provides decreased dose;
  - Allows for a decrease in systemic symptoms. It can be utilized as an elective course to tranquilize collaboration when at least two drugs are utilized simultaneously;
  - Reduces extracellular catalyst levels contrasted with GI tract because of the huge alveolar surface zone;

- Reduces avoidance of first-pass hepatic digestion by administration of active pharmaceutical ingredient;
- Best approach for the delivery of active moieties.

➢ **Disadvantages:**

- The duration of action is regularly short-lived because of the fast expulsion of medication from the lungs or because of medication digestion;
- Necessitates incessant dosing

7.2.3 BARRIERS OF PULMONARY DRUG DELIVERY SYSTEM

There are different barriers which affect the pulmonary drug delivery system. These different barrier parameters are listed below:

1. **Less Bioavailability:** Due to the less permeability of the membrane there is a decrease in bioavailability of compounds having large molecular weight.

   For example, proteins and peptides (insulin).
2. **Lung Clearance Mechanism:** Active moieties deposited in the airways will possess less residence interval time. Subsequently, the active moiety were quickly removed through alveolar and mucociliary clearance.

   For example, bacteria and macrophages.
3. **Restrictions of Traditional Dosage Forms:** The conventional device used routinely delivered drug to the target site up to 10–40%. It results into the loss of the drug being incorporated in the formulation.

   For example, the extent of salbutamol drug deposited by venture nebulizer is always more as compared to conventional device, i.e., disposable jet nebulizer (Jadhav et al., 2007).

7.2.4 RESPIRATORY PATTERNS

At the time of aerosol spraying the drug deposition can be influenced due to respiration it is due to tidal volume and recurrence decide the mean stream rates in every area of the respiratory tract, which, thus, impact the viability of every deposition mechanism. Turbulence will in general improve molecule deposition. Quick breathing is frequently connected with existence of bigger particles at the site of nasal cavity, while moderate, unfaltering inward breath builds the quantity of particles that infiltrate to the peripheral portions of the lungs, with or without breath-holding, confirmed almost the excessive conglomeration in the hairy pulmonary routes (trachea-bronchial area). Fast inward breath demonstrated comparative patterns: the trachea-bronchial regions most extremely falls and moves amongst 3 μm and 6 μm. At the point when the above contemplations are considered, the perfect situation for airborne would be:

- Size is smaller than 5 μm in case of aerosol AD;
- Slow, consistent inward breath; and
- At the time of breath-hold on completion of inhalation.

7.3 INHALATION THERAPY

There are a few favorable circumstances in delivery of active moiety to the pulmonary system comprised of non-invasive technique for conveyance. Furthermore, throughout the pulmonary tract, the average thickness of the alveolar epithelium is in between 0.1 μm and 0.2 μm. The absolute separation is around 0.5 μm and 1.0 μm in between epithelial surface and blood in the alveolar zone which are considerably a smaller in size if we compare to the bronchial region (remove between bodily fluid surface and blood: 30 μm–40 μm). Therefore, it creates the impression that pharmaceuticals after profound inward breath and statement in the peripheral (e.g., alveolar) region of the lung can be quickly assimilated. Pulmonary drug delivery accordingly has the bit of leeway, contrasted with nasal conveyance, that it is conceivable to acquire an adequately high assimilation without the need of enhancers. Then again, the human lung has a distinctive defense mechanism to counteract aerosol particles entering into the profound lung. Fundamentally, the bronchial tract and oropharyngeal passage are superb channels to take out airborne particles from the breathed in air and particles kept on ciliated epithelium are liable to mucociliary clearance to the GIT.

In this way, to convey a medication into the profound lung, one needs to surmount these channels. In any case, still drug deposited in the alveolar area of the lung, various components hinder the ingestion of breathed in pharmaceuticals. There are multiple ingestion barriers (such as bodily liquid layer, alveolar liquid layer, macrophages, and different cells, alveolar epithelium, and cellar film) which act to changing degrees by repressing medication penetration into the course. This exists contending cell take-up pathways (e.g., molecule phagocytosis by macrophages) and obviously proteolytic debasement can restrict the measure of unblemished medication available for ingestion. The capacity of these boundaries can be disabled by altogether different substances, and therefore, the ingestion of medications can be expanded, for instance, by the utilization of absorbance enhancers (for example, cyclodextrins, cleansers, and bile acids). Moreover, protease inhibitors (for example, aprotinin) and phagocytosis can restrain proteolytic debasement by macrophages decreased by bundling of substances into permeable particles. On a basic level, ingestion energy of breathed in substances rely upon their atomic weight (little particles are more quickly retained than bigger ones), pH-esteem, electrical charge, dissolvability, and security of the breathed in substance (Scheuch et al., 2006).

The other target areas inside the lung for inhalable medications are the enormous and few bronchial pulmonary routes. The most pertinent are: asthma, chronic obstructive pulmonary disorder, and tumors in the bronchus. To treat such diseases locally, one needs to convey the medications explicitly. Rather than the inward breath of medications for fundamental treatment, the inhalative treatment of asthma and COPD by methods using various pressurized metered-dose inhaler (MDI) and nebulizer are clinically settled for a long time and the medicines include commonly low sub-atomic weight particles in plans free of stabilizers and assimilation enhancers.

7.4 AERODYNAMIC BEHAVIOR OF DRUG PARTICLES

The particle size is a basic factor influencing the site of their deposition, since it decides working systems and degree of infiltration into the lungs. Particles that have higher than unit thickness will have genuine distances across smaller than their aerodynamic particle diameter (AD). On the other hand, particles with small unit thickness will have geometric measurements larger than their AD. Vaporized size circulations might be portrayed as, for all intents and purposes, monodispersed powder (uniform-sized particles or particles) or polydispersed powder (non-uniform sized powder or particles).

The upper respiratory tract and the next region of the tracheobronchial tree go about as a progression of channels for breathed in particles. Along these lines, vaporized particles greater than 100 μm are remained blocked in the upper tract of the nose and pharynx. The extremely fine powder, for instance, having a streamlined distance under 10 μm can be allowed into deeper region. Particles having streamlined distances across more noteworthy than 10 μm are probably going to affect the dividers of the throat. Particles of size 5 μm to 0.5 μm will be pass towards in the respiratory bronchioles. Particles in the surrounding air are transported by various physical systems. The important mechanism for restorative vaporizers are dispersion by Brownian movement (particles in the size scope of <0.5 μm), sedimentation by the gravitational power (particles in the size scope of >0.5 μm) and impaction (measure extend >3 μm).

7.5 DRUG DEPOSITION MECHANISM

The mechanism for deposition of drug particles into the respiratory tract includes impaction, sedimentation (gravitational statement), Brownian dissemination and electrostatic precipitation. All mechanisms act simultaneously, however, the initial two components leads to excessive deposition of large particle in the airway passage (1 mm, AD, 10 mm). Diffusion, nonetheless, is the primary determinant of statement of littler particles in the peripheral region of the lung. Impaction happens when a molecule's force keeps it from changing course in a region where there is an adjustment toward mass airstream current. It is the primary testimony component in the upper aviation routes and at or closes bronchial branching points. The likelihood for impaction increases with expanding air speed, breathing recurrence and molecule measure. Sedimentation results when the gravitational power following up on a molecule conquers the all out power of the air obstruction. The probability of sedimentation is corresponding to residence time in the aviation route and to molecule size and diminishes with increasing breathing rate. The impact of tiny aerosol particles with the gas atom applies discrete non-uniform weights at the particle surfaces, leading to diffusion and also bringing about irregular Brownian movement. The adequacy of Brownian movement in the deposition of particles is contrarily corresponding to molecule of size 0.5 μm, and is significant in deeper region of lungs up to alveoli. Bigger size particles of limited size will deposit into the upper respiratory tract up to the trachea by the process of diffusion.

Parameters affecting particle deposition in the lung:

- Various physiochemical properties of the drug will influence the regional drug deposition into respiratory tract such as particle size, its density, the charge onto the particle, shape, and size as well as the tidal volume;
- Different breathing rate in various patients (e.g., air flow rate, tidal volume);
- Pulse rate and respiratory rate, capacity of lungs to inhale and exhale also influence;
- Basic structure and physiology of the lungs and whole respiratory tract;

Apart from that, aerosol particle size and size appropriation are the most powerful influence on aerosol deposition.

7.6 CLASSIFICATION OF PULMONARY DRUG DELIVERY SYSTEM

The retention of drug moiety in the lung or its clearance is depended upon the process of design of formulation or particulate matter for inhalation. According to mechanism of drug release, pulmonary drug delivery system is categorized into three broad classes:

- Immediate release pulmonary drug delivery systems (lactose as carrier-drug mixtures);
- Controlled release pulmonary drug delivery systems (novel approaches vesicular systems, i.e., liposomes, surfactant-based micelles, small-sized nanoparticles, and polymers based microspheres);
- Sustained release pulmonic medicines (polymers-based microparticulate systems).

The merit of pulmonic medicines delivered in the form of controlled release or sustained release are as follows:

- Longer duration of action;
- Decrease in dose size;
- Improved therapeutic efficacy;
- Better patient compliance;
- Decrease in adverse reaction;
- Cost-effective, economical especially applies for sustained-type of pulmonary drug delivery system.

Such type of drug delivery system comprise of carriers like vesicular systems liposomes, nanoparticles, nano-suspensions, genetically engineered particulate carrier, microparticles, etc. Design of such special type of modified release pulmonic drug delivery system requires in-depth knowledge about the etiology of lung diseases as well as structure and function of respiratory organs. Moreover, the selection of the appropriate delivery device, optimization of the manufacturing process of the matrix or reservoir system and the type of polymer or carrier play an important role in the design of effective targeted drug delivery system.

7.7 DEVICES FOR PULMONARY DELIVERY

In pulmonic drug delivery system, the devices will play a significant role in the accomplishment of this framework. In the ongoing years, incredible steps have been made in the improvement of devices for delivery. Actually, devices acclimatized for targeting drugs to the respiratory system are significantly few times examined if we compare to dry dosage forms, for example, powder. The determination of devices for delivery of medications to the lungs is a significant factor. In the event that the medication is intended to a particular part of the lungs, at that point they chose device must be skilled enough to create and convey the particles/beads of explicit streamlined distance across (O'Callaghan, 1994; Labiris et al., 2006). The gadgets most regularly utilized for respiratory delivery incorporates nebulizers, pMDI (pressurized metered-dose inhalers) and DPI (dry powder inhalers). DPI are devices used to convey drugs, particularly proteins to the lungs. Some nebulizers are manufactured by companies which based on the mechanism of vibration, compressed air or ultrasound waves are called ultrasonic nebulizers, vibrating nebulizers, and jet nebulizers. The selection of special devise will be based upon a few considerations such as the type of the actives and its mechanism of action, the site of activity, and pathophysiology of the lung. Fluid suspensions and arrangements are nebulized viably. The general impact of airflow as well as mechanical impact in order to control the powder movement in the respiratory tract is yet unidentified (Chan, 2003; Geller, 2002; Geller et al., 2003). Different devices routinely employed for delivery to pulmonic system are as in subsections.

7.7.1 METERED DOSE INHALER (MDI)

The medicament is delivered in the form of fine powder by utilization of metered-dose inhaler (MDI) type of drug delivery system. The particle size of aerodynamic drug moiety is smaller than 5 μm, which is meant for direct inhalation through the pulmonary system. It is successfully employed in the management of asthma and COPD. These devices can be categorized into two different classes in particular (a) the precise metering device, for example, pressurized MDIs, aerosol spray pump, etc. (b) the poorly metering devices. The example of MDIs available in the pharmaceutical market are enlisted in Table 7.1.

TABLE 7.1 Example of MDI Available in the Market

Devices	Company
Autohaler	3M Pharmaceutical, St Paul, Minnesota
K-haler	Clinical Designs, Aldsworth, United Kingdom
MD Turbo	Respirics, Raleigh, North Carolina
Tempo	Map Pharmaceuticals, Mountain View, California
Bronchoair	Bronchoair Medizintechnik, Munich, Germany

7.7.2 DRY POWDER INHALER (DPI)

Assuming the dose size as the parameter for classification of DPI, the three conventional types of devices are marketed by the pharmaceutical industries. They are listed in subsections.

7.7.2.1 UNIT DOSE DEVICE

The unit dose powder inhaler comprised of a capsule filled with the medicament. The capsule is properly located in the holder. The powdered form of the drug will be accessible to the upper respiratory tract after inhalation as soon as the capsule opens up in the device. The residue of the empty capsule is removed and the new capsule is relocated in its position for delivery of the next dose.

7.7.2.2 MULTI-DOSE DEVICE

In an isolated aluminum blister reservoir, a single rounded disk comprised of four powder doses or eight powder doses is placed to design a device meant for multiple doses. It will attend the purpose of the dosage regimen for the next one to two days.

7.7.2.3 MULTI-UNIT DOSE DEVICE

Multi-unit dose devices especially consist of several units of a single dose of drug formulated and sealed as an individual unit.

Additionally, in recent times, the two types of new technology-based DPI devices presented in the pharmaceutical market are as follows:

7.7.2.4 AIR CLASSIFIER TECHNOLOGY IN DEVICES

The principle behind the air classifier is de-agglomeration of moving particles by application of large inertial forces. This type of technology-based air classifier device consists of cyclone chamber.

7.7.2.5 MULTIPLE AIR CLASSIFIER TECHNOLOGY

This type of device is the revised and research-based form of air classifier technology, which involves a parallel array of multiple classifier chambers to enhance the quantity for aerosol. The list of DPI existing in the pharmaceutical market are enlisted in Table 7.2.

TABLE 7.2 Example of DPI Available in the Market

Devices	Company
Eclipse (unit dose)	Aventis
Rotahaler™ (single dose)	GlaxoSmithKline
Aerolizers™ (single dose)	Novartis
Accuhaler™ (multi-unit dose)	GlaxoSmithKline, UK
Meghaler (multi-dose)	GGU

7.7.3 NEBULIZER

Formerly, the three different categories of nebulizer systems are commonly known and marketed, for example, piezo mesh, ultrasonic, and air-jet (pneumatic). Pneumatic nebulizers are again classified into two distinct types one is concentric and another is cross-flow nebulizers. At present the recent development launches the new nebulizer based on technology known as AAD (adaptive aerosol delivery), which replaces the old nebulizer. The example of nebulizers available in the pharmaceutical market is enlisted in Table 7.3.

TABLE 7.3 Example of Nebulizers Available in the Market

Devices	Company
DeVilbiss Pulmo-aide nebulizer	DeVilbiss Health Care
AIRIAL Mini Jet adult nebulizer machine	Venture Respiratory Inc.
Aeromist epic nebulizer	Medline Industries
Aeromist mini nebulizer	Medline industries

7.8 LUNG COMPATIBILITY OF DIFFERENT FORMULATIONS COMPRISING POLYMERS

During the development of PDDS the properties of polymers need to be investigated to check its compatibility with the lungs. The well-being of these polymers must be first decided and their similarity with lung liquid is of incredible concern. The polymers used to drag out the discharge rate for unending use may gather in the lung, particularly in the lung periphery, which isn't served by mucociliary clearance. In this manner, handling methods and definition segments must be altogether screened so as to keep away from the dangerous results. The carriers incorporated in the formulation of dry powder leads to several complications, for example, sugars, and cyclodextrins can cause bronchi constriction in huge numbers of the patients. More use of proteins and different carriers, for example, ingestion enhancers and compound inhibitors, can deliver immunogenicity, local irritation, and toxicity (Perez-Gil, 2008; Hickey and Garcia-Contreras, 2001; Heinemann et al., 2000).

7.9 METHODS FOR PREPARATION OF DOSAGE FORM FOR PULMONARY DELIVERY

A lot of traditional methods are available to prepare the DPI. These techniques have a number of confinements, for example, drug particle shape and size, average particle size or normal size distribution pattern and reduced control for drug powder crystallinity. These issues can be overcome by specialized milling techniques. Several methods commonly utilized to fabricate the particulate matter for ling delivery are as follows:

7.9.1 SPRAY DRYING METHOD

It is an advanced pharmaceutical method used to proficiently deliver the small-sized particles of colloidal range into the respiratory tract in solid form (Mosen et al., 2004; Duddu et al., 2002). Spray drying was investigated during the 1980s, as an elective method for the preparation of very fine particles of less micron size around 2 to 5 micrometers for the administration into lungs. In this procedure, the feed solution is kept at room temperature, it is pumped towards the nozzle and later it is converted into the atomized form. Then this atomized solution is dried in a special chamber using preheated drying gas. This strategy is all the more encouraging in delivering the particles of over 2-μm size. This procedure is additionally appropriate for thermolabile materials, such as drugs of category protein and peptide. More significantly, the uniform particle morphology is obtained using this technique (Gilani et al., 2005).

7.9.2 SPRAY FREEZE DRYING METHOD

This technique was investigated for pharmaceutical application in mid-1990s. It includes processing steps of both spray drying and freeze-drying techniques (Rogers et al., 2001; Maa and Prestrelski, 2000). Light and porous particles are produced by this method. This method is more specifically used for the preparation of DPI products of thermolabile drugs such as insulin (proteins and peptides) and plasmid DNA. However, this technique is costly, which is mostly used for expensive drug (Yuet al., 2004; Kuo and Hwang, 2004).

7.9.3 SUPERCRITICAL FLUID TECHNOLOGY

The essential element of this procedure is the controlled crystallization of medications from dispersion in supercritical liquids, carbon dioxide. This technique has been utilized in the pharmaceutical field for formulation of microparticles, nanoparticles, and liposomes. To improve the formulation properties, this method can be used in the case of some drugs (Chattopadhyay et al., 2007; Rehman et al., 2004).

7.9.4 SOLVENT PRECIPITATION METHOD

The method particularly encompasses of micro-precipitation and sono-crystallization. Direct controlled crystallization method is used to prepare the narrow and uniform size crystalline powder, while rapid precipitation is used for inhalable particles using antisolvent. Different antiasthmatic drugs were formulated utilizing the sonocrystallization method (Rasenack et al., 2003).

7.9.5 DOUBLE EMULSION OR SOLVENT EVAPORATION TECHNIQUE

This technique includes formulation of oil/water (o/w) emulsion. In this, the organic solvent diffuses into the aqueous phase and out of the polymer phase. This solvent is then evaporated to form drug-loaded polymeric nanoparticles. By this technique, biodegradable polymers have been explored as carrier for solid drug nanoparticles to target into respiratory tract.

7.10 INNOVATION IN THE CHARACTERIZATION OF PULMONARY DRUG DELIVERY SYSTEM

The significant progress is reported in the field of pulmonic drug delivery systems, more specifically in relation to the various evaluation parameters. These strategies can be delegated as *ex-vivo, in-vivo, and in-vitro* techniques. Different *in vitro* strategies like physicochemical parameters of particles, assessment of aerosol performance, cell culture study and molecule disintegration study have been developed. Drug administration-related devices and systems, drug in powder form and its deposition into the respiratory tract and *in vivo* pharmacokinetic studies are the developed newly. *Ex vivo* strategies are additionally created to check the viability and security parameters of medications and their conveyance components. Several evaluation procedures applies for the pulmonary drug delivery systems are given below:

Evaluation parameters:

1. *In-vitro* evaluation:
   i. Physicochemical characterization:
      a. Morphological studies:
         - TEM;
         - SEM;
         - AFM;
         - Laser light scattering.
      b. Powder flow properties;
      c. Moisture content test;
      d. Aerosol triboelectric characterization;
      e. Particles interparticulate forces measurement;
      f. Solid state characterization:
         - Thermal analysis;
         - X-ray powder diffraction;
         - Infrared spectroscopy, NMR, and Raman spectroscopy;
         - Qualitative surface analysis.
      g. Particle size and zeta potential measurement.
      ii. Evaluation of aerosol performance:
      a. Aerodynamic size distribution by TOF analysis;
      b. Dose uniformity test.
   iii. Particles dissolution studies;
   iv. Cell cultures study.
2. In-vivo evaluation:
   i. Drug administration systems:
      a. Whole body exposure system;
      b. Head-only or nose-only exposure systems;
      c. Direct intra tracheal administration;
      d. Intranasal administration.
   ii. Drug deposition studies;
   iii. Pharmacokinetic studies.
3. Ex-vivo evaluation.

7.10.1 MORPHOLOGICAL EVALUATION

The morphological evaluation of nanoparticles should be possible by using TEM; it uses various strategies for the examination of particles, such as cryogenic transmission techniques, negative stain method and freeze-fracture electron microscopy. The drug particles size, its structure, shape, and arrangement can be recognized by translation of the morphological information of TEM examination. Likewise, molecule size can be estimated. Scanning electron microscopy (SEM) is utilized to assess the molecule size and particle surface morphology (Bunker et al., 2005; Chew et al., 2005; Tang et al., 2003, 2009).

7.10.2 POWDER FLOW CHARACTERISTICS

Powder flow properties are performed to examine the flow properties and its association with the aerodynamics properties. The density (bulk and tapped), angle of repose and porosity is flow properties which need to be measured for quality control (El-Gendy et al., 2011; Maretti et al., 2014; Chan and Chew, 2003; Chan et al., 2013).

7.10.3 AEROSOL TURBOELECTRIC CHARACTERIZATION

It has been reported that electrostatic charge created during the handling of powders and their aerosolization will in general control the deposition of medication particles in the respiratory tract influencing the aerosolization qualities of inhaled medication systems. Electrostatic charge in the aerosol has appeared improved conveyance into the deeper region of the lungs and it can be utilized to target the drug in the specified area in the respiratory tract. The aerosol electrostatic properties are measured by direct and indirect measurement methods. The examples of devices used in direct measurement method are eNGI: electrical next generation impactor; TSI: twin-stage impinger; and ELPI: electrical low pressure impactor. In the indirect measurement method, the device available is E-SPART (Telko et al., 2007; Young et al., 2007; Hoe et al., 2009; Zhu et al., 2008).

7.10.4 SOLID STATE CHARACTERIZATIONS

Differential filtering calorimetry (DSC) is a standard thermal analytical method utilized to study the polymorphic changes occurs in a drug-carrier matrix. This technique is used for evaluation of drug encapsulation within the matrix. Parameters such as stability, melting point, and recrystallization are included in solid-state characterization. Bunjes et al. described the thermal analytical method to study the type of matrix pulmonary system (Begat et al., 2004). Solid-state characterization is also done by using XRPD (X-ray powder diffraction technology) FTIR and TGA (thermogravimetric analysis).

7.10.5 ZETA POTENTIAL AND PARTICLE SIZE ANALYSIS

Drug particle size and its zeta potential of colloidal solution area significant variable which decides the efficacy of formulation and its stability. Particle size in general controls the drug deposition in the lung whereas zeta potential shows the stability of the system. LD (laser diffraction) and PCS (photon correlation spectroscopy) were frequently used for estimating particle size and polydispersibility index.

7.10.6 PARTICLE IN-VITRO DISSOLUTION STUDIES

In contrast to the conventional drug delivery system, the *in-vitro* dissolution study differs for the drug product meant for inhalation. In this, dissolution profile which actually based on the emitted amount of drug or dose of drug emitted from the device and its deposition in the respiratory system. Till now, the inhaled drug products do not have any recommended and set up dissolution technique due to their distinctive nature of drug and its deposition into respiratory tract. Davis et al. investigated the use of Anderson cascade impactor in determination of the aerosol particles which are deposited on a filter membrane. *In-vitro* dissolution was further studied by placing these filter membranes into a flow through cell apparatus or Franz diffusion cell (Forbes et al., 2003; Sakagami, 2006).

7.10.7 CELL CULTURE STUDY

The formulations must be outperformed through the cell culture testing techniques before the *ex-vivo* and *in-vivo* testing. The most recent two decades have seen a various cell models got from human being and rodent tissues of aspiratory epithelium. Cells from human are less accessible than cells from different vertebrates and are agents of the clinical conditions. The deposition of medication and its disintegration held in a little volume of cell lining liquid, imitates the *in vivo* condition. Stephanie et al. prepared a device for administration of drug into respiratory tract and named it as PADDOCC (Pharmaceutical Aerosol Deposition Device on Cell cultures). It is used to simulate *in-vivo* conditions in which aerosol dose transmitted onto the epithelial cells are presented to an air-fluid interface. PADDOCC limits testing of creatures and aides in the foundation of early clinical trials.

In this system, a closed plastic compartment is connected to a dry powder spray converter and tissue kept into chamber. Pulmonary administration is less unpleasant when contrasted with another system; however, some assimilation of medication through the skin, from the nasal course and from the GIT. It is the basic technique for pneumonic medication delivery and noteworthy modification at the place of deposition of drug inside the lung can be accomplished by little adjustment in the method. The powder insufflators are utilized to convey different medications to the lungs in various respiratory and non-respiratory maladies like COPD, asthma, lung cancer, tuberculosis, diabetes, osteoporosis, and cancer.

7.10.8 DRUG DEPOSITION STUDIES

The measure of medication saved in the lung and its appropriation are key parameters in the assessment of the plan and execution of devices for pharmaceutical aerosol. Two different experimental animal methods are used to investigate the deposition of drug, one is invasive and another is noninvasive method. Dolovich studied the various imaging techniques in animal models, i.e., emission computed tomography (SPECT), gamma scintilography and tomography (positron emission) (PET). These techniques are effectively utilized in pneumonic medication delivery and inhalation toxicology. Obtrusive procedures, for example, bronchoalveolar lavage (BAL) technique can be utilized for insoluble, idle particles when the non-intrusive strategies are deficient to gauge the medication statement.

7.10.9 PHARMACOKINETIC STUDIES

Compartmental or non-compartmental techniques are utilized to estimate the pharmacokinetic parameters after administration of the drug. A mathematical model portraying the disposition of a given medication can be attained by adding the physiologically based pharmacokinetic (PBPK). PBPK comprises of various compartments with each speaking to a specific organ. Byron and Gonda addressed the lung residence time of aerosol drug by different models, the numerical and compartmental.

7.10.10 EX-VIVO TESTS

The systems of medication transport crosswise over lung tissue can be studied by various *ex-vivo* model (lung) which are helpful to predict the *in-vitro-in vivo* relationship. Sakagami and Beck-Broichsitter et al. depicted IPL *ex vivo* model appearing of drug particles. PCLS model is used for toxicity testing.

7.11 FORMULATIONS OF PULMONARY DRUG DELIVERY

7.11.1 NANO AEROSOLS

It is a pulmonary targeted drug delivery system used to deal with respiratory or systemic infections or lung diseases such as cancer and in the management of cystic fibrosis. It is also helpful to deliver the drug particulate actives more successfully to the respiratory deeper track, which is due to the intrinsic features of the aerosol as well as specialty of the route of administration. The nanoparticles made up of chitosan-TPP polyelectrolyte complex were used to target the isoniazid, an anti-tubercular drug, to the lugs. It also increases the bioavailability, thereby reducing the frequency of dosing.

7.11.2 LIPOSOMES

Several studies on liposomal formulation, incorporated with the anti-tubercular agent, have proven its efficiency to target the drug directly in the lung. It also provides a sustained type of drug release with minimal biodistribution in the blood circulation.

7.11.3 LACTOSE CARRIER SYSTEMS

Recently, in the field of inhalation systems, d-lactose as an efficient carrier-based drug formulation has been introduced to treat respiratory issues. Lactose as a small unit is used as a carrier, evaluated for drug-carrier adhesion, variation in drug-carrier interaction, and impact on the drug-aerosol show before its actual use. Drug particles mixed and blended with the coarser size lactose monohydrate (63–90 μm) to prepare DPI is well recommended due to its favorable toxicologic information and quick accessibility. Additionally, the excellent flow properties, minimum particle size, and spherical shape increase its popularity as a carrier system.

7.11.4 LARGE SIZED POROUS PARTICLES

Pulmosphere is the new porous and hollow aerosol system. It consists of phosphatidylcholine, which is a basic primary element of surfactant is observed in human lung. The residence time of this new system is more due to the big size of particles that allow it to stay inside the lungs than that other non-porous system. It also avoids the clearance of particles by phagocytic mechanism. For example, testosterone-loaded large porous particles express more bioavailability in contrast to its conventional dosage form.

7.11.5 BIODEGRADABLE POLYMER

Recently, polylactic acid and polyglycolic acid are used to fabricate microsphere belongs to biodegradable category is considered as sustained-release carriers for drug delivery into pulmonary region.

7.11.6 REVERSE MICELLES

These are the micro-droplets formed by adding little quantity of water to a non-polar solvent hexane. It represents a surfactant molecule, which forms a coat around a small water droplet. An active molecule of protein previously added in water, self-orientation works to form to micelle. Recently, this method has been utilized to deliver protein and peptide as well as biotechnologically developed product into the pulmonary tract. The list of commonly targeted active ingredients and its category for pulmonary system are mentioned in Table 7.4.

TABLE 7.4 Some Pulmonary Targeting Drugs and Their Categories

Category of Drugs	Examples
Antibiotics	Streptomycin, amphotericin B, isoniazid
Analgesics	Narcotics, benzodiazepines
Antiemetics	Phenergen, Zofran
Peptides and proteins	Peroxidase, antitrypsin, albumin, IgG
Anticholinergics	Ipratropium bromide, Tiotrop

7.12 APPLICATIONS OF PULMONARY DRUG DELIVERY

7.12.1 ASTHMA AND COPD

Asthma is a chronic disease, and the patient suffers for a long time due to the inflammation and tapering of respiratory airways. Asthma resulted into coughing, breathlessness, chest snugness and repetitively shortness of breath. The individual belonging to all age groups are affected by asthma; however, it regularly begins in the teenage years. Statistical survey gives information about more than 22 million individuals suffered from asthma out of 22 million around 6 million are youngsters.

Aerosol system that convey bronchodilators to loosen up aviation routes and corticosteroids to control aggravation in asthma and COPD are broadly utilized today and convey a demonstrated reputation. The present inhaled drug delivery market is ruled by the three primary classes of medication, for example, bronchodilators, corticosteroids, and anticholinergic. For the management of asthma, advances in medications available such as levos-albutamol inhalers which have better adequacy than salbutamol. COPD is associated to smoking, and contains emphysema and incessant bronchitis, which result into 1 million passing annually. Tiotropiuminhalers are available in market to nipple COPD (Shaikh et al., 2011; Pilcer et al., 2009).

7.12.2 IN PATIENTS ON VENTILATORS

Aerosols are the best approach to deliver the drug to a patient on ventilators. Nowadays, many companies are engaged in the development and marketing of nebulizers. Day-by-day, recent developments give new techniques for the administration of drugs for the various ages patients, like for baby's small baby masks are launched with good inhalation.

7.12.3 IN CYSTIC FIBROSIS

The pulmonic drug delivery system brings a revolution in the treatment of cystic fibrosis for a considerable length of time. The basic objective of aerosol is to convey medications to infants and children.

7.12.4 IN DIABETES

Diabetes is a disorder in which appropriate hyperglycemia results from the inadequacy of insulin secretion. Peptides, as well as proteins, are winding up progressively significant in prescription. A few organizations are dealing with insulin inhalers than some other insulin delivery options. Insulin inhalers are also going to launch just like other inhalers. The two products, i.e., dry powder and solutions, can be administered through different inhaler systems. For example, novel pMDI definitions for pulmonary delivery of proteins (Shaikh et al., 2011; Pilcer et al., 2009).

7.12.5 IN TREATMENT OF MIGRAINE

The olfactory region is a layer that separates the nose and brain. The nasal route is the best approach for targeted drug delivery to the brain. In the treatment of migraine drug of choice, ergotamine can be administered. Since decades, ergotamine administered via MDIs.

7.12.6 IN VACCINATION

While there was moderate enthusiasm for aerosol immunization 15–20 years back, advance toward application has been seen. Nearly 100 antibodies are affirmed in the U. S. About half portion of these anticipates respiratory illnesses, up till now, all are presently injected. In contrast with injectables, they expel needle fears, particularly in youngsters, yet they could likewise create a more potent local immune response than when they are infused into muscles. Measles contaminates 30 to 40 million youngsters in immature countries.

7.12.7 IN ARTERIAL HYPERTENSION

FDA affirmed Ventavis (iloprost), a treatment for arterial hypertension (serious limitation of veins results in early passing) via pulmonary route, manufactured by CoTherix (San Francisco, USA). This is the novel application attributes to the pulmonary route of administration.

7.12.8 IN ACUTE LUNG INJURY

Medication given by pulmonary route assumes a significant role in acute lung damage. Main difficulty during acute lung injury is hypoxemia and it difficult to manage. Choice of treatment is use of nitric oxide and prostacyclin, which helps to improve the oxygenation by expanding blood course through ventilated regions. Prostaglandin E1 (PGE1) is also administered by continuous aerosol device through a ventilator has additionally been appeared to improve oxygenation.

7.12.9 SURFACTANT AEROSOL

In premature infants and neonates, the issue of respiratory suffering occurs frequently where surfactant aerosol plays a significant job. While in adult patients, the loss of surfactant is likely a contributing element which results in acute respiratory distress syndrome. An ongoing report demonstrated no distinction in result for spontaneously breathing newborns who breathed in either surfactant or placebo by means of a CPAP mask.

7.12.10 GENE THERAPY VIA PULMONARY ROUTE

This is a new research of pulmonary drug delivery. Main aim of Gene treatment given by pulmonary route is for the cure of cystic fibrosis. Cationic-lipid-interceded CFTR gene transfer can altogether impact the underlying chloride deformity in the lungs of patients. Lot of issues to be defeated before clinical applications is viable. A portion of these are well-being, fruitful exchange of adequate hereditary material to suitable tissue, satisfactory quality expression, maintenance of articulation after some time and viability of articulation (Shaikh et al., 2011; Pilcer et al., 2009).

7.12.11 APPLICATION IN CANCER TREATMENT

Cancer is a very dangerous disease in which death rate is very high. The number of cancer patients increased in worldwide and one of the major causes is the growth of lung cancer and breathed in chemotherapy appears a sensible way to deal with treat lung malignancy. Various companies involved in clinical trials to assess doxorubicin HCI inward breathe arrangement in lung malignant growth patients. Upwards of 400,000 lung disease patients could profit by breathed in chemotherapy; mice associated with lung tumors are assessed for efficacy of aerosolized paclitaxel product. The management fundamentally diminished lung tumors and prolonged survival.

7.12.12 DIAGNOSTIC APPLICATION

Pulmonary drug delivery system is useful in therapeutic drug delivery as well for the diagnosis purpose. In diagnostics, some radioactive and nonradioactive materials are used size specificity. For example, aerosol of methacholine and histamine is utilized to check nonspecific bronchial response asthma.

7.12.13 IN THE CONTROLLED DRUG DELIVERY TO LUNGS

Recently some investigations are dealt with the use of aerosols for controlled delivery of drugs to the lungs. The basic aim behind these investigations is to reduce the dosing frequency of drug so vesicular approaches are tired, such as liposomes, some solid dosage forms with mucoadhesion property also available, i.e., mucoadhesive microspheres. Nanoparticulate carrier systems are also used for the sustained release of drug from a pressurized package system aerosol which results in extending the residence time of drug of in the pulmonary region, e.g., ribavirin aerosol.

7.12.14 MOST CURRENT USES OF PULMONARY DRUG DELIVERY FOR BONE DISORDERS

Some bone-related diseases can be treated by pulmonary drug delivery, like osteoporosis and Paget's disease. Many patients consult doctors with bone-related issues, and this number has increased in the last few decades. For such patients, it will be a new approach in order to treat such conditions in a better way. It is proven on the basis of clinical trials that the pulmonary route is safe, effective, and well-accepted for the delivery of peptides and proteins. The pulmonary route is the best choice to treat bone-related issues.

7.13 CONCLUSION

A wide assortment of operators has been regulated to the lung by means of oral inward breath for the treatment of various infection states. Likewise, it would be an option in contrast to parenteral medication delivery, most strikingly for the delivery of breathed-in insulin and furthermore for peptide and protein drugs. PDDS gives a sidestepping approach to hepatic first-pass digestion of numerous powerful medications and lessens medication initiated poisonous quality or unfavorable medication responses. In this manner, it presumes that further research is fundamental for the pulmonic system as a most encouraging, progressed, and alluring financial medication delivery. The viability, safety, and nature of the medication and medication products are evaluated and set up by the current compendia and state-sanctioned testing techniques, which are the needs developing with time. A better comprehension of the item testing and portrayal strategies can prompt the effective improvement of the pulmonary drug delivery system.

KEYWORDS

- **aerodynamically particulate matter**
- **broncho alveolar lavage**
- **chronic respiratory distress**
- **pulmonary devices**
- **pulmonic drug delivery system**
- **scanning electron microscopy**

REFERENCES

Begat, P.; Morton, D. A.; Staniforth, J. N.; Price, R. The cohesive-adhesive balances in dry powder inhaler formulations I: Direct quantification by atomic force microscopy. *Pharmaceutical Research,* **2004**, *21*(9), 1591–1597.

Bunker, M.; Davies, M.; Roberts, C. Towards screening of inhalation formulations: Measuring interactions with atomic force microscopy. *Expert Opinion on Drug Delivery,* **2005**, *2*(4), 613–624.

Chan, H. K. Inhalation drug delivery devices and emerging technologies. *Expert Opinion on Therapeutic Patents,* **2003**, *13*(9), 1333–1343.

Chan, H. K.; Chew, N. Y. Novel alternative methods for the delivery of drugs for the treatment of asthma. *Advanced Drug Delivery Reviews,* **2003**, *55*(7), 793–805.

Chan, J. G. Y.; Chan, H. K.; Prestidge, C. A.; Denman, J. A.; Young, P. M.; Traini, D. A novel dry powder inhalable formulation incorporating three first-line anti-tubercular antibiotics. *European Journal of Pharmaceutics and Biopharmaceutics,* **2013**, *83*(2), 285–292.

Chattopadhyay, P.; Shekunov, B. Y.; Yim, D.; Cipolla, D.; Boyd, B.; Farr, S. Production of solid lipid nanoparticle suspensions using supercritical fluid extraction of emulsions (SFEE) for pulmonary delivery using the AERx system. *Advanced Drug Delivery Reviews,* **2007**, *59*(6), 444–453.

Chew, N. Y.; Tang, P.; Chan, H. K.; Raper, J. A. How much particle surface corrugation is sufficient to improve aerosol performance of powders? *Pharmaceutical Research,* **2005**, *22*(1), 148–152.

Duddu, S. P.; Sisk, S. A.; Walter, Y. H.; Tarara, T. E.; Trimble, K. R.; Clark, A. R.; Weers, J. G. Improved lung delivery from a passive dry powder inhaler using an engineered pulmosphere® powder. *Pharmaceutical Research,* **2002**, *19*(5), 689–695.

El-Gendy, N.; Pornputtapitak, W.; Berkland, C. Nanoparticle agglomerates of fluticasone propionate in combination with albuterol sulfate as dry powder aerosols. *European Journal of Pharmaceutical Sciences.,* **2011**, *44*(4), 522–533.

Forbes, B.; Shah, A.; Martin, G. P.; Lansley, A. B. The human bronchial epithelial cell line 16hbe14o–as a model system of the airways for studying drug transport. *International Journal of Pharmaceutics,* **2003**, *257*(1, 2), 161–167.
Geller, D. E. New liquid aerosol generation devices: Systems that force pressurized liquids through nozzles. *Respiratory Care,* **2002**, *47*(12), 1392–1404.
Geller, D.; Thipphawong, J.; Otulana, B.; Caplan, D.; Ericson, D.; Milgram, L.; Bowman, C. M. Bolus inhalation of rhDNase with the AERx system in subjects with cystic fibrosis. *Journal of Aerosol Medicine,* **2003**, *16*(2), 175–182.
Gilani, K.; Najafabadi, A. R.; Barghi, M.; Rafiee-Tehrani, M. The effect of water to ethanol feed ratio on physical properties and aerosolization behavior of spray-dried cromolyn sodium particles. *Journal of Pharmaceutical Sciences,* **2005**, *94*(5), 1048–1059.
Heinemann, L.; Klappoth, W.; Rave, K.; Hompesch, B.; Linkeschowa, R.; Heise, T. Intra-individual variability of the metabolic effect of inhaled insulin together with an absorption enhancer. *Diabetes Care,* **2000**, *23*(9), 1343–1347.
Hickey, A. J.; Garcia-Contreras, L. Immunological and toxicological implications of short-term studies in animals of pharmaceutical aerosol delivery to the lungs: Relevance to humans. *Critical Reviews™ in Therapeutic Drug Carrier Systems,* **2001**, *18*(4).
Hoe, S.; Traini, D.; Chan, H. K.; Young, P. M. Measuring charge and mass distributions in dry powder inhalers using the electrical next generation impactor (eNGI). *European Journal of Pharmaceutical Sciences,* **2009**, *38*(2), 88–94.
Jadhav, K. R.; Gambhire, M. N.; Shaikh, I. M.; Kadam, V. J.; Pisal, S. S. Nasal drug delivery system-factors affecting and applications. *Current Drug Therapy,* **2007**, *2*(1), 27–38.
Jaspart, S.; Bertholet, P.; Piel, G.; Dogné, J. M.; Delattre, L.; Evrard, B. Solid lipid microparticles as a sustained release system for pulmonary drug delivery. *European Journal of Pharmaceutics and Biopharmaceutics*, **2007**, *65*(1), 47–56.
Kaur, I. P.; Singh, H. Nanostructured drug delivery for better management of tuberculosis. *Journal of Controlled Release,* **2014**, *184*, 36–50.
Kuo, J. H. S.; Hwang, R. Preparation of DNA dry powder for non-viral gene delivery by spray-freeze drying: Effect of protective agents (polyethyleneimine and sugars) on the stability of DNA. *Journal of Pharmacy and Pharmacology,* **2004**, *56*(1), 27–33.
Labiris, N. R.; Bosco, A. P.; Dolovich, M. B. Aerosols as drug carriers. In *Nanoparticulates as Drug Carriers*; Torchilin V. P., Ed.; Imperial College Press., **2006**, pp. 367–395.
Liang, Z.; Ni, R.; Zhou, J.; Mao, S. Recent advances in controlled pulmonary drug delivery. *Drug Discovery Today*, **2015**, *20*(3), 380–389.
Maa, Y. F.; Prestrelski, S. J. Biopharmaceutical powders particle formation and formulation considerations. *Current Pharmaceutical Biotechnology,* **2000**, *1*(3), 283–302.
Maretti, E.; Rossi, T.; Bondi, M.; Croce, M. A.; Hanuskova, M.; Leo, E.; Iannuccelli, V. Inhaled solid lipid microparticles to target alveolar macrophages for tuberculosis. *International Journal of Pharmaceutics,* **2014**, *462*(1, 2), 74–82.
O'Callaghan, C. Targeting drug delivery to the lungs by inhalation. *Mediators of Inflammation,* **1994**, *3*(7), S31–33.
Patton, J. S.; Byron, P. R. Inhaling medicines: Delivering drugs to the body through the lungs. *Nature Reviews Drug Discovery*, **2007**, *6*(1), 67–74.
Pilcer, G.; Vanderbist, F.; Amighi, K. Preparation and characterization of spray-dried tobramycin powders containing nanoparticles for pulmonary delivery. *International Journal of Pharmaceutics,* **2009**, *365*(1, 2), 162–169.

Rasenack, N.; Steckel, H.; Müller, B. W. Micronization of anti-inflammatory drugs for pulmonary delivery by a controlled crystallization process. *Journal of Pharmaceutical Sciences,* **2003**, *92*(1), 35–44.

Rehman, M.; Shekunov, B. Y.; York, P.; Lechuga-Ballesteros, D.; Miller, D. P.; Tan, T.; Colthorpe, P. Optimization of powders for pulmonary delivery using supercritical fluid technology. *European Journal of Pharmaceutical Sciences,* **2004**, *22*(1), 1–17.

Rogers, T. L.; Johnston, K. P.; Williams III, R. O. Solution-based particle formation of pharmaceutical powders by supercritical or compressed fluid CO_2 and cryogenic spray-freezing technologies. *Drug Development and Industrial Pharmacy,* **2001**, *27*(10), 1003–1015.

Sakagami, M. *In vivo, in vitro* and *ex vivo* models to assess pulmonary absorption and disposition of inhaled therapeutics for systemic delivery. *Advanced Drug Delivery Reviews,* **2006**, *58*(9, 10), 1030–1060.

Salama, R.; Hoe, S.; Chan, H. K.; Traini, D.; Young, P. M. Preparation and characterization of controlled release co-spray dried drug-polymer microparticles for inhalation 1: Influence of polymer concentration on physical and *in vitro* characteristics. *European Journal of Pharmaceutics and Biopharmaceutics.*, **2008**, *69*(2), 486–495.

Salama, R.; Traini, D.; Chan, H. K.; Young, P. M. Preparation and characterization of controlled release co-spray dried drug-polymer microparticles for inhalation 2: Evaluation of *in vitro* release profiling methodologies for controlled release respiratory aerosols. *European Journal of Pharmaceutics and Biopharmaceutics*, **2008**, *70*(1), 145–152.

Scheuch, G.; Kohlhaeufl, M. J.; Brand, P.; Siekmeier, R. Clinical perspectives on pulmonary systemic and macromolecular delivery. *Advanced Drug Delivery Reviews,* **2006**, *58*(9, 10), 996–1008.

Shaikh, S.; Nazim, S.; Shaikh, A.; Khan, T.; Zameeruddin, M.; Majaz, Q.; Chalikwar, S. Recent trends in applications of pulmonary drug delivery: A review. *International Journal of Pharmaceutical Research & Development*, **2011**, *2*(12), 171–180.

Tang, P.; Chan, H. K.; Raper, J. A. Validation of computation method to predict aerodynamic diameter of particles with rough surface. *Powder Technology,* **2009**, *192*(1), 74–84.

Tang, P.; Chew, N. Y.; Chan, H. K.; Raper, J. A. Limitation of determination of surface fractal dimension using N_2 adsorption isotherms and modified Frenkel− Halsey− hill theory. *Langmuir,* **2003**, *19*(7), 2632–2638.

Telko, M. J.; Kujanpää, J.; Hickey, A. J. Investigation of triboelectric charging in dry powder inhalers using electrical low pressure impactor (ELPI™). *International Journal of Pharmaceutics,* **2007**, *336*(2), 352–360.

Ugwoke, M. I.; Vereyken, I. J.; Luessen, H. Microparticlesand liposomes as pulmonary drug delivery systems: what are the recent trends? In *Pulmonary Drug Delivery Basics, Applications and Opportunities for Small Molecules and Biopharmaceutics*; Karolinc B. P., Henrik, L., 1st Eds.; Editio Cantor Verlag: Aulendorf Germany., **2007**; pp. 308–377.

Young, P. M.; Sung, A.; Traini, D.; Kwok, P.; Chiou, H.; Chan, H. K. Influence of humidity on the electrostatic charge and aerosol performance of dry powder inhaler carrier based systems. *Pharmaceutical Research,* **2007**, *24*(5), 963–970.

Yu, Z.; Garcia, A. S.; Johnston, K. P.; Williams III, R. O. Spray freezing into liquid nitrogen for highly stable protein nanostructured microparticles. *European Journal of Pharmaceutics and Biopharmaceutics,* **2004**, *58*(3), 529–537.

Zhou, Q. T.; Leung, S. S. Y.; Tang, P.; Parumasivam, T.; Loh, Z. H.; Chan, H. K. Inhaled formulations and pulmonary drug delivery systems for respiratory infections. *Advanced Drug Delivery Reviews*, **2015**, *85*, 83–99.
Zhu, K.; Ng, W. K.; Shen, S.; Tan, R. B.; Heng, P. W. Design of a device for simultaneous particle size and electrostatic charge measurement of inhalation drugs. *Pharmaceutical Research,* **2008**, *25*(11), 2488–2496.

CHAPTER 8

Nanoencapsulated Nasal Drug Delivery System

Nayan A. Gujarathi,[1] Akshada A. Bakliwal,[2] Bhushan R. Rane,[3] Vasim Pathan,[3] and Raj Kumar Keservani[4]

[1]Shri Vile Parle Kelavani Mandal's Institute of Pharmacy, Dhule–424001, Maharashtra, India, E-mail: nayang5678@gmail.com

[2]Sandip Institute of Pharmaceutical Sciences, Nashik–422213, Maharashtra, India

[3]Shri D.D. Vispute College of Pharmacy and Research Center, New Panvel–410206, Maharashtra, India

[4]Faculty of B. Pharmacy, CSM Group of Institutions, Prayagraj–212111, Uttar Pradesh, India

ABSTRACT

The interest of pharmaceutical researchers has expanded towards investigating the conceivable outcomes of nasal delivery of different medications. Intranasal drug delivery system is conventionally recognized for the cure of illnesses like cold, hack, rhinitis, and so on. Recent research activities have been made to convey different medications, particularly protein and peptide, employing rhinal path for initial use; using the standards and ideas of different nanoparticulate sedate conveyance frameworks utilizing different polymers and retention advertisers. The use of medications into nanoparticles may be a promising methodology, since colloidal plans have been appeared to shield them from the corrupting milieu in the nasal pit

Topical and Transdermal Drug Delivery Systems: Applications and Future Prospects. Nayan A. Gujarathi, PhD, Juliana Palma Abriata, PhD, Raj Kumar Keservani, MPharm, Anil K. Sharma, PhD, MPharm (Eds.)

and encourage their vehicle over the mucosal obstructions. The intranasal delivery is gone for improving medication bioavailability for fundamental medications, as assimilation diminishes with expanding molecular weight, and for medications, which are powerless to enzymatic degradation, for example, proteins and polypeptides.

8.1 BACKGROUND

Utilizing microparticles as another method for delaying the habitation period inside the adenoidal space has presented in 1987. It was suggested that microspheres of egg whites, starch, and DEAE-dextran (diethyl aminoethyl-dextran) consumed water and shaped into a semisolid viscous film, which subsequently disappeared gradually from the nasal cavity. After three hours of organization, half of the conveyed measure of egg whites, dextran microparticle (60%), and starch microparticle are present at the site of administration. It was proposed that an extended contact time could contribute to the assimilation productivity of medications. As recommended, the relative bioavailability for nasal route in comparison to subcutaneous route for human development hormone in sheep was lengthened from 0.1% for the response for 2.7% for biodegradable starch microparticles. The expansion of assimilation enhancer viz. lyso phosphatidylcholine further upgraded hormone ingestion as a general bioavailability of 14.4% was improved.

Besides, as of late strong lipid nanoparticles have additionally indicated promising outcomes and were appeared to build the cerebrum focusing of rosmarinic corrosive after nasal conveyance for potential administration of Huntington's malady.

8.2 INTRODUCTION

Nowadays, the drug administered via nasal route is progressively perceived and procured. Numerous pharmaceutical products have been created for the nasal drug delivery route that presently takes points of interest of the advantages, including accommodation and convenience, fast beginning of activity, evasion of quick hepatic digestion, potential outcomes to limit systemic presentation, higher bioavailability, and quick brain access through the nerves that puncture the cerebra form plate into the olfactory bulb (Illum,

2003). These products are generally conveyed as a nasal spray; however, some offer the choices of direct fluid or powder introduction. Recently, the issues related with this course of administration were altogether reviewed.

8.3 ANATOMY AND PHYSIOLOGY OF NASAL CAVITY

The nasal cavity is categorized into three major areas, which are discussed in subsections (Thorne and Frey, 2001).

8.3.1 VESTIBULAR REGION

It is arranged simply into the nostrils and the stratified keratinized cells and squamous epithelium layer protects this region. The sebaceous glands are also available in this region. The area of the vestibular region is about 10 to 20 cm^2 and is viewed as the least significant of the three parts in regard of medication ingestion.

8.3.2 RESPIRATORY REGION

This section contributes to the largest portion of the nasal passage, encompasses a zone of around 130 cm^2 and comprises of three turbinates in particular inferior, center, and superior, which are in charge of humidification and temperature guideline of breathed in air. The intranasal permeation of drugs is expected to eventuate fundamentally in the respiratory region since it is having highest level of vascularity.

8.3.3 OLFACTORY REGION

The olfactory region is situated at the topmost part of the nose in connection with brain over the verge of the septum. The size of the olfactory region is around 10–20 cm^2. It also comprises of the receptors directed for the feeling of smell. Human olfactory region involve lamina propria which is a connective tissue. The cilia are non-motile in this area rather than the respiratory region since they do not have Mg^{++} ATPase that produces the power for ciliary motility.

The nasal cavity additionally contains adenoidal-related lymphoid tissue (NALT). In people, the NALT is ordinarily related to the Waldeyer ring, and contains particular M-like cells, which are specific for the take-up and transcellular transport of very large molecules and microbes. The foundational resistant reaction, as shown in Figure 8.1, is induced by dendritic cells, macrophages, and T and B lymphocyte follicles. These are the composition of lymphoid tissue existing under the nasal membrane.

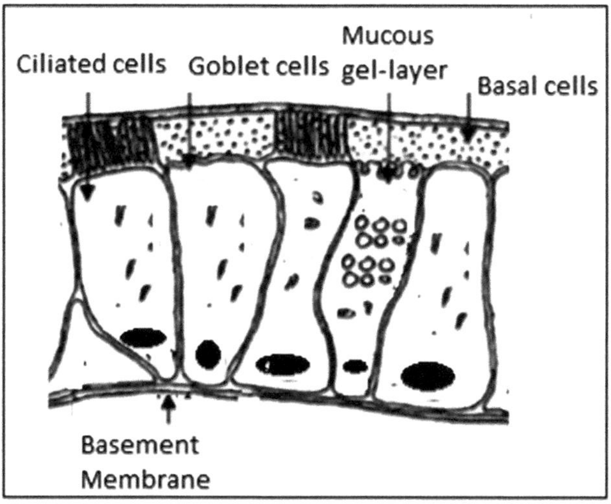

FIGURE 8.1 Nasal epithelium showing different cell types.

8.4 NANOENCAPSULATION

Nanoencapsulation is characterized as the way toward exemplifying substances with different covering materials at the nanoscale level. This method is principally utilized inside the pharmaceutical, nourishment, and restorative ventures.

8.4.1 BENEFITS OF NANOENCAPSULATION

- Preservation of active pharmaceutical moiety from decomposition;
- Delivery of drug at target site is feasible by means of conjugation or covering;

- PEGylating the nanoencapsulated particles prolongs the systemic circulation time;
- Accumulation of nanoencapsulated particles is promoted into the cell by the modification of surface charge;Fluorescent labeling for imaging.

8.5 FORMULATION CONSIDERATION IN NANOENCAPSULATED NASAL DRUG DELIVERY

Notwithstanding its physiological difficulties, the nasal route offers beneficial administration in contrast to subcutaneous or transdermal routes. Drug administration by the nasal route has been analyzed by numerous agents since this course takes into consideration high assimilation of low-atomic weight, hydrophobic medications, evasion of first-pass impacts and straight forwardness in organization by patients. Conveyance of non-biocompatible, unprotected medications and certain excipients legitimately to the fragile nasal epithelium is probably going to annoy allergic nasal stasis, and leads to intranasal discomfort. In this manner, medication details must be compatible with nasal functioning for actual application of the nasal way of administration. A suitable approach to design safe and effective nasal preparation is to utilize smaller scale or nano-encapsulation. The particle size is the basic factor that influences their collaboration and transport crosswise over mucosal surfaces.

With encapsulation and controlled release, low effective concentration of active agents can be accomplished in nasal epithelium and improve the probability of the formulation being biocompatible. The moderate discharge can result in decreased exposure of the mucosal epithelial cells, particularly if ingestion enhancers are dosed with the microcapsule, so the dynamic medication can be consumed as fast as it is discharged. Two biocompatible and biodegradable polymers, to be specific PLGA {poly (D,L-lactide-co-glycolide)} and chitosan, had been comprehensively used in embodiment of dynamic substances for nasal drug delivery.

8.5.1 CORE MATERIALS FOR NANOENCAPSULATION

The principal materials of hydrophilic and lipophilic nature (e.g., nutraceuticals compound) are most widely utilized for nanoencapsulation.

The hydrophilic substances are dissolved in water and insoluble in lipophilic components (lipids) and natural solvents. In the same manner, the lipophilic substances are insoluble in the water yet soluble in lipids and natural solvents. Some nanoencapsulated hydrophilic substances are ascorbic acid, polyphenols, etc. The nanoencapsulated lipophilic substance incorporates lycopene, beta-carotene, lutein, phytosterols, and docosahexaenoic acid (Lopez et al., 2006; Lakkis, 2007; Heyang et al., 2009; Zimet and Livney, 2009).

8.5.2 COATING MATERIALS FOR NANOENCAPSULATION

1. **Polymers Utilized in Formulation of Nanoparticles:** The biopolymers are compatible with the human body in terms of versatility (non-toxicity, non-antigenicity) and ought to be biodegradable and biocompatible (Leong et al., 2011).
2. **Natural Polymers:** Most routinely utilized natural polymers in the fabrication of polymeric nanoparticles are sodium alginate, chitosan, gelatin, and albumin (Ghosh, 2000).
3. **Synthetic Polymers:** A lot of synthetic polymers are available and commercially employed in the fabrication of nanoparticles, such as polylactides (PLA), poly(ethylene glycol), polycyanoacrylates, poly(lactide co-glycolides) (PLGA), polyanhydrides, polyglycolide (PGA), polyorthoesters, poly glutamic acid, polycaprolactone, PMA, PVP, carbomer, PMMA, PVA, polymethacrylate, polyacrylamide, etc. (Nagavarma et al., 2012).

8.6 REASON FOR DEVELOPMENT OF NASAL DELIVERY

Nasal drug delivery is an appropriate delivery technique for medications that are effective in small dosage and demonstrate no or insignificant oral bioavailability. The nasal route excuses hepatic first-pass disposal related with the oral delivery. Right now, two classes of nasally conveyed therapeutics are available. The first class of actives has low molecular weight and others contain hydrophobic moieties. Both of these actives are used for the illnesses of the nasal mucosa and sinus. List of active drugs administered via nasal route are steroids, decongestants, topical agents, anti-toxins, and other (OTC) products. The below-average envelops a couple of medications, which have adequate nasal retention for showing fundamental impacts. Significant candidates are the compounds, largely regulated by injection and hardly absorbed after oral organization, because of their insecurity in the gastrointestinal tract, poor absorption properties, and their quick and broad biotransformation.

8.7 MECHANISM OF DRUG ABSORPTION

8.7.1 FIRST MECHANISM

It includes a watery course of transport, which is otherwise called the paracellular course yet moderate and aloof. There is an opposite log-log relationship between intranasal retention and the atomic load of water-solvent compounds. The medications having atomic weight more noteworthy than 1,000 Daltons indicate poor bioavailability.

8.7.2 SECOND MECHANISM

It includes transport through a lipoidal course and it is otherwise called the transcellular procedure. It is in charge of the vehicle of lipophilic medications that demonstrate a rate reliance on their lipophilicity. Active transport is one of the mechanisms for the transport of drug across the nasal cell membrane by means of bearer intervened way or transportation across the aperture in the impervious junction.

8.8 NANOENCAPSULATION TECHNIQUES

Nanoencapsulation strategies use either top-down or bottom-up methodologies for the progression of nanomaterial (Augustin and Sanguansri, 2009).

8.8.1 TOP-DOWN APPROACH

A top-down technique includes the use of precise devices that permit the size drop and development leads to utilization of the nanomaterials being created. Example: emulsification method and solvent evaporation method:

1. **Emulsification Method:** The principle of emulsification includes center material which is busted down into polymerization arrangement. The monomers are polymerized to frame containers in a fluid arrangement.
    i. **Advantage:** Micro or nanocapsule formulated possess marginal particle size distribution.
    ii. **Disadvantage:** Difficult to control capsule formation.
2. **Solvent Evaporation Method:** The solvent evaporation method is proceeded through a liquid assembling vehicle. The microcapsule coating is dissolved in the volatile solvent, which is used here and must be immiscible with other vehicle phases wherein other components are added. A material (core) under microencapsulation is distributed in the covering polymer arrangement. After agitation, the core coating material blend is distributed in the solution of vehicle phase to get the proper size nanocapsule. The blend is then warmed (if important) to evaporate the dissolvable for the polymer. For the situation wherein the core material is scattered in the polymer arrangement, polymer recoils around the core. For the situation where the core component is bust down inside the polymeric reservoir, a lattice-type nanocapsule is formed. When all the dissolvable for the polymer is dissipated, the fluid vehicle temperature is decreased to encompassing temperature (whenever required) with proceeded with unsettling. At this stage, the nanocapsules can be utilized in suspension structure, covered on to substrates or separate out as powders. The dissolvable dissipation procedure to create nanocapsule is pertinent to a wide assortment of solid and liquid principal material. The principal material might be either water–solvent or water-insoluble components. A range of coating polymers can be utilized as film-forming components.

8.8.2 BOTTOM-UP APPROACH

In the bottom-up methodology, materials are built without anyone else get together and self-association of molecules, which were impacted by numerous elements such as temperature, pH, ionic strength, and concentration. Supercritical liquid method, consideration nanoprecipitation, coacervation, and complexation categorized under the bottom-up approach.

1. **Complexation:** In this, a specific polar particle is entangled through a hydrophobic connection inside the β-Cyclodextrin cavity supplanting water particles.
   i. **Advantage:** Efficient to secure fragile and high included esteem a polar mix such as flavor.
   ii. **Disadvantages:**
      a. Encapsulation limited to a polar mix with some appropriate subatomic measurements;
      b. β-Cyclodextrin is costly.
2. **Coacervation:** In this, the capture is because of the statement of a fluid covering material around the center material by electrostatic fascination.
   i. **Advantage:** Used to encapsulate heat-sensitive ingredients.
   ii. **Disadvantages:** The complex coacervates are very delicate, expensive, and complex technique.

8.9 TASK AND SCOPE FOR NASAL DELIVERY SYSTEMS

Existing nasal delivery devices, for example, spray pumps and pipettes can't completely exploit the depicted potential preferences of nasal delivery. An enormous portion is kept on the front segment lined by skin, which isn't the objective for either topical drugs or systemic drugs. Drugs transported near the nostrils may cause bad taste and aggravation and decrease persistent acknowledgment. At long last, lacking, and variable confirmation in the remote area lodging the openings to the sinuses and center ears, just as the olfactory locale, speaks to a genuine test for broadened utilization of nasal administration of drugs and vaccines. This applies specifically to the new progressed and costly medications requesting blend of solid dosing, high patient consistency and reproducible bio-accessibility to guarantee their adequacy and security. With respect to definition, most nasal products are now formulated as a solution and administered through metered-dose spray pumps. Fluid definitions can be constrained by the solvency, security, and portion volume. Powders, then again, are progressively steady and it is simpler to squeeze the size and surface properties. A few investigations show reduced local irritation and more rapid absorption of powders. Bioactis Ltd. ("Bioactis;" Chief, Ryoichi Nagata, MD, PhD) has been creating nasal devices, which are therapeutic devices that deliver drugs to the nose.

8.10 CURRENT APPROACHES FOR INCREASING NASAL BIOAVAILABILITY

In most of the drugs those administered via nasal route showed less bioavailability because of drug solubility problem, fast degradation of drug due to enzymes available in nostrils and poor membrane penetration. Thus, some of approaches have been suggested to overcome these limitations.

8.10.1 NASAL ENZYME INHIBITORS

The biotransformation of drug moiety in the nasal cavity can be disposed of by utilizing the catalyst inhibitors. Principally for the formulation of nasal nanoencapsulated system which contains peptide groups (e.g., proteins, and peptides), enzyme inhibitors for peptidases and proteases were utilized. The absorption enhancer such as fusidic acid subsidiaries and salts likewise indicates protein hindrance movement to build the retention and bio-accessibility of the medication. The other catalyst inhibitors usually utilized for the enzymatic action are boroleucin, amastatin, bestatin, trypsin, borovaline, and aprotinin inhibitor.

8.10.2 PENETRATION ENHANCER

The predominantly utilized agents for enhancing the absorption of the active medicament are known as permeation enhancer. In general, the absorption enhancer's act by means of one of the mechanisms mentioned below:

- Inhibit enzyme activity;
- Reduce bodily fluid consistency or flexibility;
- Decrease mucociliary clearance;
- Open tight intersections; and
- Solubilized or balance out the medication.

The mechanism of permeation enhancer is to expand the rate at which medication goes through the nasal mucosa. Numerous enhancers accomplish it by adjusting the structure of epithelial cells somehow or another, yet they ought to achieve this while making no harm or lasting change to nasal mucosa.

Ideal characteristics of penetration enhancer:

- For drug, it should be a compelling increment;
- It ought not make permanent harm or modification to the tissues;
- It ought to be non-irritant and nontoxic;
- It ought to be functional in slight amount;
- At the time of absorption, the enhancing effect should occur;
- The impact ought to be brief and reversible;
- It ought to be compatible with different excipients.

Different kinds of penetration enhancers have been assessed for natural medications, including surfactants, bile salts, chelators, phospholipid, and diethylene glycol, salt of unsaturated fatty acid, β-cyclodextrin and glycyrrhetinic acid derivatives.

Chemical penetration enhancer – classification:

- Surface active agent: Saponin, polyoxyethylene-9-lauryl ether (Laureth-9);
- Glycyrrhetinic acid derivates: Carbenozolone, glycyrrhizinate;
- Bile salt;
- Fatty acid salt;
- Chelators: Salicylates, ethylene diamine tetra-acetic acid (EDTA);
- Glycols: n-glycofurols and n-ethyleneglycols;
- Phospholipids: Lyso phosphatidylcholine (lyso-PC), Di-decanoyl –PC;
- Derivatives of cyclodextrin.

8.10.3 PRODRUG APPROACH

Prodrug approach is primarily implied for optimization of physicochemical properties, for example, solubility, taste, odor, stability, and so forth. Prodrug is normally used as pro-moiety; it contains various functional groups with undesired effects. This prodrug approach is essentially applied for improving the nasal bioavailability, particularly for the proteins and peptides to increase the permeability through the nasal membrane and also enzymatic stability. The prodrug experiences enzymatic change to discharge the active moiety, when it undergoes the process of enzymatic degradation and film destruction. By formulating the enamine subsidiaries of calcitonin, bradykinin, enkephalin, vasopressin, carnosine, and angiotensin II the absorption of these peptides is enhanced cauline with prodrug approach.

8.10.4 STRUCTURAL MODIFICATION

Alteration of medication structure without modifying pharmacological movement is one of the worthwhile approaches to improve the nasal retention. The alteration of active pharmaceutical ingredient is regularly used to adjust its physicochemical characteristics of a system, for example, sub-atomic size, sub-atomic weight, pKa, and dissolvability to improve the nasal ingestion of medication. For example, change of calcitonin-salmon to ecatonin (C-N bond substitutes the S-S bond) demonstrated preferable bioavailability over salmon calcitonin.

8.10.5 PARTICULATE DRUG DELIVERY

Molecule configuration is an inevitably play a significant role in improvement of absorption. Microspheres, nanoparticles, and liposomes are used as transporters for drug moiety entrapped within matrix or reservoir system. The characteristics of such a system might be changed to boost remedial adequacy. In general, this can result in expanded assimilation adequacy and security and diminished poisonous quality of the therapeutic efficacy. Systems can be intended to be mucoadhesive to expand the maintenance time and encourage supported discharge. Microspheres will fundamentally amplify the ingestion and in turns, bioavailability by adhering to the mucosal layer in the nose and thereby improve the nasal residence time of medication. The microspheres formulated by utilizing polymers like dextran, chitosan, and biodegradable starch effectively advances the bioavailability of different medications. Liposomes are amphiphilic in nature are recognized to promote penetration of medications through the physiological membrane. Cationic liposomes were having great permeation capacity than adversely charged anionic liposomes.

8.11 EVALUATION OF NANOENCAPSULATED NASAL DRUG DELIVERY

8.11.1 NANOPARTICLE SIZE DETERMINATION

Dynamic light scattering (DLS) technique is utilized to estimate the particle size and polydispersity index of nanoparticulate system meant for nasal administration. Most of the research work utilizes Malvern Zetasizer to determine particle size of nano-formulation. In DLS estimation, the formulation was diluted with refined water separated at 0.45 μm to keep away from various dispersing. The temperature is maintained at 25°C and the scattering angle is adjusted to 90° for the investigation of nanoparticle.

8.11.2 SURFACE CHARGE DETERMINATION

The nanoparticle surface charge was revealed by utilizing phase analysis light dispersing method. Similar methods were exploited exclusively for the determination of particle size and zeta potential assurance, with the same instrument. In the process of estimating surface charge for given test materials, the nanoparticles were diluted with refined water sifted at 0.45 μm to accomplish an ideal conductance without changing the surface charge properties of nanoparticles.

8.11.3 NANOPARTICLE TRACKING ANALYSIS (NTA)

The experiment of nanoparticle tracking analysis (NTA) was performed to determine as well as further affirm the particle size distribution and the concentration of particle in the suspension. It is carried out on NanoSight NS300 (Malvern Instruments Ltd.) outfitted with a laser light source of 480 nm and magnifying lens of 20× amplification. The tools were utilized to complete the tracking analysis along with the field of viewpoint on around 100×80×10 μm.

8.11.4 IMAGING OF NANOPARTICLE USING SCANNING ELECTRON MICROSCOPY (SEM)

The morphology of raw material and formulation was scrutinized with the help of scanning electron microscopy (SEM).

8.11.5 SYNCHROTRON SMALL-ANGLE X-RAY SCATTERING (SAXS) MODEL

Nanoparticle structure determination and its contact with a nasal mucosa study were determined by means of synchrotron small-angle X-ray scattering (SAXS) model. The inner configuration of the nanoparticulate nasal system and modification in its morphology arises due to its interaction with a replica of nasal body fluid, which is examined by technique synchrotron small-angle X-ray scattering (SAXS). The X-beam cross-segment was 200×400 μm with λ=0.1 nm. All determinations were performed at a temperature 25°C. Plastic capillaries of 2 mm internal diameter mounted on a level plane onto a thermostatic test holder were used to place test samples.

8.11.6 ENCAPSULATION EFFICIENCY (EE)

Indirect method is employed to estimate the encapsulation efficiency (EE) of drug in nanoparticle means for nasal drug delivery. The amount of active ingredient in free and precipitated form was quantified and subtracted from the total quantity of the drug moiety weighed and incorporated in the method of preparation. The amount of drug present in the formulation is revealed by HPLC. Thereafter it is expressed as a percentage of total amounts of drug in the formulation.

8.11.7 IN VITRO DRUG RELEASE

In vitro release from nanocapsule was studied using same physiological fluid present in the nasal cavity, i.e., simulated nasal electrolytic solution (SNES) at biologic human convergences of the nasal liquid. During dialysis study, 1 ml of nanoemulsion was accurately measured which is further diluted with 1 mL of SNES of pH 6.5 and lastly kept into dialysis bag. The sealed bags were soaked into a release medium (100 ml) containing 0.5% BSA maintained at 37°C and magnetically mixed at required rpm. At predetermined time point's aliquots of the disintegration medium were withdrawn and replaced with a proportional measure of fresh release fluid. Aliquot samples were subject to analysis by HPLC to estimate the amount of drug release from dosage form.

8.12 APPLICATIONS OF NANOENCAPSULATION

8.12.1 INTRANASAL ADMINISTRATION FOR VACCINE DELIVERY

Nanocarriers have several qualities that make them one of the recently used categories as an immunization adjuvant. Initially, these nanocarriers are one of the main adjuvants that can successfully build the measure of antigen that achieves the immune system. Besides, nanocarriers can control the arrival of the antigen over a prolonged period of time. In continuous with this, it also impacts the antigen accessibility after some time, and through that, likewise results in safe reaction. Thirdly, the nanocarriers can join their main role as antigen delivery system with different impacts, for example, immune-stimulation, and immunomodulation. So as to accomplish a methodical arrangement, the nanocarriers are categorized into two types – lipid-based nanocarriers and polymeric nanocarriers. A few examples have been cited for the administration of nanocarriers by means of intranasal administration like PLA-PEG nanoparticles which have prospective for the delivery of nucleic corrosive antibodies via nasal route. Numerous difficulties stay ahead, the expanding learning about the immune system, the developing field of nano-drug and the use of new nanotechnologies together with a progressively positive view towards the advantages of mucosal immunization by administrative experts imagine the future landing of another age of intranasal antibodies (Peek et al., 2008; Panyam and Labhasetwar, 2003; Vila et al., 2002; Csaba et al., 2004).

The readers may have a review on licenses for intranasal conveyance by Misra et al. (2008). There are additional licenses which depict that how the nanoparticulate medicate conveyance is affecting the specialists and pharmaceutical enterprises to create definitions by means of intranasal organization.

Polymeric nanoparticles in view of their size are specially taken up by the mucosa related lymphoid tissue. They are broadly assessed for nasal organization. Hsing-Wen et al. (2009) displayed a development which is identified with therapeutic utilization of nanoparticles having a pharmaceutical organization of chitosan and polyglutamic corrosive with bioactive operators and conveyance methods for nasal organization for upgraded penetrability.

Soane et al. (1999) have arranged nasal flu immunization utilizing chitosan. Nasal flu immunization may demonstrate to be a decent option in contrast to parenteral infusion in view of the upgrade of the mucosal resistant reaction and the simplicity of antibody organization. This examination researched the utilization of chitosan, a bioadhesive polymer, as a nasal conveyance framework with inactivated, subunit flu immunization. The marketed product available in the form of a nanocarrier system is given in Table 8.1.

TABLE 8.1 Intranasal Vaccines Formulated in Nanocarrier System

Antigen	Nano-Carrier Formulation	Conclusions
Cholera toxin	Multiple emulsion	Higher titer both qualitatively and quantitatively in mucosal membranes and systemic circulation
Tetanus toxoid	Liposomes	Intranasal administration was found effective for inducing mucosal immunity
Toxoplasma gondii, tachyzoites	Micro/nanoparticles	Improved levels and elevated mucosal and systemic immunity
Plasmid DNA	Cationic nanoparticles	25–30 fold higher beta-galactosidase response

8.12.2 INTRANASAL ADMINISTRATION FOR DIRECT BRAIN DELIVERY (DAHLIN ET AL., 2001; WESTIN ET AL., 2005)

This segment reports examination of the transport pathways of medications from the nasal route to the CNS, perceptions were made on transport route of therapeutic drugs from nasal administration. It is revealed from the animal studies that increasing the drug atomic weight, hydrophilicity, and level of ionization can diminish medication transport into the CNS after intranasal administration. Along these lines, transport of a medication legitimately into the CSF, as a measure for CNS conveyance is controlled by a blend of atomic and natural properties of the medication which are at this stage hard to anticipate. At last, various examinations have demonstrated that CNS bioavailability of low sub-atomic weight tranquilizes after intranasal installation is extremely low. Direct nose-to-brain conveyance of various medications will be reliant on different assimilation pathways. The lipophilic medication will be retained into the circulation system through passive transport mechanism. The medications additionally reach to the cerebrum from blood by intersection BBB; however, can likewise be wiped out from the CSF into the blood. The medication gets ingested from the nose by means of the olfactory locale into the CSF and conceivably further to the cerebrum. The medication retention or medication lost by means of various pathways is very subject to the qualities of medication. When medications are managed nasally, the medication will typically be cleared by the mucociliary clearance system at the same time, the above assimilation pathways for medications restrain leeway by mucociliary clearance system. The utilization of nanoparticles may offer an improvement to nose-to-cerebrum medicate conveyance since they can shield the epitomized medication from organic or potentially synthetic debasement, and extracellular transport by P-gp efflux proteins. This would build CNS accessibility of the medication.

8.12.3 INTRANASAL ROUTE FOR SYSTEMIC DELIVERY (ILLUM, 2003; STOKE ET AL., 2008)

The intranasal delivery is a compelling method to systemically convey drugs as a substitute to intravascular and oral routes as it offers the potential for quick ingestion and quick beginning of activity, while maintaining a strategic distance from hepatic first-pass digestion. Potential applications incorporate the treatment of acute pain, nausea, and vomiting. Some systemic medications, for example, calcitonin, desmopressin, painkillers as morphine and against headache mixes are now promoted through nasal route. As of recently no nanoparticulate intranasal detailing is accessible in market as, the analysts have concentrated on the delivery system to make it for explicit focusing in sick conditions. The clinical utilization of macromolecular medications connected nasally is expanding bit by bit and these medications are planned for non-parenteral application thinking about their clinical favorable position and cost of the improvement. The atomic weight of these medications is between 1,000 and 3,400 Daltons, and bioavailability is roughly 10% which causes these medications as a decent contender to create into the nanoparticulate formulation for nasal delivery. The marketed intranasal formulation were given in Table 8.2 for systemic delivery of drug.

TABLE 8.2 Marketed Intranasal Formulation

Drug Molecule	Brand	Company	Indications
Salmon calcitonin	Miacalcin®	Novartis, USA	Postmenopausal osteoporosis treatment
Estradiol	Aerodiol®	Servier Laboratories, France	Hormone replacement therapy (HRT)
Desmopressin	Desmospray®	Ferring Pharmaceuticals, US	Control of dehydration in diabetes insipidus
Nicotine	Nicotrol NS®	Pfizer, USA	Smoking cessation
Sumatriptan	Imigran®	GlaxoSmithKline, USA	Migraine and cluster
Zolmitriptan	Zomig Nasal®	AstraZeneca, USA	Migraine and cluster headache

8.12.4 EFFICACY AND TRANSPORT OF NANOPARTICLES/ NANOEMULSION/MICROEMULSION (ZHANG ET AL., 2006; BETBEDER, 2000)

The drug delivery perspective has shown application of nanoparticles composed of polymers and this shows measurably more prominent capacity, than a simple solution of the medication, to convey model medications, for example, nimodipine to the olfactory bulb or to upgrade the pharmacological movement of morphine, when these little particles are connected intranasally in mix with nanoparticles. Consequently, noteworthy upgrade of nimodipine conveyance to the CSF and olfactory bulb has been accounted for when intranasally connected to rodents in methoxy-PEG-PLA (MPEG-PLA) nanoparticles (76.5 nm width) contrasted with medication in arrangement. The examination demonstrated a reliance of the vehicle on the size and surface characteristics of the nanoparticles and the kind of cell culture utilized. Generally, extremely low amounts of particles were transported through both of the cell cultures, the little particles were transported to a higher degree than the bigger particles and laminated particles were transported through co-culture systems superior to carboxylated particles. The commercial formulation available in the pharmaceutical market are given in Table 8.3.

TABLE 8.3 Intranasally Administered Microemulsion and Nanoemulsion Formulations

Drug	Type of Formulation	Results
Clonazepam	Microemulsion	Accumulation and transport of drug in brain
Risperidone	Nanoemulsion	Rapid and efficient delivery to brain by intranasal administration
Sumatriptan	Microemulsion	Accumulation of drug in brain via nasal administration

8.13 RECENT DEVELOPMENTS IN NANOENCAPSULATED NASAL DRUG DELIVERY SYSTEM

- Agustin (2010) investigated nanoencapsulation of drug insulin by using zirconium phosphate for nasal applications. In this examination *in vitro* release profile of the hormone after the intercalation was resolved and round about dichroism was utilized to think about the hormone stability upon intercalation and release. The insulin remained stable after encapsulation at room temperature for an extended period of time and finally shell life of the peptide hormone is increased. This sort of material represents to a strong candidate to building up a noninvasive insulin transporter for the management of diabetes mellitus.
- Adryana (2016) studied nanoencapsulated statins for brain delivery via nasal route. The SVT-LCNs represent the absolute most desirable qualities for mucosal conveyance, such as small particle size, positive surface charge, long-term stability, high EE, and mucoadhesion. Furthermore, they showed two energizing highlights: First was their biodegradability by chemicals present in the mucous layer, for example, lysozyme. This demonstrates another trojan-horse system which may improve medication discharge in the nearness of the mucosa. Second was their capacity to upgrade the nose-to-mind transport as prove by fundamental gamma scintigraphy studies.

- Fabio (2018) prepared surface-modified nanocarriers to target brain via nasal route: From Bioadhesion to Brain Targeting. In this survey, delivery of nanomedicine depend on mucoadhesive polymers, particle building surface electrostatic charges, or chemical moieties focusing on the nasal epithelium will be examined and basically assessed in connection to nose-to-brain conveyance.
- Salatin et al. studied hydrogel nanoparticulate system and nanocomposites of drug for delivery via nasal route. The examination endeavors have extensively been coordinated towards the advancement of hydrogel nanosystems which have mucoadhesive properties so as to expand the living arrangement time, and subsequently increment the time of contact with the nasal mucosa and improve the medication retention. It is almost sure that the high consistency of hydrogel-based nanosystems can effectively offer this mucoadhesive property. This survey discuss about the potential advantages of utilizing hydrogel nanocomposites and hydrogel polymer-based nanoparticles for drug/antibody delivery via the intranasal administration.
- Csaba et al. (2006) used PLGA: Poloxamer and PLGA: Poloxamine mix nanosystems as gene delivery carriers via nasal route. In this work arrangement of another sort of nanoparticles comprising of blends of poly (lactic-co-glycolic corrosive) (PLGA) and polyethylene oxide (PEO) derivatives, which show the ability to associate and release plasmid DNA in a controlled way. The objective of the study is to examine the capacity of these nanoparticles to overcome cell and mucosal boundaries (for example, nasal mucosa) and, therefore, to function as quality conveyance bearers. The outcomes propose that these novel nanoparticles have a potential as transporters for the passage of DNA over the nasal mucosa.

8.14 CONCLUSION

Bioavailability of drugs administered via nasal route is particularly limited due to low solubility of drug, rapid enzymatic degradation in nasal cavity, and poor membrane penetration. Nasal drug delivery is a helpful technique for medications that are active in low portions and demonstrate no or insignificant oral bioavailability. The nasal delivery excuses hepatic first-pass transfer associated with the oral delivery. Nanoencapsulation offers protection to API from degradation. It confers targeted delivery of drug via surface coating or conjugation. Although some brain-targeted nanoparticles encapsulated with statins are being projected by several researchers. The research work recommends that the nanoparticles administered nasally were successfully served as a carrier to deliver statins directly to the brain. The nanoparticles were prepared to optimize the loading of lipophilic moieties and to give various features beneficial for nasal delivery, such as chemical and physical stability, biocompatibility, mucoadhesion, and a quick medication release. Although further studies are important to explain the uptake of the nanoparticle by the nasal epithelium. It essentially support the absorption of drug devoid of permeating the mucosa and subsequently reveal the pharmacokinetics (PK) as well as efficacy of the nano-formulation of statins delivered through the nasal route. Nanomedicines appears to be an essential tool to enable the brain delivery of potent drug unable to cross the BBB and when utilized all things considered they will satisfy the potential showed in the numerous scientific studies conducted so far.

KEYWORDS

- **encapsulation efficiency**
- **nanocarriers**
- **nanoencapsulation**
- **nasal cavity**
- **nasal drug absorption**
- **nasal drug delivery**
- **simulated nasal electrolytic solution**

REFERENCES

Adryana, C. The nasal delivery of nanoencapsulated statins–an approach for brain delivery. *International Journal of Nanomedicine,* **2016**, *11*, 6575–6590.

Agustin, D. Nanoencapsulation of insulin into zirconium phosphate for oral delivery applications. *Biomacromolecules,* **2010**, *11*, 2465–2470.
Augustin, M.; Sanguansri, P. Nanostructured materials in the food industry. *Advances in Food and Nutrition Research,* **2009**, *58*(4), 183–213.
Betbeder, D.; Sperandio, S.; Latapie, J.; Nadai, J.; Etienne, A.; Zajac, J. Biovector (tm) nanoparticles improve antinociceptive efficacy of nasal morphine. *Pharmaceutical Research,* **2000**, *17*, 743–748.
Csaba, N.; Sánchez, A.; Alonso, M. PLGA: Poloxamer and PLGA: Poloxamine blend nanostructures as carriers for nasal gene delivery. *Journal of Controlled Release,* **2006**, *113*, 164–172.
Csaba, N.; Sanchez, A.; Fernandez-Megia, E.; Novoa-Carballal, R.; Alonso, M. Chitosan nanoparticles for the delivery of plasmid DNA: Preparation and characterization. *European Journal of Pharmaceutical Sciences.,* **2004**, *23*, 54.
Dahlin, M.; Jansson, B.; Bjork, E. Levels of dopamine in blood and brain following nasal administration to rats. *European Journal of Pharmaceutical Sciences,* **2001**, *14*, 75–80.
Fabio, S. Surface-Modified nanocarriers for nose-to-brain delivery: From bioadhesion to targeting. *Pharmaceutics,* **2018**, *10*, 34.
Ghosh, P. Hydrophilic polymeric nanoparticles as drug carriers. *Indian Journal of Biochemistry and Biophysics,* **2000**, *37*, 273–282.
Heyang, J.; Fei, X.; Cuilan, J.; Yaping, Z.; Lin, H. Nanoencapsulation of lutein with hydroxy propyl methyl cellulose phthalate by supercritical antisolvent. *Chinese Journal of Chemical Engineering,* **2009**, *17*(4), 672–677.
Illum, L. Nasal drug delivery: Possibilities, problems and solutions. *Journal of Controlled Release,* **2003**, *87*, 187–198.
Lakkis, J. *Encapsulation and Controlled Release Technologies in Food Systems.* Blackwell Publishing in Ames., **2007**, 113–133.
Leong, W.; Lai, O.; Long, K.; Yaakob, B.; Mana, C.; Misran, M.; Tan, C. Preparation and characterization of water-soluble phytosterol nanodispersions. *Food Chemistry,* **2011**, *129*(1), 77–83.
Lopez, A.; Gavara, R.; Lagaron, J. Bioactive packaging: Turning foods into healthier foods through biomaterials. *Trends in Food Science and Technology,* **2006**, *17*(10), 567–575.
Nagavarma, B.; Hemant, K.; Yadav, S.; Ayaz, A.; Vasudha, L.; Shivakumar, H. Different techniques for preparation of polymeric nanoparticles: A review. *Asian Journal of Pharmaceutical and Clinical Research,* **2012**, *5*(3), 16–23.
Panyam, J.; Labhasetwar, V. Biodegradable nanoparticles for drug and gene delivery to cells and tissue. *Advanced Drug Delivery Reviews,* **2003**, *55*, 329–347.
Peek, L.; Middaugh, C.; Berkland, C. Nanotechnology in vaccine delivery. *Advanced Drug Delivery Reviews,* **2008**, *60*, 915–928.
Stoke, D.; Reber, K.; Waltzman, L.; Erns, C.; Hamilton, D.; Gawareck, D. Analgesic efficacy and safety of morphine chitosan nasal solution in patients with moderate to severe pain following orthopedic surgery. *Pain Medicine,* **2008**, *9*, 3–12.
Thorne, R.; Frey, W. Delivery of neurotrophic factors to the central nervous system. *Clinical Pharmacokinetics,* **2001**, *40*, 907–946.
Vila, A.; Sanchez, A.; Perez, C.; Alonso, M. PLA-PEG nanospheres: New carriers for transmucosal delivery of proteins and plasmid DNA. *Polymer Advance Technology,* **2002**, *13*, 851–858.

Westin, U.; Piras, E.; Jansson, B.; Bergstrom, U.; Dahlin, M.; Brittebo, E. Transfer of morphine along the olfactory pathway to the central nervous system after nasal administration to rodents. *European Journal of Pharmaceutical Sciences,* **2005**, *24*, 565–573.

Zhang, Q.; Zha, L.; Zhang, Y.; Jiang, W.; Lu, W.; Shi, Z. The brain targeting efficiency following nasally applied MPEG-PLA nanoparticles in rats. *Journal of Drug Targeting,* **2006**, *14*, 281–290.

Zimet, P.; Livney, Y. Beta-lactoglobulin and its nanocomplexes with pectin as vehicles for Ω-3 polyunsaturated fatty acids. *Food Hydrocolloid,* **2009**, *23*(4), 1120–1126.

PART III
Miscellaneous

CHAPTER 9

Skin Sensitivity and Irritation Testing for Transposing Transdermal Drug Delivery System

Mahesh P. More,[1,2] Prashant K. Deshmukh,[2] Pravin O. Patil,[2] and Nayan A. Gujarathi[3]

[1]*SRES Sanjivani College of Pharmaceutical Education and Research, Kopargaon–423603, Maharashtra, India,*
E-mail: mahesh.more@sanjivani.org.in

[2]*H.R. Patel Institute of Pharmaceutical Education and Research, Shirpur–425405, Maharashtra, India*

[3]*Shri Vile Parle Kelavani Mandal's Institute of Pharmacy, Dhule–424001, Maharashtra, India*

ABSTRACT

Skin irritation and sensitivity evaluation is a major hurdle in current development, wherever newer formulations are come in market with differential composition and components. Each excipient or chemical must be tested for its irritant potential. Guidelines asserted by various regulating agencies, need to be considered before proceeding for test protocols. As per the current scenario adapted by regulatory authorities, the researcher selected an appropriate testing protocol like an in-vitro test methodology for determining irritant potential. Various underlying techniques are covered within this chapter to understand the methodologies used to test the irritant effect of chemicals. Researcher continuously working on simulating or

Topical and Transdermal Drug Delivery Systems: Applications and Future Prospects. Nayan A. Gujarathi, PhD, Juliana Palma Abriata, PhD, Raj Kumar Keservani, MPharm, Anil K. Sharma, PhD, MPharm (Eds.)

transforming in-vivo test protocols into *in-vitro* methods to minimize the animal exposure. This chapter will cover almost recent developments and fundamental of implementing skin sensitivity and irritation effects.

9.1 INTRODUCTION

The semipermeable membrane covering the body as a shield on living organisms proposes to be offering protection from the exterior environment, UV-irradiation, endogenous water loss. The impermeable barrier did not penetrate microorganisms, but selectively allow certain types of chemicals with constricting molecular structure and weight. Molecular passage available for absorption is very high as the larger surface area is available, the substantial risk might be involved for permeating chemicals through the skin surface. Skin affording better drug delivery surface, but it might also possibly if hazardous substances might be absorbed resulting in an unusual systemic effect. Skin regulates body temperature and important excretory organs. Take a part in detoxification and helps in the elimination of certain heavy metals from the body. The skin maintains homeostatic balance in living animals.

The anatomy of the skin consists of dermis, epidermis (EP), and SC as an outer layer and deep subcutaneous layer. Squamous epithelium contains cornified keratinocytes. Keratinocytes are dead skin cells reach in keratin corneocytes. A Tight junction forms between desmosomes and intercellular space filled with lipid cement. The lipid cement consists of lipids, ceramides, neutral lipids such as free cholesterol, cholesterol ester and free fatty acids (Schurer and Elias, 1991). The composition of a thick dead cell layer on the surface of skin around 10-40 μm is in the form of desiccated keratinized cells act as a rate-limiting barrier and play a major role in the regulation of irritation. Keratinocytes also interprets immune surveillance mechanisms related to the initiation and modulation of irritation at the EP. Inflammatory responses were generated upon stimulation by external chemical or irritant causing molecules. The dermis layer consists of fibroblasts, immune cells, endothelial cells. Sugar like highly polar compounds as well as molecular weight greater than 5,000 Dalton (proteins or polymeric structure) is impermeable through the keratinized barrier. If some irritant chemical comes in contact with skin barrier, it shows an allergic reaction or adverse effect.

For a systemic effect, the transdermal drug delivery approach provides potential implications over the oral route. Initially, the transdermal way of transporting drugs across the skin shows effective treatment in hypertension, female menopause, angina, nicotine dependence, urinary incontinence, and motion sickness, etc. The therapeutic benefits of transdermal routes are steady-state plasma profile, useful for delivery of drug with a short half-life, probably avoid the first-pass effect, reduces repeated dosing based on patch type (weekly delivery or monthly delivery), increases patient compliance. It is a highly acceptable option for a wide range of patients and conditions (Hiremath, 2007).

9.2 SKIN IRRITATION

Sensitivity to the skin is the most important criterion as per new regulatory guidelines in relation to topical formulations. The presence or absence of any allergic reaction with skin and components of the formulation is an eager area of research. To identify the components or concentrations of ingredients which are responsible for any type of allergic reaction observed on the skin. Allergic reactions cause discomfort for the patient. It is the responsibility of the manufacturer to determine the sensitivity for a particular component or patch for patients either to provide some sort of testing methods/strip or test prior to marketing. The prophetic skin testing analogy was utilized to determine the allergic conditions before or after the application of the formulation. The current patch test causes irritation or inflammation after application. In this type of measurement, the patient cannot reach upto the conclusion patch is safe for use during its application time. Although the validated methods are not available yet, it could not possible to rectify the specific type of allergen that produces inflammation. For liquid-based formulation use to delivered through transdermal methods, absorbent materials used to cover the liquid formulation. After application liquid migrates towards the periphery of the patch due to capillary action and porous nature of absorbent material. Another factor while testing is available skin surface area, the patch which is in contact with skin surfaces must be insufficient if more than one allergen causing inflammation at the site of application. Hereto the analyst would not get confirmation about at what concentration and which type of allergen causing the inflammation to the skin. False-positive allergic reactions might be predicted in some of the cases, where liquid test substance reaches the periphery of absorbent materials. The liquid substance may act as a solvent for adhesive material, chemical ingredients within adhesive taps leaches at the site of application. The ingredient may react chemically to skin surfaces and shows a false positive allergic reaction (Osbourn and Tusing, 1965).

Skin Irritation is a cutaneous inflammatory reaction that shows some physicochemical changes reflected visually and some can be quantified using invasive and noninvasive methods. The physiological reaction that occurs in contact with the stimulus might be a biomolecular or chemical substance as represented in Figure 9.1.

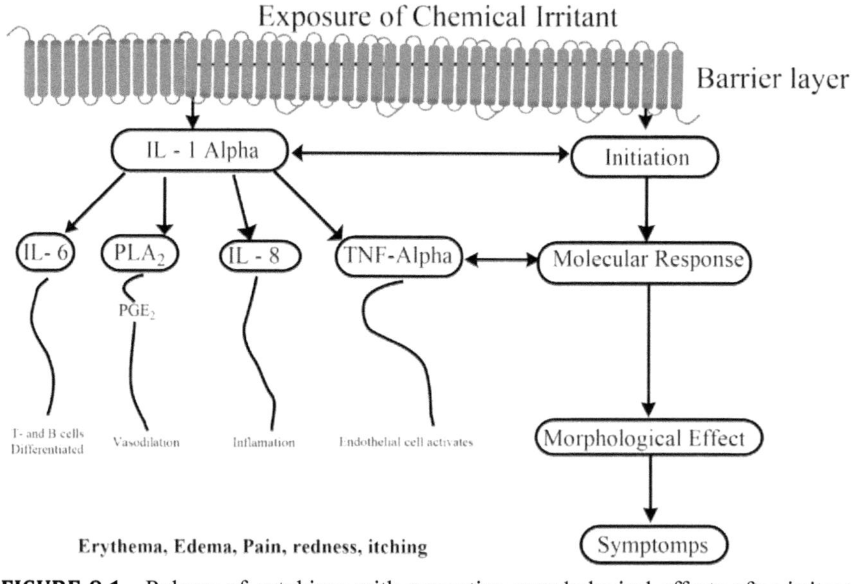

FIGURE 9.1 Release of cytokines with respective morphological effects after irritant substance exposure to the skin surface.

An observable reaction results from the local inflammatory process and visually detect by Erythryma, edema, dryness of skin, fissure, desquamation, itching, and pain-like reactions are considered as a characteristic inflammatory reaction which altered homeostasis. The clinically the physiological response from the body can be considered as rash, redness, inflammation, swelling, abnormal tissue growth, scaling, etc. (Fulzele et al., 2007).

It is very difficult to differentiate between contact dermatitis and skin irritation. It is initiated after direct inflammation on the skin surface. Harvell in 1995 defines skin irritation as a locally arising, non-immunogenic inflammatory reaction, which appears shortly after stimulation and usually disappears during a few days.

Chemically induced skin irritation can be classified as four clinically distinguished subtypes:

1. **Acute Skin Irritation:** Single exposure to powerful chemical;
2. **Cumulative Skin Irritation:** Repetitive exposure to a mild irritant;
3. **Delayed Acute Skin Irritation:** Repetitive work in a wet environment and have sensitive skin diseases;
4. **Chronic Skin Irritation (Hardening of the Epidermis):**
   i. Hyperproliferation;
   ii. Hyperkeratosis.

9.3 MECHANISM OF SKIN IRRITATION

Every substance has a small amount of irritation potential even though water might show irritation effect upon long exposure. Chemical has doubled sited effect, it contributes to impart stable physicochemical properties to formulation helps to enhance the efficiency of the product while some of them might react with skin components like lipids, proteins, etc., and shows an adverse cutaneous inflammatory reaction. If we see the example of surfactant, possibly used in all types of pharmaceuticals, cosmeceuticals. The surfactant resembles bilayer lipid membranes and contains a hydrophobic head and a hydrophilic tail. While for enhancing miscibility or reducing surface tension it forms micelle-like structure. But to some extent, few ingredients might interact with surface lipids of the skin layer and lipids get precipitated and show inflammatory response.

Initiation and modulation of skin irritation as two major modular pathways considered for the study. After chemical exposure, it damages the SC and loses the barrier function or secondly it directly acts on cells of the skin. The barrier composition containing dead keratinocytes prevents or controls action exogenous water loss and endogenous inflammation caused by chemicals. There are a few factors that need to be addressed while assessing skin irritation caused by chemicals are like, the concentration of the irritant chemical, exposure time, exposed surface, genetic variability, molecular structure, etc. (Hall-Manning et al., 1998). Delipidation and denaturation of proteins are two major reactions, while in delipidation, the structure of lipids gets disturbed and lipid cement will lead to loss of barrier function. During delipidation, it causes loss of epidermal water content and upon repeated exposure, if it gets increased, irritant penetration increases. Upon enhancing penetration, deeper damage to living keratinocytes.

Surfactant based mechanism could possibly elucidate the exact effect of the chemical on skin and promotes skin irritation. Cytoplasm tends to release as a surfactant disrupts the cell membrane and cell components were directly exposed to the chemical irritant (Osborne and Perkins, 1994). The main inflammatory cascade starts from the stimulation of pro-inflammatory cytokine interleukin-1α. While IL-1α is present in the cytoplasm. Upon stimulation of IL1α, it also starts the expression of cytokines, like IL-6, IL-8 (Nickoloff and Naidu, 1994). IL-8 is a major cytokine representing a strong inflammatory response. In some instances, researchers have recorded, activation of phospholipase A2 (PLA2) after overexpression of IL-8. PLA 2 has a major function in the arachidonic acid cascade (Terry et al., 1999).

Some chemicals may act by altering the physiology of keratinocytes. As a normal physiological function of keratinocytes, it regulates the level of protein kinases and phosphorylated proteins via the production of reactive oxygen species (ROS). Irritant exposure, keratinocytes increases the production of ROS, which oxidizes nucleic acids, proteins, and membrane lipids. It further causes cellular damage and leads to gene alteration or cytotoxicity (Corsini et al., 2002). At some extent it might be possible that, irritant may be superimposed on cell membrane e via modulation of lipid bilayer. The stratum corneum (SC) containing lipid cement might get disturbed which changes membrane fluidity. If cell membrane fluidity is affected, it loses receptor-mediated transduction mechanism. In presence of irritant response, membrane bound proteins get disturbed which directly or indirectly changes the pH environment (Welss et al., 2004).

In literature various methods have been used by researchers, it not possible to cover up all methodology in a single chapter, we had highlighted most cited and repeated methods of skin irritation represented in Table 9.1.

TABLE 9.1 In-Vitro Skin Irritation Study (Bason, Gordon, and Maibach, 1991)

Method	System	Principle and Observation
Testskin	Synthetic human epithelium	Structurally similar skin, inflammatory mediators as a response
Microphysiometer	Cell culture	Changes in pH conditions related to metabolism indicator response
Neutral red dye assay	Cell culture	Dye uptake by viable cell and percent viability as a response
Tetrahymena thermophile	Protozoa	Motility, increase or decrease motility as a response
Chorioallantoic membrane	Fertilized egg membrane vasculature	Vascular damage as a response
Skintex	Keratin collagen matrix linked to the dye	Dye release and turbidity measure by spectrophotometer related to membrane disruption

9.4 *IN-VITRO* METHODS OF SKIN IRRITATION

According to the 6th amendment of European Union cosmetic directives, dermal skin equivalents can be used to determine phototoxicity, corrosivity, and irritation potential to replace the use of animals. Till date, methodologies used for skin irritation and sensitization test are based on animal studies. The test causes pain and discomfort to animals and sometimes does not correlate to human studies. In European Union, COLIPA represents cosmetic and toiletry industry, helps to replace and animal-based test protocol and implement in-vitro method for prediction of irritant potential. For in-vitro method analysis for irritation testing, reconstructed organotypic skin samples are used which have similar morphology to human skin.

An *in-vitro* test method, determination of biological response is the simplest strategy. During determination of the potency of irritant need to find out the specific interaction either physical or chemical and based on that, the conclusion remains dependent. A following number of factors might be responsible for giving exact biological response such as concentration of chemical, duration, and frequency of exposure of chemical. Total available skin surface and exposed area can be easily determined from the *in-vitro* data types. For prediction of toxic effect of substance and penetration rate of irritant has to be determined from experimental setup. Draize test accurately predict the acute toxicity of chemicals developed by John

Draize and Jacob Spines in 1944. Till date few commercial companies using Draize test to determine the irritation potential on skin and eyes. The Draize test is the gold standard for assessing skin and ocular irritation potential (Draize et al., 1944). Various in-vitro models are elaborated as in subsections.

9.4.1 TISSUE MODELS

Several tissue models were represented by various researchers to determine the irritation and sensitization effect on skin. Mainly keratinocytes and fibroblast extracted and air-exposed reconstructed skin equivalents are used for study. Although few researchers had explored the monolayer keratinocytes for determining skin toxicity, but limitation for using monolayer keratinocytes is that concentration dependent irritation potential is difficult to predict and correlated with in-vivo response.

In-vitro tissue models are come with two variants, synthetic matrices containing human keratinocytes in multi-layered composition and fibroblast culture contains collagen as substrate with full skin equivalents in multilayer. At the interfaces of each layer air spaces available which helps to organize keratinocytes and act as barrier in between the layers. The base component is either biological or non-biological on which layer of keratinocytes are seeded. It forms a reconstituted EP after full differentiation at the dermal base components. The permeability of dermal skin equivalent is around 10–30 fold higher than normal skin. Due higher penetration rate over predicted results might be reported in literature. Composition of cell is different in comparison to in-vitro dermal skin equivalents and normal human skin. For prediction of interaction with, nerve cells, blood components, endothelium, inflammatory cells need to be analyzed using other sophisticated analytical methods.

9.4.1.1 CELL TYPES

The major barrier in selecting an appropriate type of cell types. Few of cell/tissue types are explored in this heading.

- NIKS keratinocytes are consistent and possess normal epidermal differentiation like human EP. NIKS keratinocytes generates fully stratified epithelium are basically pathogen-free, non-tumorigenic, and p53 wild type.

- SV40 large T antigen-transformed human epidermal keratinocyte 2 cells (Kim et al., 2017).

9.4.2 NON-INVASIVE BIOENGINEERED RESPONSE METHODOLOGY

Transdermal patch irritation study was elaborated by Serup and Agner (1990), revealed that the irritant response was measured non-invasively after application of patch to human volunteers. The researcher selected 12 healthy volunteers for study with ages between 21 and 45 yr. Irritant response quantified by various bioengineering techniques as follows: Measurement of trans epidermal water loss via evaporimeter, skin color from colorimeter and superficial blood flow from laser doppler flowmetry and edema measured using ultrasound A-scan. The upper arm from the flexor side was chosen for application of patch in men and women volunteers. Test sample containing filter disc were applied using large Finn chambers on scan pore tape. The evaluation parameters were assessed after 1 h of patch removal and reevaluated after 48 h (parameters assessed from Table 9.2, observations recorded and response quantified).

TABLE 9.2 Parameter Scoring based on Colorimeter Reading for Irritant Effect Determination

Scale	Reaction Observation and Scoring
0	No reaction
½	Slightly scaling or very weak erythema
1	Weak erythema possibly slight infiltration
2	Marked erythema, infiltration possibly vesicle and crusting
3	Pronounced erythema, infiltration, possibly vesicle bullae, pustule or pronounced crusting

Non-invasive bioengineering methods employed by researcher is as follows. For measurement of trans epidermal water loss by using evaporimeter. At the skin surface water vapor gradient forms a boundary layer between skin and ambient air. A pair of hygrosensors and thermistors type sensor assembled with evaporimeter helps to determine vapor pressure gradient of boundary layer formed at the skin interface. Water vapor gradient at the skin interface helps to quantify trans epidermal water loss. Insulating gloves use to avoid direct contact with the probe during measurement, also helps to minimize the measurement error. The skin read time is 30–45 sec., pen recorder measure and display the values. The protecting cover uses to avoid convection of air during measurement and fluctuation from the read out value minimizes.

9.4.3 SUPERFICIAL BLOOD MEASUREMENT BY LASER DOPPLER FLOWMETRY

Superficial blood flow measurement gives important insights and measured using laser Doppler flowmeter. The instrument uses a bandwidth of 20 Hz–12 kHz with stable gain around 10. For stable measurement a plastic block (2.5 × 10 × 1.5 cm) is used, an adjustment is given for probe (hole) at the center of the block. Care should be taken, and Velcro straps are provided, to avoid skin compression during measurement.

9.4.4 MICRO COLOR METHOD

During the irritation, sensory organ like skin shows differentiation of color by accumulating inflammatory response. Visual observation did not get exact color differentiation. Micro color containing illuminating xenon flash lamp measures the accurate color differentiation (Serup and Anger, 1990). Guidelines notified by International Commission of Lighting System utilizes nonlinear perception method. The color expressed in 3 Dimensional coordinate system on X-axis coordinates green and red color, Y-axis coordinate blue and yellow color and z-axis reflects brightness. For final calculation values from X-axis which reflects green and red color used which represents erythema (Serup and Agner, 1990).

9.4.5 PULSE MODE ULTRASOUND SCANNER

Skin thickness is again an important parameter in bioengineering method, measured by using A-mode pulse ultrasound scanner (20 MHz). This measures the distance between acoustic echoes to SC and dermis to subcutaneous interface. The values expressed in mm and acoustic velocity of

1,580 m/sec is used. During measurement of normal skin thickness 2.2% coefficient of variation is considered (Hoffmann et al., 1991).

9.4.6 ZEIN TEST

Surfactants mostly interact with insoluble proteins such as keratin, SC. The interaction gives valuable information such an acute or serious damage like desquamation, redness, burn, etc., SC swelling with or without protein denaturation effect. Zein is a vegetable protein and model experiments for identification of irritant can be possible. Chemical irritant specifically surfactants interact with proteins, changes in proteins structure or morphology correlated with irritant effect of surfactant. At industrial scale the zein test use to determine mildness of surfactant, results can be interpreted above the critical micelle concentration (CMC) of surfactant.

To test the irritant effect of surfactant, simulated zein test resembles the skin surfaces and zein superficial film effect was studied. On dilute solution of surfactant zein forms monolayer, time of relaxation studies using Langmuir balance technique at constant pressure. From the technique, it helps to correlate the protein conformation in the presence of surfactant. Industrial surfactant like alkylpolyglucoside, lauryl ether sulfate and sodium dodecyl sulfate tested experimentally by Pezron et al. (1996). Corn contains polypeptide molecule called zein, which is slightly water-soluble. Tertiary structure contains 9 repeated units while secondary structure consists of polypeptide chains of polar amino acids. The average molecular weight is around 113 g and polypeptide composition is glutamine, leucine, alanine, prolin, and serine. It contains 60% of non-polar polypeptide chains.

The method determines the solubilized protein by surfactant concentrations. The detailed method is as follows: in Erlenmeyer flask, 1% solution of surfactant (40 mL) is taken and neutralize the surfactant solution using alkali if necessary to pH 7. Weigh accurately 2 g of zein and raise the temperature up to 30°C for 1 h. Centrifuge the surfactant solution containing protein at 2,000 g for 20 min, supernatant was analyzed for solubilized protein using Kjeldahl method. With slight modification in the Kjedhl method for the determination of organic nitrogen in mg from 100 g of surfactant solution gives zein number. At liquid water interface bidirectional behavior of monolayer formed by protein is determined using the Langmuir Blodgett system (Pezron et al., 1996). Standardized correlation gives interaction between zein protein and surfactant categorized irritation effect as given in Table 9.3.

TABLE 9.3 Classification Criteria for Determination of Skin Irritation Using Zein Test

Zein Number	Observation for Irritation Potency of Surfactant
0–200	No skin inflammation
200–400	Slight Skin Inflammation
400–500	Skin inflammation, slight desquamation
500–600	Strong skin inflammation, desquamation
Above 600	Very strong skin inflammation and desquamation

9.4.7 STRUCTURAL ACTIVITY RELATIONSHIP-BASED SKIN IRRITATION MODELLING

Chemicals with particular effect could correlate the SAR irritation potential of chemicals. SAR-based model screen the irritation potential from data set available, but data set did not represent the chemicals with particular effect. Based on specific effect or physicochemical interaction SAR model randomly select group of chemicals. Few researchers have reported the acute irritant potential of esters, from physicochemical effect. The model categorized irritant and nonirritant chemicals based on dataset available and screen the same derivatives for potential effects. The physicochemical parameters considered during model are water solubility (lower in irritants than nonirritant), density (lower in irritants), Hansen dispersion (higher in irritants), Hansen hydrogen bonding (higher in irritants) and sum of partial positive charges (lower in irritants).

The irritant response could be categorized with changes in the chemical composition of EP. As experiment devised by Smith et al. categorized different types of esters studied using Draize test. Most commonly used fragrance contains ester, that might possible responsible for irritant effect on the skin. Around 34 esters were studies on rabbit using the Draize test, and irritant response were analyzed. The esters showing irritant effect is within predictive space of SAR models. Out of 34 ester samples 16 samples shows positive effect while 18 samples represent negative.

The test results were incorporated in the database and refine SAR subsets. After linear discriminate analysis 10 irritant and 50 non-irritant randomly selected to predict the exact effect (Smith et al., 2000).

9.4.8 CONTINUOUS ELECTRON PARAMAGNETIC RESONANCE

Redox reaction can be identified beneath skin by utilizing continuous electron paramagnetic resonance, formation of free radical, oxygen/nitrogen species, barrier function, drug penetration, oxygen concentration, barrier function, etc., can be assessed. Skin traps identification gives irritant effect of chemicals. For details experiments verified by Fuchs et al. (2000) described as follows: Human skin biopsies from Healthy skin can be obtained from patients undergoing skin surgery. Excess tissue trimmed and fat removed from the skin. Human skin attached to Teflon tubes with epidermal surface at the uppermost side and connected to "O" ring. The Teflon tube interface sealed with soft paraffin wax to minimize the external contact. The Teflon tube placed in somewhat larger size diameter plastic tube containing electrolyte solution. The spin traps applied on the epidermal surface; chemicals tested in triplicate as three epidermal discs were used per chemical. The 50 mg of test substance was applied on Vaseline site of the epidermal surface and spread evenly to cover the surface of epidermal disc. Vaseline was used as negative control and 20% SLS with Vaseline shows positive control during the experiment. After 24 h of exposure with irritant, skin disc was hydrated with electrolyte solution and transcutaneous electrical resistance (TER) was recorded using Wheatstone bridge. A Wheatstone bridge with a low voltage alternating current used for determination of TER. TER values below 10 kOhms indicate a corrosive action, and 10 kOhms is used as a threshold value to distinguish between corrosive and non-corrosive skin irritants (as shown in Table 9.4) (Fuchs et al., 2000).

TABLE 9.4 Four-Point Grading Scale for Identification of Irritant Effect

Grading	Description of Response
0	No reaction
1+	Mild erythema
2+	Distinct erythema spreading beyond treatment site
3+	Strong, spreading erythema with edema

9.4.9 HUMAN REPEATED INSULT PATCH TEST (RIPT)

Weight of evidence analysis providing comparative data for evaluation between animal patch test and human repeated insult patch test (RIPT).

Quantitative and numerical response were recorded from observed skin surfaces (irritant score as shown in Table 9.5). The response and histological data recorded from animal and human could be compared properly. The primary skin irritancy effect predicted in animal have comparatively same dermatological response in human or if less response predicted in both cases.

TABLE 9.5 Skin Reaction Score to Test the Primary Irritant Substance in Laboratory Animals

Skin Reaction		Scale
Erythema and Eschar Formation	**Edema Formation**	
No erythema	No edema	0
Very slight erythema	Very slight edema	1
Well defined erythema	Slight edema	2
Moderate to severe erythema	Moderate edema	3
Sever erythema	Severe edema	4

According to Consumer Products Safety Commission (CPSC), primary skin irritation is defined as any substance that is not corrosive and that human experience data indicate is a primary irritant and/or means a substance that results in an empirical score of 5 or more in laboratory animal studies

9.4.9.1 METHODOLOGY

Primary irritation to the skin is measured by the patch test technique on both abraded and intact skin using albino rabbits. After 24 h of exposure, the patches are removed and the resulting reactions are graded on the basis of the values in Table 9.5. Additional readings are taken after 72 h. The values for erythema and eschar formation at 24 and 72 h are added for both intact and abraded skin. Similarly, the values for edema formation at 24 and 72 h for intact and abraded skin are also added. This yields a sum of eight scores that is divided by four to yield a primary irritation score. The primary irritation score (after division by 4) can lie in the range 0–8. A substance is defined by the CPSC as a primary skin irritant only when yielding a score of $\geq$5 using this scoring method (Tardiff et al., 2003).

9.4.10 IN-VITRO CELL CULTURE EXPERIMENTS (SIMULTANEOUS PHOTOTOXIC AND CYTOTOXIC ESTIMATION)

The cytotoxicity directly correlated with the irritant effect of chemical or surfactant. Benavides et al. in 2004 experimented the irritant as well as phototoxic effect of range surfactants on incubated cells. Murine Swiss albino 3T6 (ATCC CCL96), murine Swiss albino 3T3 (ATCC CCL92) fibroblasts and HaCaT keratinocyte cell line was used for designing study. The cell culture was incubated in growth media and harvested after 80% confluent with trypsin/EDTA. The suspension cell density of 2×10^4 to 5×10^4 mL^{-1} were seeded in 96 well plates. For analyzing irritant potential 3T6 fibroblast cells were incubated with variable concentrations of surfactant and response measured via neutral red uptake assay. The incubation time around 3 h at 37°C. The cells were rinsed with phosphate buffer saline (250 μL) in twice. Add 100 μL per well neutral red desorb solution was added, and plates were shaken for 10 min. The spectrophotometric absorbance was recorded at 550 nm. Non surfactant treated well (Control Well) represent 100% viability. Negative control, Positive control and test sample in variable concentrations were assessed.

9.4.10.1 PHOTOTOXIC EFFECT

For phototoxic experiments, Benavides et al. developed 3T3 fibroblast cells and HaCaT human keratinocyte cell lines to analyze the effect in the presence or absence of irradiation. The plates seeded with cultured cells after 80% confluence density. The resazurin dye use for analyzing the irritant performance. The surfactant/chemicals are added into each well via serial dilution method. The one plate exposed to UVA-visible light and the other one wrapped with foil for non-irradiated effect. The SOL 500 lamp providing exposure of 5 J/cm^2 and examined under an inverted microscope for morphological changes. The surfactant solution was removed and washed with normal culture medium and test with combined neutral red dye and resazurin assay. The reduction of resazurin to resorufin takes place via electronic transport chain reaction between oxygen and cytochrome oxidase enzyme. The reduction of resazurin correlated with mitochondrial activity. The spectrophotometric absorbance was recorded at 530 (excitation) and 590 (emission). Comparative percent cell viability between UVA and non UVA samples were analyzed (Benavides et al., 2004).

9.4.11 THREE DIMENSION (3D) CULTURED HUMAN SKIN MODELS

Human skin equivalents are provided by many commercial industries, such as Toyobo Co., Ltd., Osaka, Japan is providing living skin equivalents-high (LSE-High) for commercial applications. The human skin equivalents are used to form a 3-dimension arrangement which has similar morphology and structure alike human skin. The 3 D model consist of two layers epidermal and dermal layer. On collagen gel keratinocytes as differentiated epidermal layer and fibroblast as dermal layer embedded to constitute the human skin structure. As like human skin, the 3D culture model contains basal layer, stratum spinosum, granular layer, SC within the epidermal layer. Upon modification of culture medium, the barrier function of SC can be improved and modulated. 3D dimensional skin model is used for alternative strategy over human and animal models, but the limitation of using skin model is that, in contact with irritant, it does not show edema or erythema like visual differentiation. In 3D Skin model, cytotoxicity test was assessed via MTT assay or neutral red uptake study. The metabolic reduction of this soluble tetrazolium salt to a blue formazan precipitate is dependent on the presence of viable cells with intact mitochondrial function.

Kinetic and dynamic analysis of skin irritant can also be predicted from these models. It was hypothesized that; chemical irritant follows first-order kinetics after application on skin surfaces. If barrier function from skin model decreases it is directly correlated with skin irritant potential and conclusion can be nearer to damage of SC by chemical (Kano et al., 2006).

Kinetics within intact skin for representing irritant effect was expressed from two equations as follows:

$$V = (100 - V_\infty) \exp(-k_1 t) + V_\infty \; when \; 0 \leq t < T \quad (1)$$

$$V = (100 - V_\infty) \exp(-k_1 t) + V_\infty \; when \; 0 \leq t < T \quad (2)$$

whereas; V is the percent cell viability; t is the incubation time; T is the inflection time; k_1 and k_2 are the first order decreasing rate constant early and late phases, respectively; V_∞ is the percent skin viability at infinite time after irritant application.

9.5 EPISKIN

Episkin is commercial reconstructed EP developed by L'Oréal Paris for testing of chemicals for their irritant potency. The reconstructed EP consists of dermis layer formed by human collagen matrix (Type I), further it is covered with Type IV collagen. The EPISKIN consists of fully developed SC and stratified keratinocytes. The reconstructed EP kept in 12 well plates and medium changed after each 24 h period of time.

For testing of chemicals for their irritant potential, chemicals solutions or viscous liquids get exposed to episkin culture medium. About 10 μL of solution added into each well separately. Highly viscous chemical or solid material (10 mg) applied with micro spatula carefully on skin surface and spread it. For better contact between epidermal layer 10 μL PBS was added into each well. The reconstructed tissue incubated with chemicals at 37°C/5% CO_2 for 15 min. After incubation epidermal layer washed with 25 mL PBS. After washing reconstructed tissue further incubated at 37°C/5% CO_2 for 42 h, intensification of effect to applied chemicals. After incubation with chemicals, MTT assay uses to rectify the viability of cells. The reconstructed episkin incubated with MTT for 3 h 37°C/5% CO_2 in dark. To extract the resulting formazan product, each EP was separated from the collagen matrix and both were placed in acidic isopropanol (0.04 N) for 72 h at 4 ◦C. Extracts (200 μl) were transferred into a 96-well plate and the optical density was measured at 570 nm against a blank of acidic isopropanol. When extracts were not transparent, turbidity was eliminated by centrifugation (5,000 rpm, 4 min) prior determination of absorbance. If cell viability fell under 50% of the control, the chemical was considered as an irritant compound (Borlon et al., 2007).

9.6 STRAT TEST

Three dimensional structure consist dermal and epidermal layers formed by pathogen-free NIKS keratinocytes progenitor cells (primary human cell). Intact basement membrane separates dermal and epidermal layers. The composition containing EP, dermis, and epithelial layer possess barrier function and irritation potential determined through electrical impedance. EP contains stratified keratinocytes; dermis contains fibroblast cell and epithelial-mesenchymal cell regulates signal. Strat test is a replicated alternative to human skin, developed for determining toxicity potential of chemical substances. Simulate architecture and biological response to human skin. In contact with chemical irritant, certain types of signaling mechanisms were started, like paracrine signaling. Signaling mechanism between keratinocytes and dermal fibroblast possibly starts with expression of proteases, inhibitor, cytoskeletal proteins, chemokine, cytokine, etc. Strat test consist of primary human cell, and simulates architecture as well as biological response similar to human skin.

9.6.1 SKIN MODEL PREPARATION

All the components are provided by Stratatech Corporation along with Strata Life culture media for production process. Human skin model developed in 0.6 cm^2 surface area with individual cell culture inserts provided by StrataTech to product full-thickness skin. Optimized concentrations of culture components are provided by StrataTest to get epidermal and barrier response for electrical impedance. The epidermal stratification and barrier function was desirably achieved after stratified keratinocytes produced. In tissue culture insert within area provided 0.6 cm^2, on type I collagen normal human fibroblast cells forms dermal layer. On the dermal layer NIKS keratinocytes are seeded and for creating terminal differentiation lifted towards air interface. Before test, the multilayer model gone through panel test to confirm the tissue morphology, viability of cell within dermis and EP and barrier function. Rasmussen et al. suggest chemical irritant were tested up to 7 days with refrigerated storage. Environmental irritants were tested like ozone, cigarettes, smoke, or UV radiation via detection of relative oxygen species generation (ROS). A non-fluorescent indicator dye used for detection after oxidation becomes fluorescent (Rasmussen et al., 2010).

9.7 SKINETHIC™ RECONSTRUCTED HUMAN EPIDERMIS (RHE)

Three-dimensional epidermal skin model is developed on the air-liquid interface, and barrier function controls the signal transport. Due to histological similarity with human EP, irritant response consigned from permeability barrier function. Standard criteria and detailed methodology provided by manufacturer along with cell culture kit from Episkin explored the potential use of skin model for testing commercial samples. From test protocols evaluation, the researcher evaluated 85 common chemicals enlisted in OECD data and provided classification of irritant, toxic, and corrosive substances (Alépée et al., 2015).

9.8 *IN-VIVO* METHODS

9.8.1 MODIFIED DRAIZE TEST

Rats weighing 150–200 g were selected for dermal irritancy study as elaborated by Cargil et al. (1986). Circular gauze pad with 1 mm thick and 16 mm diameter was applied on the dorsal surface of the animal using adhesive tape. About 50 mg of test sample applied on gauze pad and retained for 3 or 7 days. After application of gauze pad skin surface evaluated and scored according to modified Draize scoring method, reflects irritation index and potential of irritant. For analyzing the irritancy potency Primary and secondary irritation index (SII) was categorized. The primary irritation index (PII) was predicted after immediate removal of the patch from the skin surface and after 72 h. While in case of SII, determined after 7 days of dressing removal. The maximum score for PII or SII was 8, as per modified Draize possible score for erythema or edema is 4 each. The score equal to 2 reflects mild irritant potency while greater than or equal to 6 shows moderate irritant effect. More than 6 score indication of potential irritant effect (Cargill et al., 1986).

9.8.2 PATCH ABSORPTION METHOD

On human volunteer, 5 test sites were selected on paravertebral skin. Filter paper disk (20 mm) fitted on extra-large Finn chambers for fixing samples on skin surfaces. The normal control consists of 0.2 mL 2% SLS fixed for 24 h, after removing chambers rinse with water and allow for air drying. The other volunteers for test sample analysis were treated for 1 h and rinsed with water and tested for trans epidermal water loss as well as erythema observations recorded. Negative control considered for one volunteer and applied one empty patch and use for comparison with test and positive control volunteer. The trans epidermal water loss was measured using Tewameter TM 210 (Courage-Khazaka, Cologne, Germany). The all samples assess from the exposure point, i.e., 0 h to 7 h or till patch removal. The values are expressed as $g/m^2/h$ is any marks or irritation response observed. The erythema index was measured using Mexameter MX 16 (Courage-Khazaka, Cologne, Germany). The method is based on absorption of chemical irritant at skin surface, while mexameter use to measure the melanin and hemoglobin in the skin. The device specifically measures the content of melanin and hemoglobin (erythema) in the skin and the method is based on the absorption principle. The irritant potential of chemicals on human volunteer recorded and classified as per standard score of erythema or edema observed (Katsarou et al., 2000).

9.8.3 OCCLUSIVE PATCH TEST

The filter paper is fixed in extra-large Finn chamber with 18 mm diameter. Test substances are dispersed in distilled water and filter papers are dipped into it. SDS (0.5%, 17 mM) is used as positive control and distilled water as vehicle control during the study.

Finn chambers are applied on the upper back of subjects and fixed using opposite adhesive tapes. Before application, the skin subjects clean and rested on the bed for 15 min, helps to reduce thermal as well as nervous sweating effect. By use of evaporimeter EP2 trans epidermal water loss measured. The laser Doppler perfusion imager (LDPI) PIM 1.0 (Lisca Development AB, Linköping, Sweden) was used to measure the superficial tissue perfusion. The tissue perfusion is reflected through color-coded image on incidence of He-Ne laser source (670 nm) on skin surface containing suitable fluorescence dyes encapsulated with formulation. During measurement scanner head distance and skin surface was fixed to 15 cm. The amount of light scattered by erythrocytes is displayed in units of volts (V). The measurement was performed in a dark room to reduce interference of bright light and sensitivity of the system was maintained (Bodin et al., 2000).

9.8.4 PATCH TEST

A simple patch test was described by Hiremath SSR in 2007 as follows: around 30 animals were selected and the monolithic matrix transdermal patch was applied to a shaved area on the dorsal side of mice. The test was evaluated for 21 days after application of the patch on the shaved skin surface and observed up to 23 h on a daily basis (removing and applying of same patch). The skin reactions are evaluated as erythema, edema, and other type of indicative irritation forms a scale. On the basis of score patches are classified as irritant or sensitive to dermal surface. If serious reaction observed at the site of application, test patch is discontinued and reaction score was recorded. After each 23 h, patch were removed, and the dermal surface was evaluated for reaction via visual observations from clinical experts (Hiremath, 2007). Observation scale is provided in Table 9.6.

TABLE 9.6 Scale Provided for Recording the Response

Scale	Response
0	No evidence of irritation
1	Minimal erythema, barely perceptible
2	Definite erythema, readily visible; minimal edema or minimal popular response
3	Erythema and papules
4	Definite edema
5	Erythema, edema, and papules
6	Vesicular eruption
7	Strong reaction spreading beyond test patch

9.8.5 4 H PATCH TEST

The test protocol consists of Webril pad with plain hill top chamber (25 mm diameter) where 0.2 g or 0.2 mL sample can be held. The test conducted for 4 h on 30 human volunteers on the upper outer arm as provided in protocol by Basketer et al. (2004). The progressive application of test samples is desirable or recommended from 15, 30 min followed by 1, 2, 3, and 4 h of interval with incremental concentration. This incremental time points reduces unacceptable strong immediate reaction of applied material on skin surface. Although each progressive patch applied on new skin site to avoid interference. Based on satisfaction of result analyzer can select shorter exposure period or longer exposure period. If the irritant concentration is lower and irritation is time increases then, the analyst can omit the shorter exposure time from study. The treated sites are evaluated after 24, 48, and 72 h of patch removal. If irritant effect observed on volunteer, then causative substance test is discontinued on that person. If analyst found, irritant effect within 4 h of test then, substance considered as stronger irritant. Substances scoring can be classified according to irritant effect observed on human volunteer (Basketter et al., 2004).

9.8.6 DERMAL MICRO DIALYSIS METHOD

This method developed under minimally invasive steps can be used for measuring endogenous and exogenous solute concentrations (Plock and Kloft, 2005). By using dermal microdialysis method, we can also predict PKPD, disease monitoring and estimation of neurotransmitter. The thin semipermeable dialysis membrane is inserted in between dermis layer which mimics the small blood vessel function. The thin dialysis membrane allows transport of biomolecules or substances across the membrane which helps to maintain concentration gradient across it and diffusion function can be correlated. The dermal microdialysis helps to deliver the biomolecules or substances towards skin tissue. There are a few advantages and limitations of methods available. The dermal microdialysis method can be used for:

- Measuring cellular response through transport of inflammatory mediators;
- Skin metabolism pharmacokinetics and pharmacodynamics (PD) can be analyzed;
- Absorption of drug as well as biomolecules can be determined;
- Skin dermal microdialysis can be use as alternative microdialysis method;
- Skin irritation or inflammation can be determined through microdialysis by measuring the release of neuropeptides or inflammatory marker.

The dermal microdialysis setup method as shown in Figure 9.2. By minimally invasive method 10 mm dialysis membrane was implanted beneath the skin in between dermis and EP. The dialysis membrane is used as a linear probe for analysis. Outer side of microdialysis membrane is connected to fraction collector and inner end of the membrane is connected to microinjection pump. Before the start of the experiment, 50% ethanol was perfused for 30 min., and deionized water subsequently for 15 min. The after complete prewashing steps, Krebs-Ringer Solution is perfused continuously for 30 min. During experimentation, Krebs-ringer is perfused at a rate of 2 μL/min. During in-vitro and *in vivo* analysis, dialysate sample collected after every 30 and 60 min. of time intervals. The microdialysis probe implanted in dorsal region of rat, before implantation skin was punctured with 19-gauge intravenous needle. The guiding needle cannula is used for insertion of 10 mm dialysis membrane probe beneath the epidermal layer. The needle removed slowly leaving membrane in the dermis. The outlet end connected to micro fraction collector via tubing connector. Around 120 min. equilibration time was set as the initial washout period within 60 min of interval two baseline of dialysate samples plotted. During equilibration test samples applied of the skin surface and potential biomarkers were quantified from probe connected to outer section (Fulzele et al., 2007).

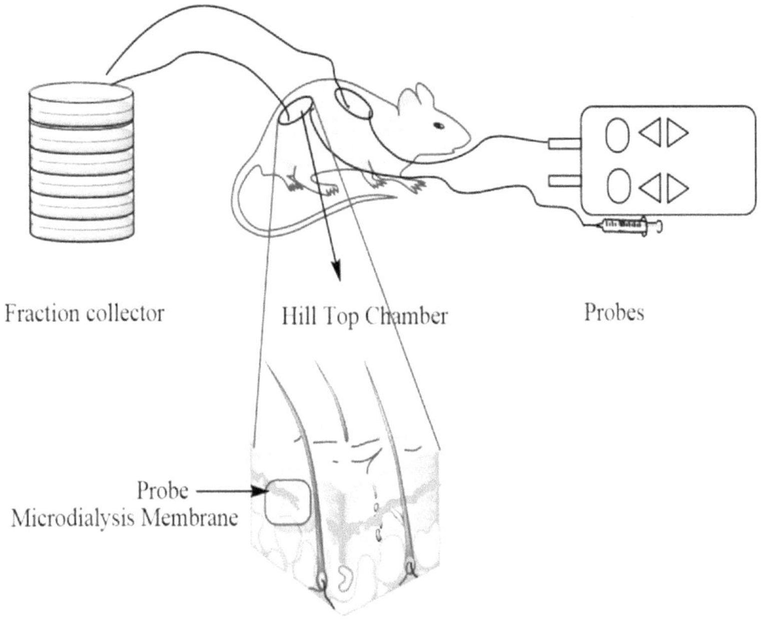

FIGURE 9.2 Dermal microdialysis model.

9.9 SUMMARY AND FUTURE PROSPECTS

The transdermal drug delivery system (TDDS) is an alternative to systemic delivery. This route has a specialized category in a pharmaceutical formulation based on formulation and effect. Various challenges faced by formulation scientists during development, the above test gives a glimpse of knowledge for the selection of appropriate testing methods for in-process quality checks. The formulator can analyze the individual components or combination of components from transdermal patches for skin sensitivity and irritant effect. The in-vitro methodologies described herein can be helpful for analyzing the irritant effect of polymeric ingredients before proceeding with actual formulation preparation. Certain advanced in-vivo techniques are also elaborated for analyzing the final formulation based on regulatory aspects. From the above methods, researchers can develop alternative methods of skin sensitivity and irritant testing with special emphasis on In-vitro methods. The desirable scaling values and their description are given in the tabulated form, which will be helpful for designing new methods. Cell-based assays are being currently preferred by industries for ease of testing, facilities available, and regulatory compliance.

KEYWORDS

- **episkin**
- **laser doppler perfusion imager**
- **patch test**
- **reactive oxygen species**
- **skin irritation test**
- **skin sensitivity test**
- **strait test**

REFERENCES

Alépée, N.; Grandidier, M. H.; Tornier, C.; Cotovio, J. An integrated testing strategy for *In vitro* skin corrosion and irritation assessment using skinethic™ reconstructed human epidermis. *Toxicology In Vitro,* **2015**, *29*(7), 1779–1792.

Basketter, D. A.; York, M.; McFadden, J. P.; Robinson, M. K. Determination of skin irritation potential in the human 4-H patch test. *Contact Dermatitis,* **2004**, *51*(1), 1–4.

Bason, M. M.; Gordon, V.; Maibach, H. I. Skin irritation: *In Vitro* assays. *International Journal of Dermatology,* **1991**, *30*(9), 623–626.

Benavides, T.; Mitjans, M.; Martínez, V.; Clapes, P.; Infante, M. R.; Clothier, R.; Vinardell, M. P. Assessment of primary eye and skin irritants by *in vitro* cytotoxicity and phototoxicity models: An *in vitro* approach of new arginine-based surfactant-induced irritation. *Toxicology,* **2004**, *197*(3), 229–237.

Bodin, A.; Fischer, T.; Bergh, M.; Nilsson, J. L. G.; Karlberg, A. T. Skin irritation from air-oxidized ethoxylated surfactants. *Contact Dermatitis,* **2000**, *43*(2), 82–89.

Borlon, C.; Godard, P.; Eskes, C.; Hartung, T.; Zuang, V.; Toussaint, O. The usefulness of toxicogenomics for predicting acute skin irritation on *in vitro* reconstructed human epidermis. *Toxicology,* **2007**, *241*(3), 157–166.

Cargill, R.; Engle, K.; Rork, G.; Caldwell, L. J. Systemic delivery of timolol after dermal application: Transdermal flux and skin irritation potential in the rat and dog. *Pharmaceutical Research,* **1986**, *3*(4), 225–229.

Corsini, E.; Sheasgreen, J.; Marinovich, M.; Galli, C. Use of differential display-polymerase chain reaction to identify genes selectively modulated by chemical allergens in reconstituted human epidermis. *Toxicology In Vitro,* **2002**, *16*(4), 427–431.

Draize, J. H.; Woodard, G.; Calvery, H. O. Methods for the study of irritation and toxicity of substances applied topically to the skin and mucous membranes. *Journal of Pharmacology and Experimental Therapeutics,* **1944**, *82*(3), 377–390.

Fuchs, J.; Groth, N.; Herrling, T. *In vitro* and *in vivo* assessment of the irritation potential of different spin traps in human skin. *Toxicology,* **2000**, *151*(1–3), 55–63.

Fulzele, S. V.; Babu, R.; Ahaghotu, E.; Singh, M. Estimation of proinflammatory biomarkers of skin irritation by dermal microdialysis following exposure with irritant chemicals. *Toxicology,* **2007**, *237*(1–3), 77–88.

Hall-Manning, T.; Holland, G.; Rennie, G.; Revell, P.; Hines, J.; Barratt, M.; Basketter, D. Skin irritation potential of mixed surfactant systems. *Food and Chemical Toxicology,* **1998**, *36*(3), 233–238.

Hiremath, S. R. R. Cytotoxicity, anti-tumor activity, cumulative skin irritation and sensitization study of 5-fluorouracil from a transdermal patch for Dalton's lymphoma ascites cells. *Journal of Health Science,* **2007**, *53*(3), 275–281.

Hoffmann, K.; Gerbaulet, U.; El-Gammal, S.; Altmeyer, P. 20-MHz B-mode ultrasound in monitoring the course of localized scleroderma (Morphea). *Acta Dermato-Venereologica Supplementum,* **1991**, *164*, 3–16.

Kano, S.; Sugibayashi, K. Kinetic analysis on the skin disposition of cytotoxicity as an index of skin irritation produced by cetylpyridinium chloride: Comparison of *in vitro* data using a three-dimensional cultured human skin model with *in vivo* results in hairless mice. *Pharmaceutical Research,* **2006**, *23*(2), 329–335.

Katsarou, A.; Davoy, E.; Xenos, K.; Armenaka, M.; Theoharides, T. Effect of an antioxidant (quercetin) on sodium-lauryl-sulfate-induced skin irritation. *Contact Dermatitis,* **2000**, *42*(2), 85–89.

Kim, C. W.; Kim, C. D.; Choi, K. C. Establishment and evaluation of immortalized human epidermal keratinocytes for an alternative skin irritation test. *Journal of Pharmacological and Toxicological Methods,* **2017**, *88*, 130–139.

Nickoloff, B. J.; Naidu, Y. Perturbation of epidermal barrier function correlates with initiation of cytokine cascade in human skin. *Journal of the American Academy of Dermatology,* **1994**, *30*(4), 535–546.

Osborne, R.; Perkins, M. An approach for development of alternative test methods based on mechanisms of skin irritation. *Food and Chemical Toxicology,* **1994**, *32*(2), 133–142.

Osbourn, R. A.; Tusing, T. W. Sensitivity test patches. In *Google Patents, US Patent,* **1965**, *3, 212*, 495.

Pezron, I.; Galet, L.; Clausse, D. Surface interaction between a protein monolayer and surfactants and its correlation with skin irritation by surfactants. *Journal of Colloid and Interface Science,* **1996**, *180*(1), 285–289.

Plock, N.; Kloft, C. Microdialysis—theoretical background and recent implementation in applied life-sciences. *European Journal of Pharmaceutical Sciences,* **2005**, *25*(1), 1–24.

Rasmussen, C.; Gratz, K.; Liebel, F.; Southall, M.; Garay, M.; Bhattacharyya, S.; Pirnstill, J. The StrataTest® human skin model, a consistent *in vitro* alternative for toxicological testing. *Toxicology In Vitro,* **2010**, *24*(7), 2021–2029.

Schurer, N. Y.; Elias, P. M. The biochemistry and function of stratum corneum lipids. *Advances in Lipid Research,* **1991**, *24*, 27–56.

Serup, J.; Agner, T. Colorimetric quantification of erythema—a comparison of two colorimeters (Lange micro color and Minolta chroma meter Cr-200) with A clinical scoring scheme and laser-doppler flowmetry. *Clinical and Experimental Dermatology,* **1990**, *15*(4), 267–272.

Smith, J. S.; Macina, O. T.; Sussman, N. B.; Karol, M. H.; Maibach, H. I. Experimental validation of a structure-activity relationship model of skin irritation by esters. *Quantitative Structure-Activity Relationships: An International Journal Devoted to Fundamental and Practical Aspects of Electroanalysis,* **2000**, *19*(5), 467–474.

Smith, J. S.; Macina, O. T.; Sussman, N. B.; Luster, M. I.; Karol, M. H. A robust structure-activity relationship (Sar) model for esters that cause skin irritation in humans. *Toxicological Sciences,* **2000**, *55*(1), 215–222.

Tardiff, R. G.; Hubner, R. P.; Gevecker Graves, C. Harmonization of thresholds for primary skin irritation from results of human repeated insult patch tests and laboratory animal skin irritation tests. *Journal of Applied Toxicology: An International Journal,* **2003**, *23*(4), 279–281.

Terry, C. M.; Clikeman, J. A.; Hoidal, J. R.; Callahan, K. S. TNF-α and IL-1α induce heme oxygenase-1 via protein kinase C, Ca^{2+}, and phospholipase A2 in endothelial cells. *American Journal of Physiology-Heart and Circulatory Physiology,* **1999**, *276*(5), H1493–H1501.

Welss, T.; Basketter, D. A.; Schröder, K. R. *In vitro* skin irritation: Facts and future. State of the art review of mechanisms and models. *Toxicology In Vitro,* **2004**, *18*(3), 231–243.

CHAPTER 10

Regulatory Aspects of Drug Development for Dermal Products

Nayan A. Gujarathi,[1] Akshada A. Bakliwal,[2] Bhushan R. Rane,[3] Vasim Pathan,[2] and Raj Kumar Keservani[4]

[1]Shri Vile Parle Kelavani Mandal's Institute of Pharmacy, Dhule–424001, Maharashtra, India, E-mail: nayang5678@gmail.com

[2]Sandip Institute of Pharmaceutical Sciences, Nashik–422213, Maharashtra, India

[3]Shri D.D. Vispute College of Pharmacy and Research Center, New Panvel–410206, Maharashtra, India

[4]Faculty of Pharmacy, CSM Group of Institutions, Prayagraj–211002, Uttar Pradesh, India

ABSTRACT

It is essential to understand the diverse regulatory aspects for the development process of drugs for transdermal and topical application. The human skin plays a vital role for topical products. In the invention of any new pharmaceutical product, there is a series of phases as pre-clinical phase, Phase I, Phase II, and Phase III. In this chapter, we focus on regulatory aspects of drug development for dermal delivery. Throughout the development process of approving any product, it is necessary to take an advice from regulatory agencies. Therefore, the United States, Europe, and Australia implement investigational new drug (IND) and New Drug Application (NDA), European Medicines Agency (EMA) and Australian Therapeutic

Topical and Transdermal Drug Delivery Systems: Applications and Future Prospects. Nayan A. Gujarathi, PhD, Juliana Palma Abriata, PhD, Raj Kumar Keservani, MPharm, Anil K. Sharma, PhD, MPharm (Eds.)

Goods Association (TGA) guidelines, respectively being applicable for all pharmaceutical industry. This includes various regulatory aspects consisting clinical study, bioavailability, bioequivalence study, transference study, and so forth. It also focuses on various examples of topically functional moieties with their specification.

10.1 TOPICALLY APPLIED DRUGS

As per a recent study in drug development, the intact human skin functions as an effective mode of administration for drugs. Drugs to be applied topically can be delivered as dermal drug products and transdermal drug products. The drugs which are applied onto the surface of the skin in order to achieve the localized action are known as dermal or topical drugs while the drugs which are envisioned for systemic activity are called transdermal drugs.

10.1.1 MECHANISM

Drugs are capable of producing either local or systemic effect when it is applied on skin. Drugs may penetrate into the stratum corneum (SC). Then percutaneous absorption can be seen which includes several steps, i.e., the adsorption into the SC by diffusion through it, viable epidermis (EP), and papillary dermis into the microcirculation.

10.1.2 BENEFITS

- It avoids GI incompatibility;
- Prevent first-pass liver or GI metabolism;
- Drug delivery can approximate zero-order;
- Can maintain steady drug concentration;
- Provides long term therapy from single dose;
- Enhancement of therapeutic efficacy;
- Improvement of patient compliance;
- Allowance of rapid termination of drug input.

In short, dermal system used to maximize the bioavailability, optimize therapeutic efficacy, and minimize side effects.

10.2 DRUG DEVELOPMENT AND REGULATORY PROCESS

Beginning with the discovery phase, any new pharmaceutical product development is divided into a series of phases. These phases are divided into pre-clinical phase, Phase I, Phase II and Phase III. In pre-clinical phase, product's toxicological, pharmacological, pharmacokinetic, and pharmacodynamic properties are examined in experimental animals. Phase I involves an examination of pharmacokinetics (PK), pharmacodynamic, and safety in healthy volunteers. This phase also examined the new product's dose range by which one can select the minimal, optimal, and maximum effective doses for further study. In Phase III, a definitive proof of efficacy is provided and the safety is further examined so that a risk/benefit profile can be considered. The schematic representation of distinct features for clinical trial phases are given in Figure 10.1; and compound success rate for screening of new entity is given in Figure 10.2.

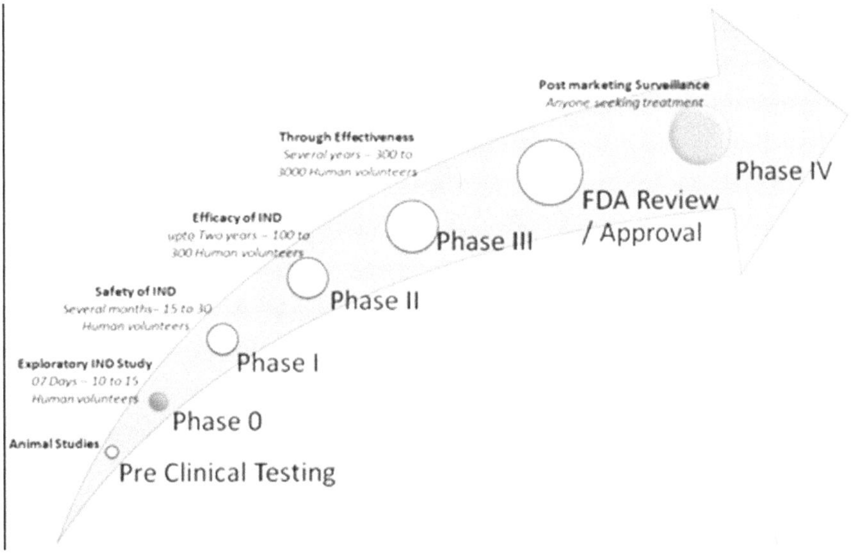

FIGURE 10.1 Overview of drug development process.

For approval of the product, advice from the regulatory agencies is useful throughout the development process. For such purpose, in the United States, advice meetings begin with a pre-IND meeting and a pre-new drug application (NDA) meeting (https://www.fda.gov/regulatory-information/

search-fda-guidance-documents#guidancesearch). In Europe, the regulatory agency (EMA) allows advice meeting as well as pre-submission meetings which serve a similar role (https://www.ema.europa.eu/en/documents/regulatory-procedural-guideline/european-medicines-agency-guidance-applicants-seeking-scientific-advice-protocol-assistance_en.pdf and https://www.ema.europa.eu/en/human-regulatory/marketing-authorization/marketing-authorization-guidance-documents#Pre-submission). The Australian TGA also has a formal procedure for meetings (https://www.tga.gov.au/sites/default/files/pm-argpm.pdf).

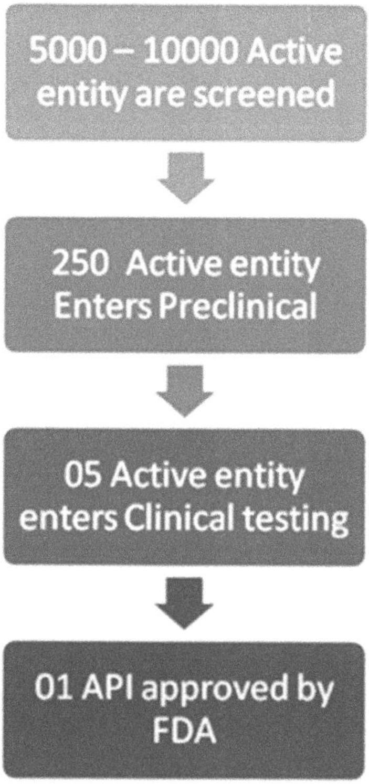

FIGURE 10.2 Compound success rate for new chemical entity.

The United States, Europe, and Australia all encourage these meetings with regulatory agencies (Figure 10.3) so that development programs can be designed to better meet the needs of regulator. (https://www.fda.gov/media/87478/download).

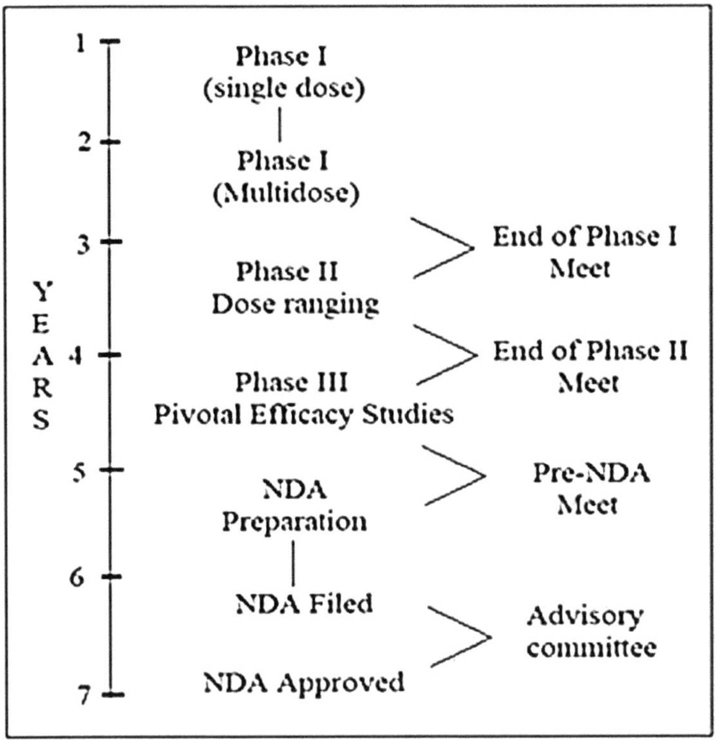

FIGURE 10.3 Typical sequence for advice meetings as illustrated for the United States.

10.3 REGULATORY REQUIREMENTS

The agency considered deliverance of drug from dermal formulation as novel for the treatment of various diseases. This includes treatment or anticipation of systemic diseases. For the approval of new drug application, the type of studies needed should be customized and are largely uttered by the following contemplations:

- The critical nature of the active pharmaceutical ingredient;
- The systemic dosage form of the identical drug existing in the sell;
- Medical and biopharmaceutical rationale;
- Safety, efficacy, and pharmacodynamics (PD) of drug entity based on the literature;
- Experience of the agency with the drug and/or drug delivery system.

It should be stressed that these drug delivery systems require careful scientific evaluation based on the pharmacological and pharmacokinetic properties of the drug and drug delivery system. Critical drugs or the drugs which require patient titration may need additional specialized studies to define the safety, efficacy, and labeling requirements.

10.3.1 PREREQUISITES FOR ACCOMPANYING CLINICAL TRIALS

Based on the geographical regions, the unapproved products have some regulations regarding the clinical trials. In US, it is obligatory to file IND before commencing the clinical trials (https://www.gmp-compliance.org/files/guidemgr/ind1.pdf.) The IND embraces information associated to the pharmacology and toxicology of the product, the chemical properties, industrialized, and controls, and the proposed clinical protocol and any clinical results obtained to date (https://www.accessdata.fda.gov/scripts/cdrh/cfdocs/cfcfr/CFRsearch.cfm?CFRPart=312). After submitting the IND, there is 30 days towards the U.S. Food and Drug Administration (FDA) to review the information and to decide whether to allow the proposed investigation to proceed. Before the investigation can proceed, it is mandatory to have approval from an institutional review board (IRB).

The clinical trial exemption (CTX) scheme and the clinical trial notification (CTN) scheme are the two schemes in Australia which conducts clinical trials (https://www.tga.gov.au/sites/default/files/clinical-trials-handbook.pdf and https://www.tga.gov.au/form/special-access-scheme). The information and protocol regarding the clinical trials comes under a CTN and the information for scientific validity, safety, and ethical concerns and approval of the conduct of the trial is submitted to a Human ethical research commission. Beneath a CTX, data about the product is acquiesced to the TGA for safety analysis and to the Human Research Ethics Committee for reviewing scientific validity and ethical concerns.

The request to a clinical trial authorization (CTA) is an essential in Europe (https://ec.europa.eu/health/sites/health/files/files/eudralex/vol-10/13_cp_and_guidance_eudract_april_04_en.pdf). This request for a CTA is made toward proficient authority for each country in which a clinical trial is to be conducted. A CTA generally consists of the application form, a clinical protocol, an investigator's brochure, and a summary of what is known about the product, in the form of an investigational medicinal product dossier

(IMPD). In the European Union, before conducting the human research approval is required by the Ethical committee.

10.3.2 BIOPHARMACEUTICAL CONSIDERATION

Biopharmaceutical considerations are vital for topical drug delivery system. Specific kinds of clinical and/or toxicological studies which need to perform can be determined by biopharmaceutical consideration. There is complete requirement to define:

- The bioavailability of drug delivery;
- The pharmacokinetics and metabolism of active therapeutic moiety;
- *In-vivo* reproducibility of the drug delivery system.

Depending on the action of the product, i.e., whether it is systemic or local, the other ingredients are added in the formulation. For example, scopolamine formulation which is used to prevent sea sickness has the systemic action while lidocaine which is used in local anesthesia has the localized action. The systemic effects are desired in the systemic action while for local action the systemic effects are not desired. There will be abundant interest to make the formulation which remains on the skin, without revelation to other body parts.

The guidance documents on the development of pharmaceutical formulation were issued by the EMA. It includes some discussion on items which is considered in favor of developing a medicinal formulation. Development of medicinal formulation includes physicochemical characteristics of the ingredients, compatibility between components, homogeneity, and performance characteristics (https://www.ema.europa.eu/en/documents/scientific-guideline/note-guidance-development-pharmaceutics_en.pdf). A short segment of this guidance reports special issues with transdermal systems, such as consideration of reservoir properties, adhesion to the skin and release testing.

With an oral dose form, the dose is generally specified in weight units, such as milligrams. This would not be suitable for many dermal products, particularly if the goal of the product is to achieve systemic absorption. It would also be difficult to compare different dermal products of the same active ingredient since the amount of active ingredient in the formulation may be the same but the relative absorption could be diverse.

Accordingly, transdermal products often express the strength as a weight unit "delivered" or absorbed per unit time, for example, micrograms per hour or milligrams per day. This approach is more useful to the prescribing physicians, predominantly when it is essential to calculate some equivalency to another formulation, whether it be another transdermal formulation or a formulation with a different route of administration. A good example of this is the transdermal delivery of fentanyl. Many of the transdermal fentanyl products express delivery in micrograms per hour, and there may be elaborate charts or formulae to aid in estimating equivalency to other dosage forms or other opioid products (https://www.accessdata.fda.gov/drugsatfda_docs/label/2005/19813s039lbl.pdf).

Another common difference observed with transdermal products is to articulate the amount of drug required with respect to the area of the patch. This is because the delivery of active ingredients across the skin is proportional to the surface area, so an augment in dosage level is often associated with an increase in the patch size. (https://www.accessdata.fda.gov/drugsatfda_docs/label/2005/19813s039lbl.pdf)

10.4 PHARMACOKINETIC EVALUATION PARAMETERS OF DERMAL DOSAGE FORMS

In evaluating the dermal dosage forms there is need to:

- Evaluate the site of drug administration in order to optimize the drug delivery;
- Consider the parameters such as optimal skin area for drug administration, blood flow at that site, etc.;
- Assess the role of skin in controlling drug delivery to systemic circulation.

As compared to other routes, pharmacokinetic evaluations of dermal dosage forms are relatively altered based on the objectives of the drug delivery.

If the product or formulation is applied to cure a dermatological disease that means they should not reach the systemic circulation. For those traditional pharmacokinetic methods of measuring the drug concentration in blood may not be useful, other than to assess the quantities which unintentionally reach the systemic circulation. For these dermal products, it may be more relevant to measure drug concentration in tissue biopsies taken from the spot of administration.

Additionally, whereas the half-life of a substance in blood will depend on rates of elimination, primarily via metabolism or excretion in the liver or kidney, the half-life of a dermal application may depend heavily on the rate of its exit from the skin, either due to absorption into the body, or washing of the application site, or more rarely metabolism within the skin. Indeed, skin can serve as a reservoir for the substance, creating a sort of sustained-release mechanism. The PK of such a process can be difficult to model (Singh and Roberts, 1994).

When the dermal application is intended to reach the systemic circulation, then blood levels of the active agent are considered the "standard" for assessing PK. However, the modeling of PK can be challenging, as the skin often acts as a reservoir. In this case, the absorption phase can be prolonged over many days.

A guidance document that discusses some pharmacokinetic issues with transdermal formulation has been issued by the EMA (https://www.ema.europa.eu/en/documents/scientific-guideline/guideline-pharmacokinetic-clinical-evaluation-modified-release-dosage-forms_en.pdf). Although it does not specifically address the difficulties of modeling pharmacokinetic data from transdermal delivery, it does mention that the effect of changing site of application should be examined if a product is intended to be applied to different areas of the body. The guidance also mentions that it is necessary to evaluate in human subjects, the potential for irritation and sensitization at the site of application, over, and above the possibility for phototoxicity or photosensitivity.

The basis of approval of dermal drug delivery systems from a biopharmaceutical standpoint consists of:

- Demonstration that the plasma concentration of active moiety is comparable to those with approved dosage form;
- Demonstration of batch to batch reproducibility;
- Defined pharmacokinetic parameters to support drug labeling.

It may not always be desirable to define the bioavailability profile of dermal drug delivery system relative to I.V. dose. The use of further mode of administrating drugs will provide additional pharmacokinetic data may be desirable so as to permit appropriate labeling.

10.5 BIOAVAILABILITY AND BIOEQUIVALENCE STUDY OF DERMAL PRODUCTS

For transdermal products which are intended to deliver an agent into the systemic circulation, the general approach to measuring bioavailability and bioequivalence can be similar to that used for products administered orally (https://www.ema.europa.eu/en/documents/scientific-guideline/draft-note-guidance-investigation-bioavailability-bioequivalence_en.pdf). However, there are some additional considerations. EMA guidance indicates that for transdermal drug delivery forms, bioequivalence should be evaluated after single and multiple dosing. The probable reason behind it is the skin can serve as a reservoir, thus bioequivalence should be studied under steady-state conditions, which require multiple dosing.

On the other hand, for certain types of topical solutions for the treatment of dermatological diseases, U.S. regulations allow bioequivalence to be waived (https://www.govinfo.gov/content/pkg/FR-1998-06-18/pdf/98-16141.pdf). It is already recognized that the absorption of an agent through skin can differ dramatically by the skin area to which it is applied, for example, abdomen compared to thigh or arm. For this reason, it is essential that when bioequivalence is being evaluated, the same body area should be used for all products being tested.

Furthermore, due to the varying characteristics of different types of transdermal devices, there is a concern about the repeatability of a demonstration of bioequivalence if the release mechanisms are different for two transdermal dosage forms being compared. In this case, the bioequivalence study should be replicated in order to demonstrate the reliability of the result (https://www.govinfo.gov/content/pkg/FR-1998-06-18/pdf/98-16141.pdf).

Finally, for demonstration of bioequivalence of two transdermal dosage forms, the EMA emphasizes that in addition to considering the PK, one should also demonstrate the comparability of other characteristics, including adherence, irritation, sensitization, phototoxicity, and the general adverse-event profile (https://www.govinfo.gov/content/pkg/FR-1998-06-18/pdf/98-16141.pdf).

The FDA does not have a general guidance document for bioequivalence of transdermal formulations. However, bioequivalence requirements are thoroughly described in the U.S. Code of Federal Regulations (https://www.govinfo.gov/content/pkg/FR-1998-06-18/pdf/98-16141.pdf).

10.6 DEMONSTRATING BIOEQUIVALENCE FOR LOCALLY ACTING DERMAL PRODUCTS

For substances not intended to reach the systemic circulation, the approach to establishing bioequivalence by measuring systemic concentrations may not be sound. In these cases, a pharmacodynamic or clinical endpoint approach may be more reliable. This concept is discussed in the EMA guidance (https://www.ema.europa.eu/en/documents/scientific-guideline/draft-note-guidance-investigation-bioavailability-bioequivalence_en.pdf) and is also discussed in the U.S. Code of Federal Regulations (https://www.accessdata.fda.gov/scripts/cdrh/cfdocs/cfcfr/CFRSearch.cfm?fr=320.21).

The FDA does have a guidance that focuses specifically on topical dermatological corticosteroids (http://www.fda.gov/downloads/Drugs/Guidance ComplianceRegulatoryInformation/Guidance's/ucm070234.pdf and https://www.fda.gov/media/70931/download). These agents can provide a vasoconstrictive effect, which has been found to be a useful PD endpoint for establishing bioequivalence. The guidance document goes into great detail on the conduct of an assay, sometimes passed on to as the skin blanching method. This assay is also discussed in an EMA publication (https://www.ema.europa.eu/en/documents/scientific-guideline/questions-answer-guideline-title-clinical-investigation-corticosteroids-intended-use-skin_en.pdf). Additionally, there is a general guidance document for bioequivalence of topical dermatological formulations, which includes a detailed discussion of a skin-stripping method for establishing comparability.

10.7 TRANSFERENCE STUDIES

Often, a key concern with a dermal product is the potential for the active ingredient to be exposed to subjects other than the intended recipient. This is not so much a concern for patch formulations but can be a serious concern for gels, creams, and sprays. Concern about such transference will be elevated for acutely hazardous substances such as fentanyl. The mechanism of transference is generally as simple as skin-to-skin contact, for example, a parent hugging a child, but can also extend to contamination of surface areas with which subjects may come in contact, including clothing, bedding, and even carpeted areas if contaminated skin cells are shed over long periods of time. For spray formulations, there may be concern about air-borne contamination. At this time, there is no specific guidance document on transference. However, the FDA and other regulatory authorities expect transference data for any dermal products for which there is a possibility of transference and where such transference could create an acute or chronic hazard to human subjects. Examples of transference data can be found in the package inserts for products such as AndroGel® (https://www.accessdata.fda.gov/drugsatfda_docs/nda/2012/022309Orig1s001.pdf) and EstroGel® (https://www.fda.gov/media/114329/download) and in a paper by Mazer et al. which compare transference potential for a patch versus a gel formulation of testosterone (Mazer et al., 2007). Transference studies are generally conducted using a pharmacokinetic endpoint. The product is applied to the intended recipient and the potential for transference is determined by having an unintended recipient make contact by rubbing the area of application. Transference potential may be examined at short and long time points after application, and in the presence and absence of clothing.

10.8 WASHING STUDIES AND INTERFERENCE FROM OTHER APPLIED PRODUCTS

Since the effectiveness of a transdermal product may be decreased if the active ingredient is washed off the skin, most regulatory agencies will expect to see some data on the effect of washing the skin at short and long time points after application. This is of most concern for gel, cream, and spray formulations. The information is generally based on pharmacokinetic studies and is often reflected in the package insert or summary of product information. For example, if washing at short time points after applications suggests that the effectiveness of the product is reduced, the patient would be instructed not to wash for some period of time after application. An example of this can be found in the U.S. EstroGel package insert. Examples of study designs can be found in the papers by Schumacher et al. (2009) and Rolf et al. (2002). Regulatory agencies may also wish to know what will be the effect of the application of other substances such as sunscreen, lotions, or even deodorant on the absorbed dose. This can readily be examined in a simple crossover study in which subjects apply the product either by itself, or in combination with some potentially interfering product. The endpoint will most likely be pharmacokinetic measures to evaluate any fluctuation either peaks or valley in absorption that is observed.

10.9 LOCAL TOLERANCE TESTING REQUIREMENTS IN EXPERIMENTAL ANIMALS AND HUMANS

With topical products, there is a particular concern about local tolerance. Accordingly, suitable studies evaluating local tolerance should be conducted in experimental animals prior to exposing human subjects. The EMA has a guidance document which outlines some of the considerations in such testing (https://www.ema.europa.eu/en/documents/scientific-guideline/guideline-non-clinical-local-tolerance-testing-medicinal-products_en.pdf). Local tolerance in experimental animals is also mentioned in International Council for Harmonization (ICH) M3 (R2), which emphasizes that such testing can be included in standard toxicology studies and should not require separate studies (https://database.ich.org/sites/default/files/M3_R2__Guideline.pdf).

Most regulatory agencies will also expect to see some evaluation of the potential for local irritancy in humans. This can often be accomplished with appropriate irritancy testing as a secondary measure in one or more pivotal studies. Under certain circumstances, where a pivotal study is not conducted for the registration of a new product, it may be necessary to conduct a special study to evaluate the potential for human irritancy. Irritancy of skin is most often evaluated by some variation of the Draize criteria (Vinardell and Mitjans, 2008) or by irritancy response scales and in addition, there are recommended scales in the FDA guidance on lidocaine products (Kirkland et al., 2006).

10.10 PHOTOSAFETY TESTING REQUIREMENTS

For active ingredients which are photo reactive, it is expected that the potential for photo irritation or photoallergy to be examined. The FDA has issued a guidance document which outlines methods of assessing this issue (https://www.fda.gov/media/71026/download). The EMA has a similar guidance document (https://www.ema.europa.

eu/en/documents/scientific-guideline/note-guidance-photosafety-testing_en.pdf). Keep in mind that inactive excipients may also be photo reactive and may require safety testing (https://www.fda.gov/media/72260/download).

10.11 ASSESSMENT OF ADHESION PROPERTIES

Patch formulations often have difficulty with adhesion characteristics due to the vast variation in skin to which patches are applied. This is made particularly difficult with the wide usage of skin creams and liquid soaps which may contain oils, fats, and waxes that interfere with adhesion. It can be quite challenging to formulate a patch that will remain on the skin for its intended duration of use but can still be removed without damaging the skin.

Regulatory agencies expect that patch adhesion will be evaluated during the development program (https://database.ich.org/sites/default/files/Q8_R2_Guideline.pdf). The FDA does not have a specific guidance which addresses the adhesion of dermal patch products; however, the concept is conferred in all aspect in draft guidance on lidocaine patches (https://www.accessdata.fda.gov/drugsatfda_docs/psg/Lidocaine_draft_Topical%20patch_RLD%20020612_RC10-18.pdf).

Generally, a measure of adhesion is included in each clinical trial conducted with the product. The question raised in the patient's daily diary to assess each day whether a patch is fully adhered, or whether the patch is loosely bound.

10.12 LABELING ASPECTS OF TRANSDERMAL PRODUCTS

Various regulatory agencies have certain expectations on labeling of transdermal products. A draft EMA guidance on content of a review of product distinctiveness explains that for transdermal patches, quantitative information should be provided about the content of active substance within each patch, the mean dose delivered per unit time, and the area of the releasing surface (https://www.ema.europa.eu/en/documents/regulatory-procedural-guideline/overview-comments-received-draft-revision-2-smpc-guideline-proposal-revision-european-commission_en.pdf).

It is also expected that the Summary of Product Characteristics list all the ingredients of the patch, including the adhesive, the release liner, and the backing material.

The FDA does not specifically include any discussion of dermal products in the general labeling guidance documents. However, dermal products may be mentioned in guidance documents for certain product classes.

The same guidance document specifies that for transdermal products, adhesion characteristics should be specified in the package insert (https://www.fda.gov/media/71359/download).

For Australia, the general guidance for labeling indicates that the product information should specify the quantity of active ingredient, along with the quantity of active ingredient released per unit of time (https://www.tga.gov.au/sites/default/files/consult-labeling-medicines-140822-guideline.pdf)

10.13 PEDIATRIC USE OF TRANSDERMAL FORMULATIONS

It is worth highlighting those transdermal formulations can be a preferred method of administration for certain pediatric populations. Transdermal formulations may be easy to apply, and may be favored in populations who are unable to swallow or who are averse to an injection. Absorption may be enhanced relative to adults because the SC, particularly in newborns, is thinner than adults, and more highly perfuse, thus favoring drug absorption. The younger pediatric groups also have a higher surface: weight ratio than adults.

- The bioavailability parameters (e.g., C_{max}, T_{max}, C_{avg}, AUC, fluctuation index, and parent/metabolite ratio) generated during the clinical pharmacology and biopharmaceutical studies.
- Data for all the anatomical application sites that will be proposed in the prescribing information.
- Dose proportionality data for the proposed dosing range.
- Tables and figures, including baseline-unadjusted levels of estradiol and metabolites. In the event that baseline-adjusted levels are more appropriate, this fact should be clearly indicated.
- The nominal mean *in vivo* delivery rate.

This is recommended data to include in U.S. package inserts for transdermal formulations of estrogen-containing products – AUC, area under the curve; C_{max}, maximum concentration; Tmax, time of maximum concentration; C_{avg}, average concentration.

However, the development of pediatric transdermal formulations can also be demanding because there may be variability of absorption by site of administration, and absorption characteristics of the skin may vary by age group, thus requiring careful evaluation of age-dependent absorption. These issues are discussed in detail in an EMA reflection paper on formulations of choice for pediatric populations (https://www.ema.europa.eu/en/documents/scientific-guideline/reflection-paper-formulations-choice-pediatric-population_en.pdf).

10.14 SPECIAL CONSIDERATION FOR TOPICAL PRODUCTS FOR TREATMENT OF PSORIASIS

The EMA has published a guidance document on the development of products for the management of psoriasis. This publication is particularly useful because it provides insights into the needs of regulators for many types of dermatological diseases. For example, it is recommended to study the absorption and PK of dermatological formulations on both healthy and diseased skin, in order to determine whether differences exist in the product performance. Additionally, the effect of frequency of applications on amount of drug retained as a reservoir in the skin should be examined. Absorption and PK from various application sites and in various age groups should be examined. As with any program, population PK is expected to play a key role in identifying outlier high and low absorbers and the potential consequences (https://www.ema.europa.eu/en/documents/scientific-guideline/draft-guideline-clinical-investigation-medicinal-products-treatment-psoriatic-arthritis_en.pdf).

10.15 SPECIAL CONSIDERATION FOR TOPICAL DERMATOLOGICAL CORTICOSTEROIDS

Prescription and over-the-counter corticosteroids are widely used for dermatological applications and their absorption through the skin has been extensively studied. Accordingly, there is considerable guidance on the development of these products. There is an EMA guidance which provides a detailed perspective on the types of data needed for approval of such products (https://www.ema.europa.eu/en/documents/scientific-guideline/clinical-investigation-corticosteroids-intended-use-skin_en.pdf). The Potential adverse reactions from topical corticosteroids are as follows:

- Skin atrophy, which becomes often irreversible, producing clinical thinning of the skin;
- Telangiectasia, purpura, striae;
- Rosacea-like and perioral dermatitis with or without skin atrophy;
- Rebound, which may lead to steroid "dependence;"
- Impairment of healing;
- Effects on the eye: increased risk of glaucoma, cataract, and exacerbation of mycosis and of herpes simplex;
- Systemic effects, resulting from suppression of the hypothalamic-pituitary–adrenal axis;
- Miscellaneous events such as depigmentation, hypertrichosis, and soon.

The EMA suggests an evaluation of the effect of several factors on the extent of dermal absorption and clinical efficacy, including the concentration of the active ingredient, the characteristics of the formulation being used to apply the active ingredient, the site of application, the condition of the skin (e.g., damaged or undamaged) and the conditions of application (e.g., occluded or not occluded).

10.16 SPECIAL CONSIDERATIONS FOR TOPICAL LIDOCAINE PRODUCTS

The FDA guidance on topical lidocaine products also provides a useful case study of the expectations of a regulatory agency for successful approval of a locally acting, topical product.

This guidance provides detailed expectations, including the number of studies required, study designs, requirements for blinding, and what to measure. There are examples of scoring systems to use for local irritation, and a discussion on how to display patch adhesion data.

10.17 SPECIAL CONSIDERATIONS FOR TOPICAL FENTANYL PRODUCTS

Fentanyl has enjoyed wide use as a transdermal product for the cure of painful conditions. However, fentanyl is also a potentially very hazardous medication, as disastrous consequences could result from overexposure or unintended exposure to healthcare workers or family members. Due to an increasing interest in the development of generic transdermal fentanyl products, the FDA has issued a guidance that outlines expectations for these products. (https://www.accessdata.fda.gov/drugsatfda_docs/psg/Fentanyl_transdermal%20extended%20release%20film_NDA%20019813_RV10-18.pdf).

Listings requested by the FDA for evaluation of topical lidocaine products.

- Center/site, number of subjects;
- Adverse events, reason for discontinuation;
- Race, age, sex;
- Analysis populations for each patch:
  - Test product per-protocol population for irritation analysis (yes/no), reason for omission;
  - Reference product per-protocol population for irritation (yes/no), reason for omission;
  - Test product per-protocol population for sensitization (yes/no), reason for omission;
  - Reference product per-protocol population for sensitization (yes/no), reason for omission;
  - Test product per-protocol population for adhesion (yes/no), reason for omission;
  - Reference product per-protocol population for adhesion (yes/no), reason for omission.
- Patch removed due to strong skin irritation reaction (yes/no);
- Time from first patch application to removal for unacceptable irritation;
- Cumulative number of patches removed for unacceptable irritation;
- Cumulative number of detached patches;
- Reinforced with tape (yes/no);
- Number of days until reinforcement with tape;
- New patch application due to detachment (yes/no);
- Date of a new patch application due to detachment;
- Time from application to detachment;

- Designation of skin sensitization (yes/no);
- Per each visit if data exist:
  - o Visit number, date of visit, days from baseline;
  - o Reason for exclusion from each per-protocol population per visit;
  - o Time from patch application to detachment for both test and reference products;
  - o Irritation scores for each product;
  - o Sensitization scores for each product;
  - o Adhesion scores for each product;
  - o Identity of the evaluator;
  - o Adverse events;
  - o Reason for discontinuation.

The guidance recommends three studies: one to assess PK, another to examine the effect of an overlay on product performance, and a third to assess the potential for irritation and sensitization. As for the lidocaine guidance, the FDA provides a detailed listing of the data analyzes they expect to receive.

10.18 SUMMARY

Dermal products may be intended to provide systemic exposure for treatment of systemic disease, for example, transdermal testosterone or estrogen for hormonal replacement, or they may be intended to provide local exposure to treat a dermatological disease, for example, post-herpetic neuralgia or psoriasis. In either case, regulatory agencies expect that such products be fully characterized with regard to their local and systemic pharmacokinetics and PD properties, and that they be evaluated for their potential toxicological effects and other adverse events. Products designed for topical administration may differ from other products in the way dose levels are expressed, since in most cases, only a portion of the administered dose is absorbed.

One particular challenge for dermal products intended for local exposure is the demonstration of bioequivalence; some regulators will expect to see a demonstration of clinical efficacy for such products, which requires larger trials, thus making development more expensive. Development of topically applied products may also require special studies, such as to evaluate transference, or the effect of washing the skin after application, or the effect of co-application with lotions and antiperspirants. If the product is a patch, the adhesion properties must be examined as an element of any pivotal studies to support registration.

Although products applied on dermal layer have their own unique challenges, there are also unique advantages, so there is no doubt that applied products will play a key role in health care for the anticipated future.

KEYWORDS

- **clinical trial authorization**
- **clinical trials**
- **drug development**
- **regulatory agencies**
- **regulatory aspects**
- **topical products**

REFERENCES

AndroGel® US Prescribing Information: Abbott Laboratories, Abbott Park, IL. March 2011. https://www.accessdata.fda.gov/drugsatfda_docs/nda/2012/022309Orig1s001.pdf (accessed on 27 May 2022).

Code of Federal Regulations. *Bioavailability and Bioequivalence Requirements*, **2020**: United States Code of Federal Regulations. 21 CFR 320, Washington. https://www.accessdata.fda.gov/scripts/cdrh/cfdocs/cfcfr/CFRSearch.cfm?fr=320.21 (accessed on 27 May 2022).

Code of Federal Regulations. *Title 21–Food and Drugs. Chapter I—Food and Drug Administration, Department of Health and Human Services: Part 312—Investigational New Drug Application,* April 1, **2020**. https://www.accessdata.fda.gov/scripts/cdrh/cfdocs/cfcfr/CFRsearch.cfm?CFRPart=312 (accessed on 27 May 2022).

Code of Federal Regulations. *Types of Evidence to Measure Bioavailability or Establish Bioequivalence,* 2020: United States 21 CFR 320.24, Washington. https://www.accessdata.fda.gov/scripts/cdrh/cfdocs/cfcfr/CFRSearch.cfm?CFRPart=320 (accessed on 27 May 2022).

Duragesic Fentanyl Transdermal System: US Prescribing Information. **2008**. https://www.accessdata.fda.gov/drugsatfda_docs/label/2005/19813s039lbl.pdf (accessed on 27 May 2022).

EstroGel® US Prescribing Information: Ascend Therapeutics, Herndon, VA. January 2008. https://www.fda.gov/media/114329/download (accessed on 27 May 2022).

European Agency for the Evaluation of Medicinal Products. *Note for Guidance on Modified Release Oral and Transdermal Dosage Forms: Section II (Pharmacokinetic and Clinical Evaluation).* https://www.ema.europa.eu/en/documents/scientific-guideline/guideline-pharmacokinetic-clinical-evaluation-modified-release-dosage-forms_en.pdf (accessed on 27 May 2022).

European Agency for the Evaluation of Medicinal Products. *Note for Guidance on the Investigation of Bioavailability and Bioequivalence: The European Agency for the Evaluation of Medicinal Products, Human Medicines Evaluation Unit.* London, England. https://www.ema.europa.eu/en/documents/scientific-guideline/draft-note-guidance-investigation-bioavailability-bioequivalence_en.pdf (accessed on 27 May 2022).

European Agency for the Evaluation of Medicinal Products. *Note for Guidance on Non-Clinical Local Tolerance Testing of Medicinal Products*, **2016**: The European Agency for the Evaluation of Medicinal Products, Evaluation of Medicines for Human Use, London, England. https://www.ema.europa.eu/en/documents/scientific-guideline/guideline-non-clinical-local-tolerance-testing-medicinal-products_en.pdf (accessed on 27 May 2022).

European Agency for the Evaluation of Medicinal Products. *Note for Guidance on Photo Safety Testing*, **2002**: The European Agency for the Evaluation of Medicinal Products, Evaluation of Medicines for Human Use, London, England. https://www.ema.europa.eu/en/documents/scientific-guideline/note-guidance-photosafety-testing_en.pdf (accessed on 27 May 2022).

European Commission. Detailed guidance on the request to the competent authorities for authorization of a clinical trial on a medicinal product for human use, the notification of substantial amendments and the declaration of the end of the trial: European Commission. Enterprise Directorate-General. Single market: Management & legislation for consumer goods Pharmaceuticals: Regulatory framework and market authorizations. Revision 3. Brussels, Belgium. https://ec.europa.eu/health/sites/health/files/files/eudralex/vol-10/13_cp_and_guidance_eudract_april_04_en.pdf (accessed on 27 May 2022).

European Medicines Agency. *Clinical Investigation of Corticosteroids Intended for Use on the* Skin – **1987**: European Medicines Agency, Committee for Medicinal Products for Human Use, London, England. https://www.ema.europa.eu/en/documents/scientific-guideline/clinical-investigation-corticosteroids-intended-use-skin_en.pdf (accessed on 27 May 2022).

European Medicines Agency. *EMEA Guidance for Companies Requesting Scientific Advice or Protocol Assistance*, **2009**. https://www.ema.europa.eu/en/documents/regulatory-procedural-guideline/european-medicines-agency-guidance-applicants-seeking-scientific-advice-protocol-assistance_en.pdf (accessed on 27 May 2022).

European Medicines Agency. *Guideline on Clinical Investigation of Medicinal Products Indicated for the Treatment of Psoriasis* – **2005**: European Medicines Agency, Committee for Medicinal Products for Human Use, London, England. https://www.ema.europa.eu/en/documents/scientific-guideline/draft-guideline-clinical-investigation-medicinal-products-treatment-psoriatic-arthritis_en.pdf (accessed on 27 May 2022).

European Medicines Agency. *Note for Guidance on Development Pharmaceutics* – January 28, **1998**: European Medicines Agency, London, England. https://www.ema.europa.eu/en/documents/scientific-guideline/note-guidance-development-pharmaceutics_en.pdf (accessed on 27 May 2022).

European Medicines Agency. *Proposal for a Revision of the European Commission Guideline on Summary of Product Characteristics (Draft).* Doc. Ref. EMEA/663087/2009, **2009**: European Medicines Agency, Committee for Medicinal Products for Human Use, London, England. https://www.ema.europa.eu/en/documents/regulatory-procedural-guideline/overview-comments-received-draft-revision-2-smpc-guideline-proposal-revision-european-commission_en.pdf (accessed on 27 May 2022).

European Medicines Agency. *Questions and Answer on Guideline Title*, 2006: Clinical Investigation of Corticosteroids Intended for Use on the Skin: European Medicines Agency, Committee for Medicinal Products for Human Use, London, England. https://www.ema.europa.eu/en/documents/scientific-guideline/questions-answer-guideline-title-clinical-investigation-corticosteroids-intended-use-skin_en.pdf (accessed on 27 May 2022).

European Medicines Agency. Reflection Paper. *Formulations of Choice for the Paediatric Population* – **2006**: European Medicines Agency, Committee for Medicinal Products for Human Use, London, England. https://www.ema.europa.eu/en/documents/scientific-guideline/reflection-paper-formulations-choice-paediatric-population_en.pdf (accessed on 27 May 2022).

European Medicines Agency: *Draft EMA Guidance on Pre-Submission for Initial Marketing Authorization Applications for Human Medicinal Products in the Centralized Procedure.* https://www.ema.europa.eu/en/human-regulatory/marketing-authorisation/marketing-authorisation-guidance-documents#Pre-submission (accessed on 27 May 2022).

Food and Drug Administration. *Draft Guidance on Fentanyl* – **2009**: United States Department of Health and Human Services, Food and Drug Administration, Office of Generic Drugs, Washington. https://www.accessdata.fda.gov/drugsatfda_docs/psg/Fentanyl_transdermal%20extended%20release%20film_NDA%20019813_RV10-18.pdf (accessed on 27 May 2022).

Food and Drug Administration. *Draft Guidance on Lidocaine* – **2007**: United States Department of Health and Human Services, Food and Drug Administration, Office of Generic Drugs, Washington. https://www.accessdata.fda.gov/drugsatfda_docs/psg/Lidocaine_draft_Topical%20patch_RLD%20020612_RC10-18.pdf (accessed on 27 May 2022).

Food and Drug Administration. Guidance for industry. *Content and Format of Investigational New Drug Applications (INDs) for Phase 1 Studies of Drugs, Including Well-Characterized, Therapeutic, Biotechnology-Derived Products*: United States Department of Health and Human Services, Food and Drug Administration, Centers for Drug and Biologics Evaluation and Research, Washington, DC. https://www.gmp-compliance.org/files/guidemgr/ind1.pdf (accessed on 27 May 2022).

Food and Drug Administration. Guidance for Industry. End-of-Phase 2A Meetings. United States Department of Health and Human Services – September **2008**: Food and Drug Administration. Centers for Drug and Biologics Evaluation and Review, Washington. https://www.fda.gov/regulatory-information/search-fda-guidance-documents#guidancesearch (accessed on 27 May 2022).

Food and Drug Administration. *Guidance for Industry*. Formal Meetings Between the FDA and Sponsors or Applicants. United States Department of Health and Human Services – May **2009**: Food and Drug Administration. Centers for Drug and Biologics Evaluation and Review, Washington. https://www.fda.gov/regulatory-information/search-fda-guidance-documents#guidancesearch (accessed on 27 May 2022).

Food and Drug Administration. *Guidance for Industry. Nonclinical Studies for the Safety Evaluation of Pharmaceutical Excipients*, **2005**: United States Department of Health and Human Services, Food and Drug Administration, Centre for Drug Evaluation and Research, Washington. https://www.fda.gov/media/72260/download (accessed on 27 May 2022).

Food and Drug Administration. Guidance for industry. *Non-contraceptive Estrogen Drug Products for the Treatment of Vasomotor Symptoms and Vulvar and Vaginal Atrophy Symptoms—Recommended Prescribing Information for Health Care Providers and Patient Labelling*–**2003**: United States Department of Health and Human Services, Food and Drug Administration, Centre for Drug Evaluation and Research, Washington. https://www.fda.gov/media/71359/download (accessed on 27 May 2022).

Food and Drug Administration. Guidance for Industry. *Topical Dermatological Drug Product NDAs and ANDAs–in vivo Bioavailability, Bioequivalence, in Vitro Release, and Associated Studies*, **1998**: United States Department of Health and Human Services, Food and Drug Administration, Centre for Drug Evaluation and Research, Washington. https://www.govinfo.gov/content/pkg/FR-1998-06-18/pdf/98-16141.pdf (accessed on 27 May 2022).

Food and Drug Administration. *Guidance for Industry: Photo Safety Testing Requirements*, **2004**. United States Department of Health and Human Services, Food and Drug Administration, Centre for Drug Evaluation and Research, Washington. https://www.fda.gov/media/71026/download (accessed on 27 May 2022).

Food and Drug Administration. Guidance. *Topical Dermatologic Corticosteroids: In vivo Bioequivalence*, **1995**: United States Department of Health and Human Services, Food and Drug Administration, Office of Generic Drugs, Division of Bioequivalence, Washington. https://www.fda.gov/media/70931/download (accessed on 27 May 2022).

Food and Drug Administration. SOPP 8001.6: Procedures for Parallel Scientific Advice with European Medicines Agency (EMA)–September 19, **2018**: United States Department of Health and Human Services. Food and Drug Administration, Centre for Drug Evaluation and Research, Washington. https://www.fda.gov/media/87478/download (accessed on 27 May 2022).

International Conference on Harmonization. ICH Harmonized tripartite guideline. *Guidance on Nonclinical Safety Studies for the Conduct of Human Clinical Trials and Marketing Authorization for Pharmaceuticals M3 (R2)*, **2009**: International Conference on Harmonization, Brussels, Belgium. https://database.ich.org/sites/default/files/M3_R2__Guideline.pdf (accessed on 27 May 2022).

International Conference on Harmonization. *ICH Harmonized Tripartite Guideline.* Pharmaceutical Development Q8 (R2) – **2009**: ICH, Brussels, Belgium. https://database.ich.org/sites/default/files/Q8_R2_Guideline.pdf (accessed on 27 May 2022).

Kirkland, C.; Yelverton, C.; Flesischer, A.; Camacho, F.; Feldman, S. Gel vehicles are not inherently more irritating than creams. *Journal of Drugs in Dermatology,* **2006**, *5*(3), 269–272.

Mazer, N.; Fisher, D.; Fisher, J.; Cosgrove, M.; Belle, D.; Eiller, B. Transfer of transdermally applied testosterone to clothing: A comparison of a testosterone patch versus a testosterone gel. *The Journal of Sexual Medicine,* **2007**, *2*(2), 227–234.

Rolf, C.; Kemper, S.; Lemmnitz, G.; Eickenberg, U.; Nieschlag, E. Pharmacokinetics of a new trans- dermal testosterone gelling on adotrophin-suppressed normal men. *European Journal of Endocrinology,* **2002**, *146*(5), 673–679.

Schumaccher, R.; Gattermeir, D.; Peterson, C.; Wisdom, C.; Day, W. The effects of skin-to-skin contact, application site washing, and sunscreen use on the pharmacokinetics of estradiol from a metered-dose transdermal spray. *Menopause,* **2009**, *16*(1), 177–183.

Singh, P.; Roberts, M. Dermal and underlying tissue pharmacokinetics of lidocaine after topical application. *Journal of Pharmaceutical Sciences,* **1994**, *83*(6), 774–782.

Therapeutic Goods Administration. *Access to Unapproved Therapeutic Goods via the Special Access Scheme: Therapeutic Goods Administration*, Department of Health and Ageing, Canberra, Australia. https://www.tga.gov.au/form/special-access-scheme (accessed on 27 May 2022).

Therapeutic Goods Administration. *Australian Regulatory Guidelines for Prescription Medicines.* Appendix 5. Conduct of meetings between TGA and sponsors – **2004**: Therapeutic Goods Administration, Department of Health and Ageing, Canberra, Australia. www.tga.gov.au/publication/australian-regulatory-guidelines-prescription-medicines-argpm (accessed on 21 July 2022).

Therapeutic Goods Administration. Guidance on Therapeutic Goods Order No. 79. *General Requirements for the Labelling of Medicines* – **2008**: Australian Government, Department of Health and Ageing, Therapeutic Goods Administration, Canberra, Australia. www.tga.gov.au/medicine-labels-guidance-tgo-91-and-tgo-92 (accessed on 21 July 2022).

Therapeutic Goods Administration. *The Australian Clinical Trial Handbook.* A simple, practical guide to the conduct of clinical trials to international standards of good clinical practice (GCP) in the Australian context–March **2006**: Therapeutic Goods Administration, Department of Health and Ageing, Canberra, Australia. www.tga.gov.au/resource/australian-clinical-trial-handbook (accessed on 21 July 2022).

Vinardell, M.; Mitjans, M. Alternative methods for eye and skin irritation tests: An overview. *Journal of Pharmaceutical Sciences,* **2008**, *97*(1), 46–59.

Index

D

I

N

T